Technical Document Basics

FOR ENGINEERING TECHNICIANS AND TECHNOLOGISTS

Technical Document Basics

FOR ENGINEERING TECHNICIANS AND TECHNOLOGISTS

DAVID W. RIGBY

North Seattle Community College

The Wordworks™ Series

Upper Saddle River, New Jersey
Columbus, Ohio

Library of Congress Cataloging-in-Publication Data

Rigby, David W.
Technical document basics for engineering technicians and technologists / David W. Rigby.
p. cm. -- (Wordworks)
Includes index.
ISBN 0-13-490137-1
1. Technical writing. I. Title.

T11 .R565 2001
808'.0666--dc21

00-034675

Vice President and Publisher: Dave Garza
Editor in Chief: Stephen Helba
Executive Editor: Debbie Yarnell
Associate Editor: Michelle Churma
Production Editor: Louise N. Sette
Production Supervision: Clarinda Publication Services
Design Coordinator: Robin G. Chukes
Text Designer: Ceri Fitzgerald
Cover Designer: Ceri Fitzgerald
Production Manager: Brian Fox
Marketing Manager: Jimmy Stephens

This book was set in Optima by The Clarinda Company. It was printed and bound by Victor Graphics, Inc. The cover was printed by Victor Graphics, Inc.

10 9 8 7 6 5 4 3 2 1
ISBN 0-13-490137-1

Welcome to Wordworks™

Wordworks™ is a series of four communication skills manuals. The manuals consist of three writers' guides for engineering and technical applications and an additional guide to in-service spoken communication. The manuals are designed to provide in-demand information in a readable fashion. They are matter-of-fact and use-oriented.

Each manual focuses on specific exit skills that are necessary for job performance. In this respect the texts are somewhat unique. They were inspired by the carefully tailored goal orientation of corporate seminar manuals. This strategy is at the heart of the streamlined manuals of corporations where specific skill outcomes are the narrow focus of company class time. For skills manuals this strategy can accelerate and focus the learning process, and the approach is particularly useful in college programs where English components are never more than a course or two of the total learning experience.

The *Wordworks™* manuals are conversational, visual, and practical so that the learning experience is accessible. Chapter-by-chapter discussions encourage a learning curve of understanding. Important concepts and practical applications are identified and explored. Models and other illustrations are also features of the texts.

The texts rely on an extensive use of graphics to conceptualize ideas, and models are provided to draw attention to desirable skills. This approach is intended to help students build a strong understanding of logical design features that they can use to construct any writing project.

Three of the texts in the series deal with the basics of the craft of writing, and the series uses a writer-to-writer strategy to expore and explain this craft. The manuals are intended to be learning tools, but because they focus on a craft, they are not intended to be overly academic. To the extent to which streamlining can be achieved in a college environment, the *Wordworks™* series provides a practical approach to skill training that is compatible with the limited time available for communications offering in technical programs. The manuals encourage curiosity and provide a learning-oriented climate, but they also simplify the path to practical knowledge and skills.

Each title in the *Wordworks™* series is intended to complement the other titles:

Basic Composition Skills for Engineering Technicians and Technologists. This first text of the series is intended to help upgrade fundamental skills in writing. It is a thorough discussion of the problems that are encountered by writers and the solutions to those problems. This book is uniquely designed to build upon existing skills that are part of every work day.

Writer's Handbook for Engineering Technicians and Technologists. The second title in the series is a writer's handbook of the rules and practices of writing. Part style guide, part grammar book, part technical writing reference, the *Writer's Handbook* is designed to be a bridge that can support both *Basic Composition Skills* and *Technical Document Basics.* The *Writer's Handbook* is specifically intended for engineering and technical students.

Technical Document Basics for Engineering Technicians and Technologists. The third title in the *Wordworks*™ series develops a concentrated focus on basic technical writing know-how. The text is designed to identify and explore the documentation standards that are used to develop and produce technical projects. The basic skills are condensed into a thoroughly illustrated and readable text.

Workplace Communications for Engineering Technicians and Technologists. An additional text supports the *Wordworks*™ concept with an exploration of spoken communication. Studies reveal that 80% or more of our work-related communcation in trade and industrial settings is handled in conversation. The absence of training tools in this area is an invitation for communication problems if only because spoken messages outnumber our memos by four to one! *Workplace Communications* helps identify and improve in-service communication skills.

About Technical Document Basics

Technical Document Basics is designed for college students in engineering technologies or engineering fields who want to get up to speed in technical writing fundamentals. The manual is intended for use in two-year engineering technology programs. Today's technicians and engineers need to understand the basic skills in technical writing in order to face writing challenges in business and industry.

Engineering technology graduates usually have dual ambitions, so they need writing skills that can be directed to two distinct types of applications. Their near-term writing needs are likely to be somewhat industry oriented. Their long-term writing needs may be academic if they continue their schooling for four-year engineering degrees. *Technical Document Basics* examines both industrial and academic applications of technical writing fundamentals.

At a practical level, this book is intended to serve both workplace and college writing tasks without any additional coursework. *Technical Document Basics* is a "first book" or entry-level text, but it was constructed with the understanding that it might be the last book for a student completing the two-year degree with the career goal of being an engineering technician. However, beyond learning fundamental writing skills, the engineering technician or technologist will also have to adapt these skills to company specifications.

On the other hand, many engineering technology graduates of two-year programs continue their studies in engineering. Four-year students are more likely to take additional courses in technical writing, in which case this volume will provide the groundwork for the further development of their writing craft.

Technical Document Basics draws a distinction between "corporate" writing and "academic" writing. Corporations use technical writing skills to serve the goals of industry. Consequently, such items as assembly instructions or maintenance manuals are major concerns for a company. Academic writing, of the sort that an upper-division college student will write for a university, is a distinctly different type of production.

The text presents a group of document types that are typical of what is popularly called "technical writing." They are the "drivers" that make up a recognizably large part of the documentation of thousands of industries from coast to coast. Assembly instructions and maintenance manuals are typical presentations in this category. They are vital links to product marketing, product uses, and product services. Engineering technology majors (two-year programs) will find that these are exactly the types of documents they need to be able to design and construct.

The text also addresses an application that is specifically useful for college studies in four-year engineering programs. The practical reality of "college papers" calls for immediate attention in a first course in technical writing. College transfer students appear to be little concerned about how to write industry-related documentation when faced with college term papers and laboratory projects.

A unique element of *Technical Document Basics* is the focus on "integrated texts," meaning a text that incorporates all its media as a finished product. Much of *Technical Document Basics* is focused on designing a package and putting the package together. Desktop publishing software programs allow students to create well-designed projects. Editing is no longer the province of a select few; the skills are in the hands of many, and writers must learn to control their craft.

The use of graphics and layouts is a focus of *Technical Document Basics*. Graphics and mathematical elements have unmistakably large roles to play in technical writing. It is fairly obvious that the need for graphics, schematics, symbologies, and mathematical systems forces writers to be *very* careful in the way they compose their projects. Authors are usually juggling three, four, or five communications systems, and the English part of the text is often fused in a medley of media. The writing is critical, but it can appear to be a minor player in a large game plan of math or physics, chemical engineering or electrical engineering, or any of many technical fields.

Technical Document Basics uses a broad assortment of samples that reflect both academic writing styles and industry practices. Students are likely to need and use both varieties of documents. All the samples are student productions. Many of the samples and models were co-op projects that were developed for genuine applications in company settings; studying them is part of the educational process of this text. As writing experiences, these exercises are of enormous value for working writers, and they serve as valuable examples to readers of the *Wordworks*™ series as well.

Technical Document Basics encourages a new look at the way writers perceive writing. Technical writing is really technical composing, which may give an author a new slant on writing. Science and engineering create unique challenges. Writers must visualize what they cannot say, and calculate what they cannot see. What authors really create is a text that is a technical composite of visual and mathematical elements. It needs to be graphic and mathematical because this is the way it can best construct its meaning. Technical writing is not just *about* engineering. It is itself engineered. It *is* engineering.

NOTE: Readers of this text do not have to be familiar with computers, even though the text frequently comments on the use of computer technology. The text assumes that most college-level readers have some word processing knowledge. If students have no computer skills, cannot afford the gear, or do not have access to a computer lab, they are still welcome to learn about the craft of technical writing. The expense will be one dollar for a glue stick and a little time at the copy machine!

This book is designed as a stand-alone text, but I would encourage readers to also use a desktop reference guide to rules for writers, preferably the *Writer's Handbook* (another volume of the *Wordworks* series) or a similar guide intended for technicians and engineers.

Contents

Work in Progress

1. A Case Study

At the beginning of each of the following chapters you will have the opportunity to read a page from the working notes of a writer in the workplace. Like you, the author is an engineering technician. This employee is working on an associate's degree in engineering technology and works for a local corporation. She will take you through the paces of a typical writing project that is representative of the documentation that is developed by her company.

I would like to welcome you to the world of Genus Specialty Systems. My name is Shirley, and I have worked at Genus for 16 years. The company has generously supported my college education, and I am in the process of completing my electronics-related degree. Although I didn't see myself as a writer, most of my time is now dedicated to documentation support; that means I do the writing and am responsible for it.

Genus is a medium-sized corporation. It is based in Seattle, Washington, and has 700 employees and a number of domestic and international sales representatives located around the world. The company manufactures a number of products and specializes in lighted assemblies for devices that include indicators, meters, gauges, interface panels, and microprocessor keyboards. Genus is a supplier to original equipment manufacturers (OEM) such as Lockheed and Boeing and is a "spares" supplier to customers such as the U.S. Government and domestic and international airlines. Genus is certified by the International Standardization Organization (ISO) to the 9001 level. This is quite an achievement, because it means a company must be able to "say what it does, do what it says, and prove it" (this is where correct and responsible documentation comes in).

My original position involved very little writing. I was hired as an assembler. I then moved into the Quality Control department as an inspector. In the last five years I have moved from coordinator to auditor within the Quality Systems Department. An auditor, as we use the term, is someone who studies production efficiency. I have found myself in increasing depths of paperwork. I think the variety of positions I have had helped make my writing "user-friendly" because I have worked at many levels of the company. I have had to build, inspect, and conform products for customers based on other people's paperwork.

My typical day is what I will describe for you at the beginning of the following chapters. Rather than go through my workday from nine to five, I will explain the development of a specific writing project I was recently assigned. For this project I was to update all aspects of an ISO 9001 Genus process regarding the handling of stock products. I will walk you through the phases of the project in brief sketches that will give you an understanding about how technical writing projects evolve from conception to completion.

S.B.

Origins: The Ruins of Science

Technical writing is a medium that developed over a period of several hundred years, although the repair manuals, maintenance manuals, site reports, and environmental impact statements of our time represent more recent trends. There are old and deep roots to the methods of engineering and science, and those roots have partly defined the mission of technical writing. A technical writing document is not difficult to produce, but imitation is not the best approach. It helps to understand the mission of a technical document. This type of writing carries a special burden: the demands of engineering and scientific thinking. If authors—particularly technicians, engineering techs, and engineers—have a basic understanding of the tradition behind tech writing, they will be in a better position to master the craft.

Much of what we know of early science and engineering is based on the surviving artifacts that reflect the technological achievements of earlier times. If the pyramids of Cheops are properly aligned in terms of astronomical bearings, and if the innermost altar of the vast temple of Abu Simbel is designed to receive sunlight on a specific ceremonial day, then we can measure the understanding of astronomy that was achieved by Egyptians and applied to the engineering and construction of monumental works. Each year, on October 20, the dawn light enters the temple, reaches the full length of two hundred feet to the rear sanctuary, and illuminates the statue of the god figure Amun and the statue of Rameses II, who ascended to power on that day—thirty centuries ago! We can imagine the awe and respect in which the priests of this temple were held. The temple still stands three thousand years later, with its seventy-foot-high guardians carved out of the rock hills of the Nile.

Like the Egyptians and other early cultures of the deserts, the Greeks and the Romans depended on engineering for providing the prosperity they enjoyed. The great Roman engineering feats of aqueduct design, for example, stretched from ten to over fifty miles and carried five million to fifty million gallons of water a day. Many were made of concrete, which the Romans formulated twenty centuries ago. One of the aqueducts built in the time of Hadrian still transports the precious water of the mountains to Athens.*

I mention these few artifacts of early Mediterranean culture to suggest both the presence and the role of scientific thought and the engineering skills that were available thousands of years ago. The ancient monuments are not simply monuments to a culture. They are also monuments to scientific and engineering understanding. They are often all we know of early scientific reasoning, because the paperwork is gone. The blueprints are gone. Even the tools of the engineer's trade are gone. We can understand the early engineers only by their achievements. Indeed, we can look at the ancient monuments with the eye of an engineer rather than the eye of the historian. The science of archeology is fast becoming the archaeology of science. An astronomer from the Smithsonian Astrophysical Observatory, Gerald Hawkins, put to rest the hundreds of years of puzzlement over Stonehenge, the ancient stone ruins of England. Was it a temple? Was it an instrument for astronomical calculations? Was it both? The solutions evolved from computer analysis, not from archaeology.

If the knowledge was there, the question remains, Where are the books? Was there a vast treasure of scientific and engineering truths that the learned passed on from generation to generation? The artifacts attest to the fact that there was such a body of knowledge for each of the major historical cultures. Could it have been oral? Possibly, but it is doubtful. You may have heard the conjecture that Homer's *Iliad* was originally passed down as an oral tradition. To some extent this could have been true of scientific inquiry except for one problem: exactitude. If you forget a detail in a tall tale, who cares if you alter the yarn just a bit? If you forget your lyrics you can always scat sing for a line or two. But among engineers, then and now, there is little room for error lest the castles tumble

** Such engineering skill is testament enough to Roman technology, but there are surviving engineering documents, such as the works of the engineer and architect Vitruvius. His ten books date from the first century B. C.*

down. The legendary work on the chalkboard of any math professor's classroom tells the story. Writing is critical to the success of scientific pursuits because writing is critical to the success of precision.

The idea of performing mental math or spoken math is alarming. It is hard enough to get through college calculus, even with every mechanical convenience available. Doing even a little banking without a pen or pencil is hardly practical, and it is sure to be costly.

The early cultures in the area of Turkey and Iran worked long and hard on written math systems. The perfect symmetry of the base-l0 system that we use—one in which all numbers are based on 0 and units from 1 to 9—is one that we might think was an easy discovery or one that is natural or true of the world. We are so conditioned to the base-l0 decimal system that we assume it to be true and perfect. Certainly it is wonderfully symmetrical, but our computers are based on binary numbering, an even simpler system. In early times math stumbled through still other systems, usually base -12, - 14, - 16, or -60 systems.

Like other languages, math can speak only if its symbols and logic structure are understood. If we are presented the simple math

$$11 + 11 = 26$$

we cannot understand the symbols and the logic because we speak and think base-10 math. The displayed calculation is based on the earlier base-12 system. We perceive the unit 11 to be a unit of 10 plus a unit of 1. In fact, it is a unit of 12 plus a unit of 1. The early math systems may not have been triumphs of logic but they doubtless proved practical. Of course, here again, the evidence is scant. Much of the paperwork is lost. But the next time you pick up a ruler, count the inches, and count the hours from midnight to noon and from noon to midnight. The next time you buy a pound of coffee, count the ounces. And there was a time when two weeks was a fortnight, a unit of 14. Count the minutes in an hour, or the seconds in a minute. How the ancient mathematicians arrived at these systems may never be clear. The base-10 system could have been a mere derivation of finger counting, physiological and dependable. In the Lascaux Cave and other sites in the south of France, the well-known cave paintings are of animals—and hands.

We know that all the great historical cultures had writing systems (except the Inca). We can presume that the measure of their success in science and engineering and mathematics depended, in measure, on their willingness and ability to write. The surviving tablets usually indicate that commercial math was commonplace, and schoolbooks were probably as common as religious texts. Learning depended not just on writing but on learning *from* writing. The much-heralded Dead Sea Scrolls were documents of religious significance. The great tablet archives such as those of Ebla in Syria (discovered in 1964) reveal the more practical side of ancient days among the Sumerians and Etruscans and other early peoples: accounting math, multilingual dictionaries, schoolbooks, and so on. Tablets, by the way, were not the 50-pound slabs of etched marble we see in Hollywood movies. They were handheld like a pocket calculator and shaped to fit the palm, usually in the size and shape of a large block of shredded

Primò, in Triang. obliq. f c b, ex datis tribus lateribus f b, f c, & c b, Distantiis nimirùm observatis, quæritur angulus f b c. Deinde in Triang. obliquang. a b c, ex datis itidem tribus lateribus, b c Dist. Arcturi à Lucid. Lyræ, & a b, atque a c complementis harum Stellarum Latitudinariis, quæritur angulus a b c: modo, quem in superioribus fusiorem dedimus. Qui angulus a b c subtractus ex angulo f b c, relinquit angulum f b i. Tum Logarithmorum f b & f b i Summa, erit Logarithmus Perpendiculi f i. Et hujus Perpendiculi Antil. subductus ex Antilogarithmo Hypothenusæ f b, relinquit Antilog. lateris b i; quod demtum ex complemento Latitud. Arcturi b a, manifestabit latus a i. Ergo in Triangulo rectang. f i a, summa Antilogarithmorum crurum a i & f i, est Antilog. Hypothenusæ f a, id est, complementi Latitudinis quæsitæ. Et complem. Latit f a Logarithmus subtractus à Logarithmo f i, relinquit Logarithmum anguli f a i, qui est Differentia Longitudinum Stellæ & Arcturi e g. Ergo non ignorabitur ipsa Longitudo Stellæ quæsitæ ♎ e. Exempl. gr.

Sit quærenda Longit. & Latit. Stellæ ante IV flexuram prioris Draconis, ex Distantiis.

```
                                   Gr.  M.  S.
     Distantia Arcturi & *         47   40  40   b f
     Distantia Lucida Lyra & *     30   32   0   f c
     Distant. Arcturi & Luc, Lyra  58   52  22   b c

               Gr. M. S.                Gr. M. S.                 Gr. M. S.
Long. Arcturi  19  29  6 ♎  Latitud. B. 31  1  0  b g   Compl.  58  59  0  b a
Long. Lyra     10  33 17 ♑  Latitud. B. 61 47 17  c d   Compl.  28  12 43  c a

              Gr. M. S.                                 Gr. M. S.
       b f    47  40 40  Log.   30197            a b  58  59  0  Log.   15430
       b c    58  52 22  Log.   15545 Add.       b c  58  52 22  Log.   15545 Add.
Differ.       11  11 42  Log.   45742 Differ.          0   6 38         30975
       f c    30  32  0                          a c  28  12 43
Summa         41  43 42                    Summa      28  19 21
Semissis      20  51 51  Log.  103238      Semissis   14   9 40  Log.  140794
Differ.       19  20 18                    Differ.    28   6  5
Semissis       9  40  9  Log.  178403 Add. Semissis   14   3  2  Log.  141561 Add.
                               281641                                  282355
                                45742 Sub.                              30975 Sub.
                        Resid. 235899                          Resid.  251380
              17  54 17 Semiss. 117949              16  31 52  Semiss. 125690
   f b c  35  48  34                         a b c  33   3 44
                                             f b c  35  48 34
                                             f b i   2  44 50

     Lat.    f b   Log.  30197                          Antilog. 39555
     Angul.  f b i Log. 303803          Gr. M. S.
     Perp.   f i   Log. 334000           2  1 51        Antilog.    63 Add.
                                  b i 47 38 42          Antilog. 39618
                                  b a 58 59  0
                                  a i 11 20 18          Antilog.  1971
                                  f i  2  1 51          Antilog.    63 Add.
                                  f a 11 31  0          Antilog.  2034
                                      89 60  0
                Latitud. quæsita. f e 78 29  0

     f a   Log. 161117 S.
     f i   Log. 334000                   Gr. M. S.
   f a i   Log. 172883                   19 13 25
                         Longit. Arct.   19 29  6 ♎
                                          9 15 41 ♎ Longitudo * quæsita.
```

wheat cereal. The clay was moistened, then written on with a reed stylus of some type and allowed to dry in the desert heat—probably a matter of minutes during most of the year.

Throughout history writing has been rather perishable. Early writing was probably lost in a matter of decades and not millennia. Tablets were readily erased with water if there was a need to use them again, and papyrus or animal skins or other organic products obviously could not last. Much was lost also due to the plundering and burning that accompanied conquests of one culture over another. Fortunately, some of the best available "documents" that lucidly illustrate the methods of early engineering in every imaginable arena of domestic and royal life are the Egyptian tomb paintings. When the tombs were robbed, the walls of the empty vaults were left to speak of life in the times of the pharaohs. It is significant that this extraordinary resource of early scientific and engineering achievements is almost entirely graphic. Endless walls of paintings depict every facet of the daily life of the times.

Perishable though books are, they allow us to keep precise records of what we know. Because writing is a fairly permanent medium for our language, we can perfect it. Math is simply another language to be rendered in writing. We assign symbols to the math and proceed to write mathematical statements. Our visual world is no different. If we draw the shape of a bison on a cave wall between 10,000 and 15,000 years ago, the bison has been "written." We use the word *draw* when we mean to transfer the representation of life to paper or canvas. The painting is no more real than the word *bison*. Both are representational or symbolic. Both are "written."

It is true that we do not have a great trove of documents of early science and engineering in the form of "books." But scientific writing is not and probably never was dependent on language in the way we are conditioned to think about writing. Writing our spoken tongue is only one of the languages available to us. The mathematician hardly needs it. The engineer *must* draw. Engineers have a kinship with artists that is seldom recognized. Illustration and modeling and design are critical expressions of knowledge. Somehow, only architects seem to be granted some esteem for their "artistic" ability. Alas, the Mona Lisa is known by everyone, while da Vinci's engineering drawings are seldom seen by anyone. We only hear about his helicopters and submarines. It is not a coincidence that the great painter was also a major engineer of his time. Fortunately, a treasure of his scientific illustrations survives.

There is, after all *some* written evidence, and there are some mathematical remnants for us to work with. However, we must not confuse writing of the ordinary kind with the writings of the scientist and engineer, because the written language is only one testament of knowledge. Math is another, visualizations yet another, and the final construct yet another. With the precision of our language, our math, and our art, engineers craft the legacy of our cultures. The monuments seem to stand silent, but they speak.

The Prodromus Astronomiae *was published in 1690, by one of our lesser known astronomers. Johannes Hevelius (1611-1687). The calculations on this page of text were used for determining the latitudes and longtitudes of stars.*

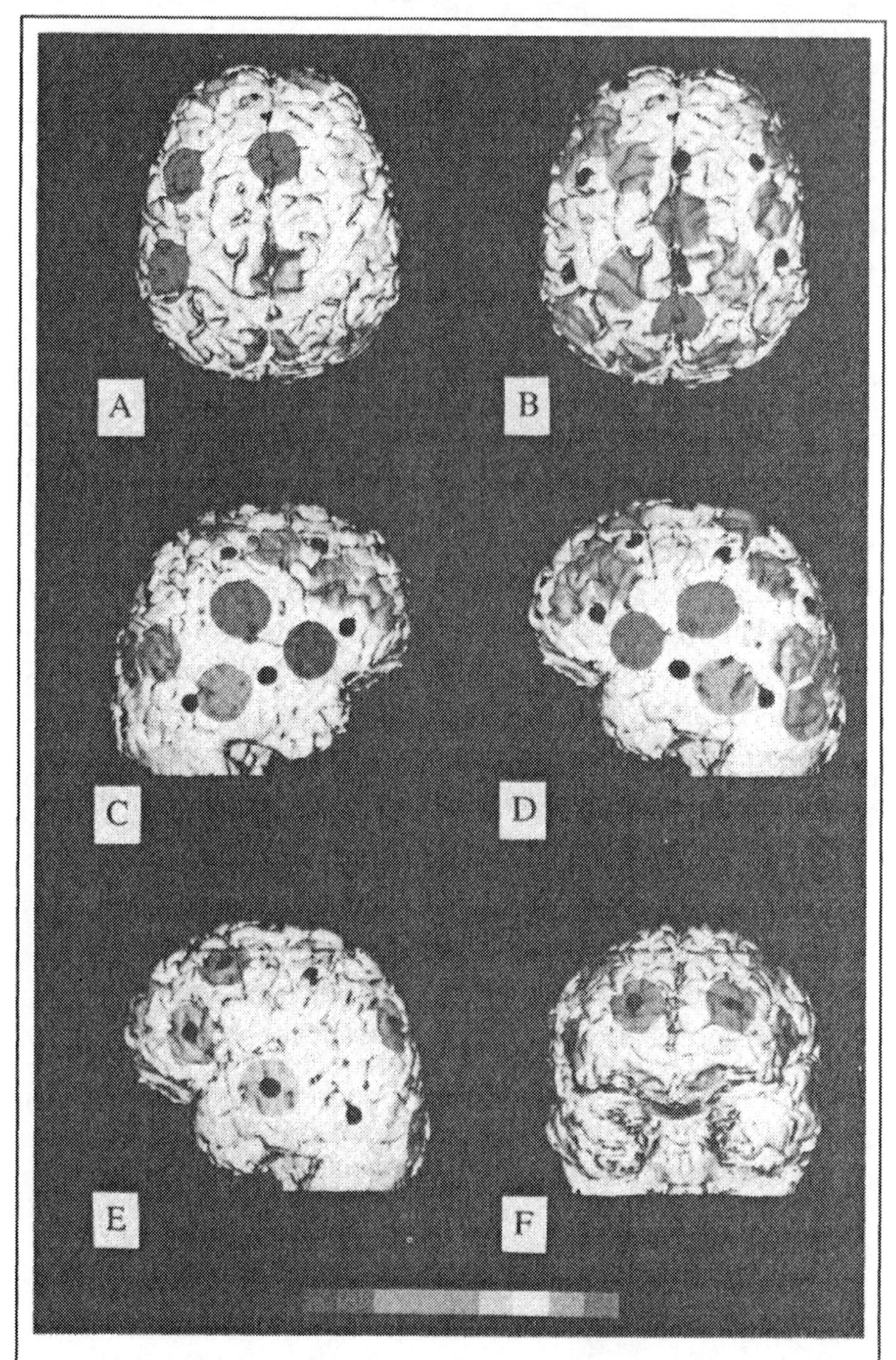

Figure 4. The average changes of amplitude (A), local and interhemispheric coherence (B, C, D and E, F respectively) in a 26 year old healthy male while doing mental arithmetic for 1 min, with respect to EEG at rest (eyes closed). The color-coded changes of amplitude are entered at the respective positions of the electrodes, the changes of coherence between the electrodes. Color code: amplitude between -5 μV (blue) and +5 μV (red); coherence between -0.25 (blue) and +0.25 (red).

The Languages: The Symbolic Structures of Science

Technical writing is not strickly a matter of writing. Writing plays a unique role when a document is perhaps 50% graphic and perhaps another 10% or more of the text is developed in some other symbolic system, typically chemistry or physics or math. Only about 40% of such a text is the word count. The true *supporting* or *adjunct* role of writing in technical documentation is often ignored.

The uniqueness of technical writing is hard to overlook: in no other type of writing does writing assume *less* of a role! The actual percentage of a technical document in written English can vary enormously. An excursion down the aisles of any periodical room in a college library will tell the tale. The more scientific the magazines, the more likely they are to look like the walls of ancient tombs. Many corporations recognize the role of graphics and graphic design in meeting their publication needs. Long before the age of *Photoshop,* an industry pacesetter in the Northwest lumber products industry put a graphics designer (a professional illustrator) in charge of the technical documentation department. He, of course, specialized in illustration. You will notice that several very famous moments in science are among the samples selected for this chapter. Observe the use of *language* and *math* and *illustration* in these old documents. In particular, notice the need for a *balanced* use of all three systems of communication.

Technical documents readily use symbolic systems other than the language system of English. Many technical and scientific specialties rely heavily on computer languages such as Visual Basic, Java, Pearl and C++. Other specialties rely on the formulas and principles of chemistry, or the laws of physics and the methods for resolving the issues of such laws, or the many areas of math from calculus to number theory to topology to differential geometry. These systems *are* languages. Mathematicians do not *perform* math; they *speak* and *write* math. They understand their world of mathematical principles in exactly the same fashion as philosophers understand philosophical principles expressed in the conventional language system.

Computer-generated modeling is a vivid new graphic tool in science and engineering. Here, researchers study the activities of the brain in models. (Courtesy of Reinhard Thaller, et al. An Approach to a Synopsis of EEG Parameters . . . . Brain Topography Journal; *Volume 4, No.1, 1991).*

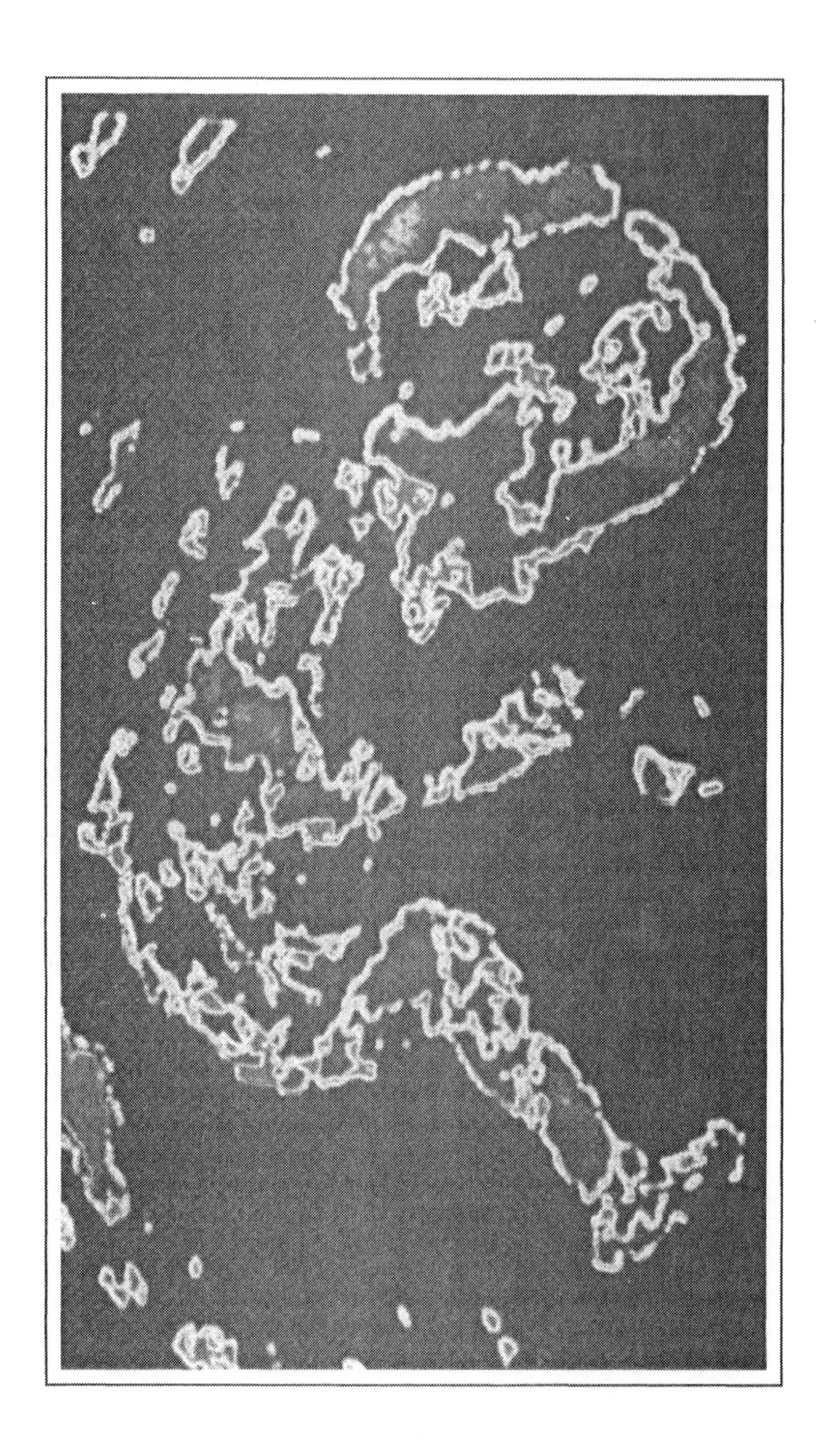

At the other extreme are visual systems that are important to scientific and engineering documentation. These graphic elements may appear to be the least symbolic of the scientific languages, at least in the case of photographs, but they can be very symbolic. What we choose to call a "graphic" in a text involves a host of possibilities from photographs to schematics. If we choose to use a photograph or a realistic drawing or an assembly drawing we are being as realistic as we can be in representing one reality or another, but the medium remains highly symbolic.

More abstract are illustration systems such as charts and graphs and schematics. These devices involve systems rather than likenesses and are more dependent on a knowledge of the logic used to design them. Everyone is likely to comprehend a photograph, but the audience begins to diminish as the graphic content becomes more symbolic and dependent on one or another system of logic. At a highly abstract level, graphics evolve into such symbologies as you will find in electronic schematics. Unless you can "read" schematics you cannot understand them. This means that a graphic reaches a codified level at which the symbols must be known in order to understand the illustration. Such a system behaves suspiciously like a language, does it not? Schematics are read top down and left to right, exactly in the fashion of Indo-European languages. What happens between supply and ground is patterned and regularized as the "language" of electronics.

What technical "writing" really involves is *systems integration*. Tech writing is a mixture—a fusion—of symbolic systems (and we will look very closely at the issue of multisystem formatting in this text). The word count in a technical project may be quite small, but the writing is critical. When we weave together the basic text and add the mathematics, the charts, the graphics, the appendices, and so on, we can make a terrific muddle of the project if we do not have the skill to *compose* the document—the skill to engineer the composite.

For example, I recently received a twenty-five-page project that was composed as a hard-copy user's guide to *Quickbooks Pro,* a dedicated software application. The composition of the project was 90% screen captures and 10% writing. The writing assumed enormous importance precisely because there was so little of it, and the writing gave continuity to an otherwise choppy project. The 10% made the project a huge success, and the mini-manual is being used in several computer labs.

Composing or engineering will be a big part of our discussions of how to assemble technical documents. First, composing the organization is important because technical subjects are inherently complicated. We might suppose that most tech writing is therefore inherently complex. This is not quite true, but it is true enough to any writer who is trying to render a complex subject on paper.

Island in a silent sea, an ultrasound image of a fetus in utero demonstrates one of many new ways to illustrate our world. Other imaging technologies in medicine include CAT scans, PET scans, and MRI scans. (Courtesy of Elaine Marieb, **Human Anatomy and Physiology,** *Addison-Wesley.)*

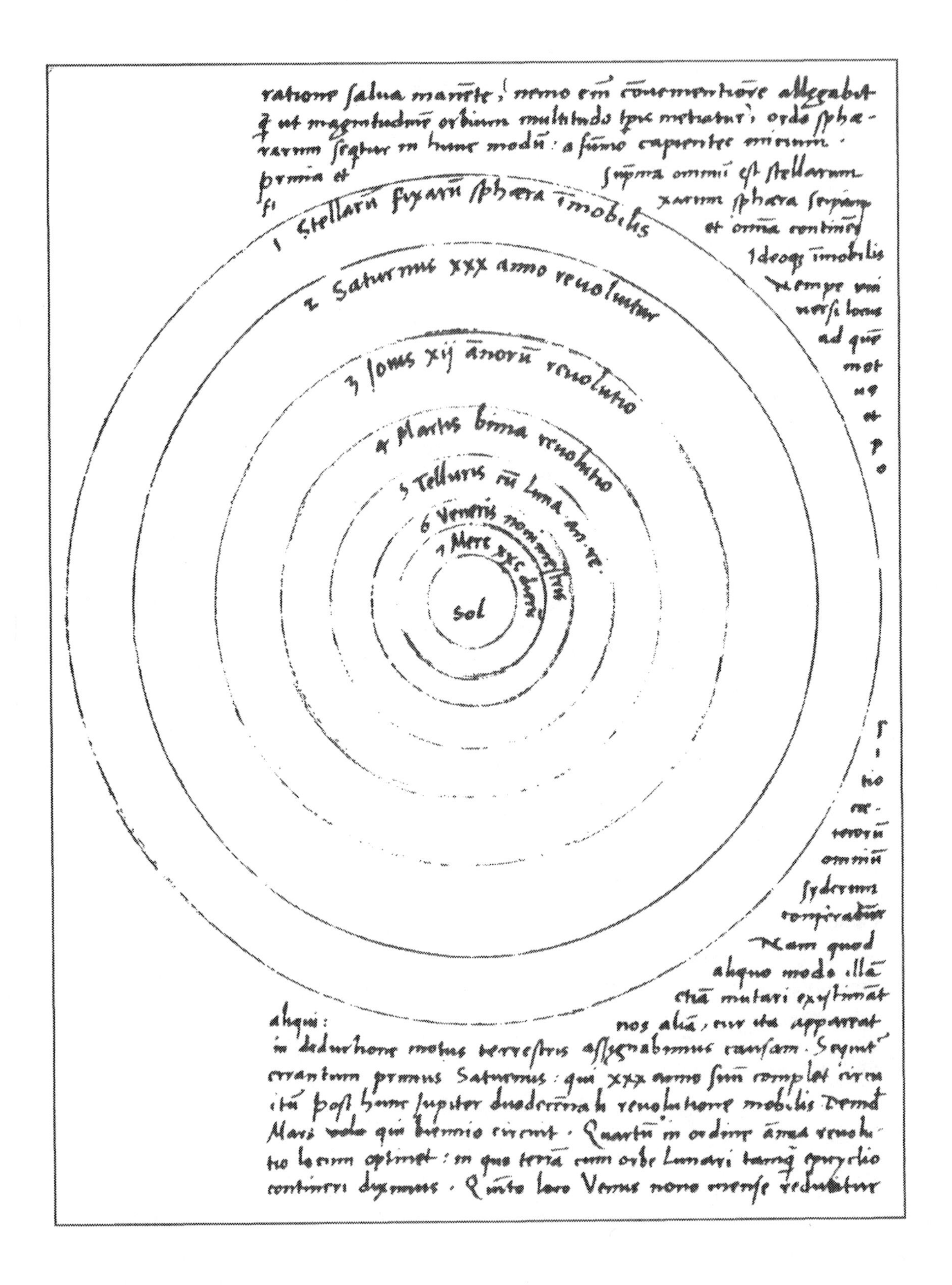
ratione salua manente, nemo enim convenientiorem allegabit
quam ut magnitudine orbium multitudo temporis metiatur, ordo sphae-
rarum sequitur in hunc modum, a summo capientes initium.
prima et suprema omnium est stellarum fi-
xarum sphaera seipsam
et omnia continens
ideoque immobilis
nempe uni-
versi locus
ad quem
mot-
us
et
po-
si-
tio
ce-
terorum
omnium
syderum
conferatur.
Nam quod
aliquo modo illam
etiam mutari existimant
aliqui: nos aliam, cur ita appareat
in deductione motus terrestris assignabimus causam. Sequitur
errantium primus Saturnus, qui xxx anno suum complet circui-
tum. Post hunc Jupiter duodecennali revolutione mobilis. Deinde
Mars, qui biennio circuit. Quartum in ordine annua revolu-
tio locum obtinet: in quo terram cum orbe Lunari tanquam epicyclio
contineri diximus. Quinto loco Venus nono mense reducitur

Second, the task of technical writing is complicated by the "multilingual" quality that emerges when a text also involves, for example, mathematics and schematics. The challenge of balancing them properly is a craft, but it is not difficult to master. In truth, of course, it is the English that often holds it all together, and so the systems integration critically depends on the function of writing.

On the Revolutions *is here seen in the original manuscript version by Nicholas Copernicus (1473–1543). A true revolution in science began with this illustrated page from the famous text.*

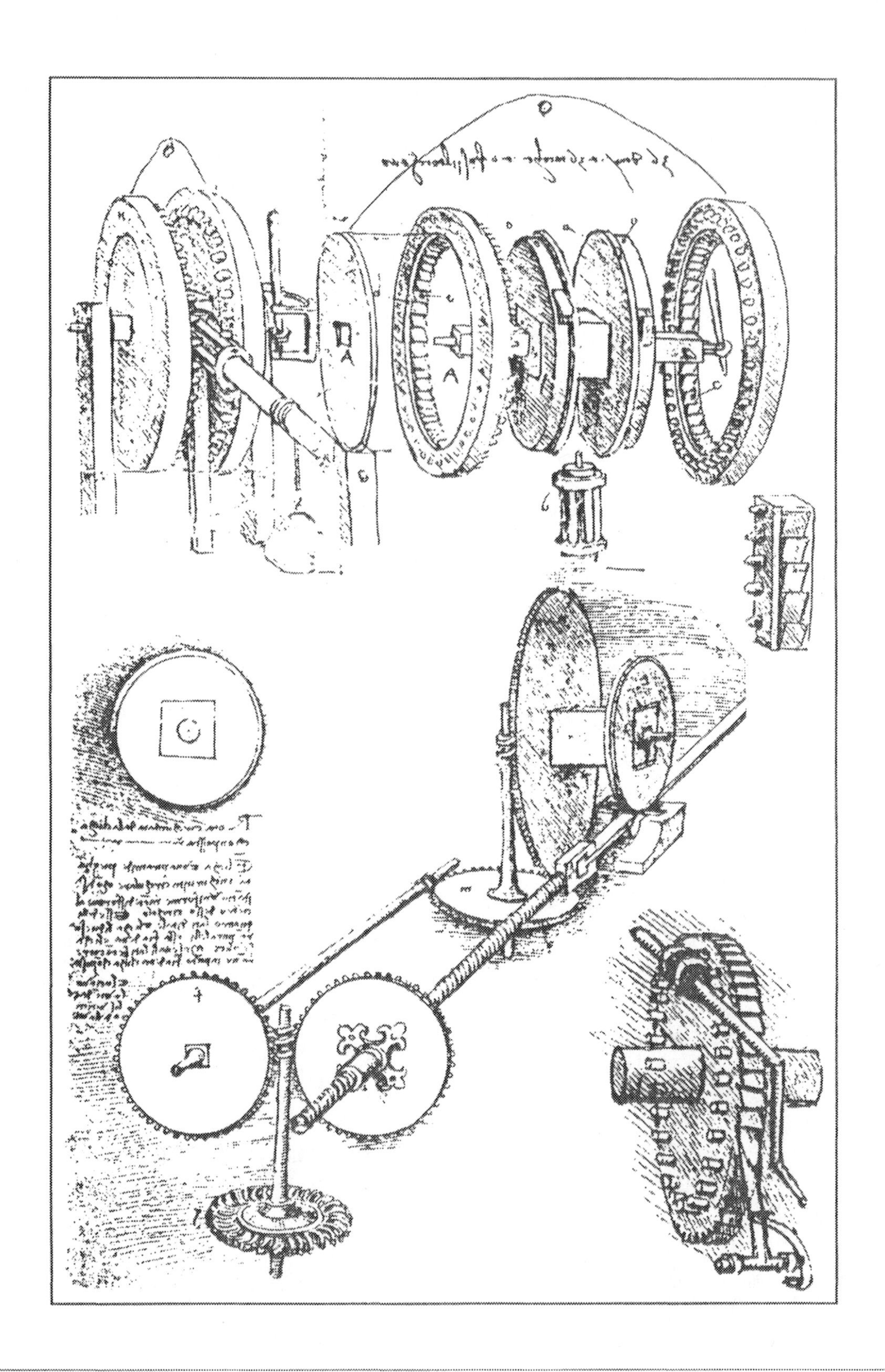

The Illustrations: Multimedia Presentations

Technical writing is a mixture of languages, yet the presence of all the additional symbolic systems does more than serve the purpose of being precise. We use these systems to do more than construct an understanding of our world. An engineer is more concerned with how we intend to use this knowledge. We use the systems to construct, period. The heavy leaning toward graphics often reflects visualizations of ideas because engineers construct applications of their ideas. In other words, blueprints, engineering drawings, and designwork of all types create a bridge between theory and reality.

A lot of technical writing is modeling. In practice, a great many technical documents become multimedia presentations as a result. The documents, and certainly the illustrations, construct the application of a model, which is the necessary way we engineer the creation of a final product, say a Titan rocket. There is no room for error, and computer simulation and laboratory modeling are similar to the math systems and the graphic systems we see in tech writing. All of them are models.

The historical necessity for both the multilingual document and the multimedia document is apparent in the samples we see in this chapter. Until very recently only a publishing company could do justice to sophisticated and complex technical documents. Only a select few had or needed the skills to control the multimedia composition involved in tech writing. Even today it is a rare document that is sent to a publisher with a fully integrated text with all the graphics in position. Instead, one package is the manuscript or disks of the English language text, and another package contains the graphics, the appendices, and so on.

Now that there is computer support for desktop publishing, we have a better opportunity to master the basics of technical writing. In other words, the layout and designwork skills, the formatting and editing techniques, and the proper application of practices and procedures are coming under our control. We now have all the software support to create camera-ready finished documents. The task of the writer is the engineering or construction of documents themselves.

Gear drives by Leonardo da Vinci (1452–1519). These drawings are a composite from his notebooks, where you will find the famous drawings for flying machines and the first bicycle ever seen on anyone's drawing boards.

De Motu
rporum

PROPOSITIO XVI. THEOREMA XIII.

Si medii denſitas in locis ſingulis ſit reciproce ut diſtantia locorum a centro immobili, ſitque vis centripeta reciproce ut dignitas quælibet ejuſdem diſtantiæ: dico quod corpus gyrari poteſt in ſpirali quæ radios omnes a centro illo ductos interſecat in angulo dato.

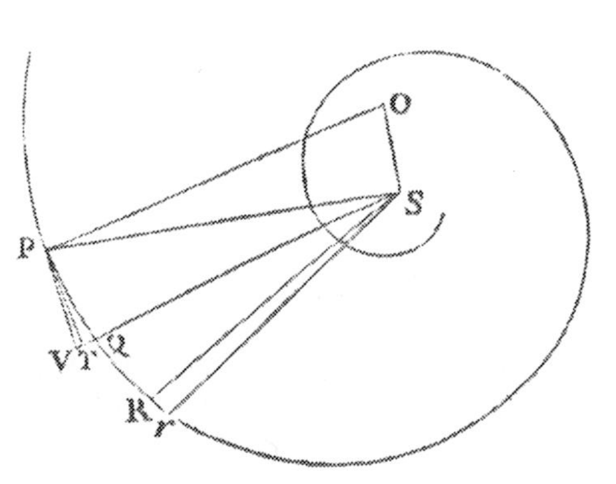

Demonſtratur eadem methodo cum propoſitione ſuperiore. Nam ſi vis centripeta in P ſit reciproce ut diſtantiæ SP, dignitas quælibet SP^{n+1} cujus index eſt $n+1$: colligetur ut ſupra, quod tempus, quo corpus deſcribit arcum quemvis PQ, erit ut $PQ \times PS^{\frac{1}{2}n}$; & reſiſtentia in P ut $\frac{Rr}{PQq \times SP^n}$, ſive ut $\frac{\overline{1-\frac{1}{2}n} \times VQ}{PQ \times SP^n \times SQ}$, ideoque ut $\frac{\overline{1-\frac{1}{2}n} \times OS}{OP \times SP^{n+1}}$, hoc eſt, ob datum $\frac{\overline{1-\frac{1}{2}n} \times OS}{OP}$, reciproce ut SP^{n+1}. Et propterea, cum velocitas ſit reciproce ut $SP^{\frac{1}{2}n}$, denſitas in P erit reciproce ut SP.

Corol. 1. Reſiſtentia eſt ad vim centripetam ut $\overline{1-\frac{1}{2}n} \times OS$ ad OP.

Corol. 2. Si vis centripeta ſit reciproce ut SP *cub.* erit $1-\frac{1}{2}n=0$; ideoque reſiſtentia & denſitas medii nulla erit, ut in propoſitione nona libri primi.

Corol. 3. Si vis centripeta ſit reciproce ut dignitas aliqua radii SP cujus index eſt major numero 3, reſiſtentia affirmativa in negativam mutabitur.

Scholium.

Cæterum hæc propoſitio & ſuperiores, quæ ad media inæqualiter denſa ſpectant, intelligendæ ſunt de motu corporum adeo parvorum, ut

The task involves logical designwork, composing, systems integration, and utilization of media. Software simply puts us in a position to do a sensational job of it. What is obvious, however, is that technical writing is not a matter of "writing." You cannot miss this basic fact if you look at the illustration of the three-hundred-year-old original edition of Newton's *Principia*. The software system, whether it is *PageMaker* or a *Word* equation editor, handle the *multilingual* and *multimedia* tasks of technical writing in addition to the "writing."

There is substantial transition work in writing a technical composition because of the multiple logic structures in the text. Transitions are important in any kind of writing, but the intersystem transitions of technical writing are unique. These are *display transitions*. A writer must always move readers along, but if the author uses mathematics and schematics, the author has the important task of moving readers in an out of logic systems other than the spoken text. These leaps into and out of, let's say, English, and Boolean algebra and a schematic of digital gates, for example, can be a challenge to write and a challenge to read. We have little choice but to master the task, since the various logic systems are of equal importance: the language, the math, and the graphics.

As we will see, graphics elements have always been fundamental features in scientific writing. The visual evidence helps demonstrate the intentions of the document. We could look at a number of historical moments to see how investigators came to rely on drawings, for example, rather than samples of the actual materials under scientific investigation. The graphics are convenient shortcuts that visualize the studies of the scientist and the undertakings of the engineer.

Graphics are popularly understood to be visual demonstrations. The illustrations represent evidence. Today, for example, the use of a photograph as a specimen is standard procedure. It is *not* the specimen, but we popularly accept the substitution process. The evidence, by proxy, is certainly adequate for reporting research in scientific papers. The illustration is usually a variation of the modeling concept. The biochemist will want to see and study the *model* of the cell that was photographed with a microscope; the engineer is likely to want to construct the model itself and show us the illustration as an embodiment of his or her ideas.

Philosophiae Naturalis Principia Mathematica *is one of the most influential scientific documents in the history of science. Isaac Newton (1642–1727) defined the laws of gravity, which redefined the laws of our universe.*

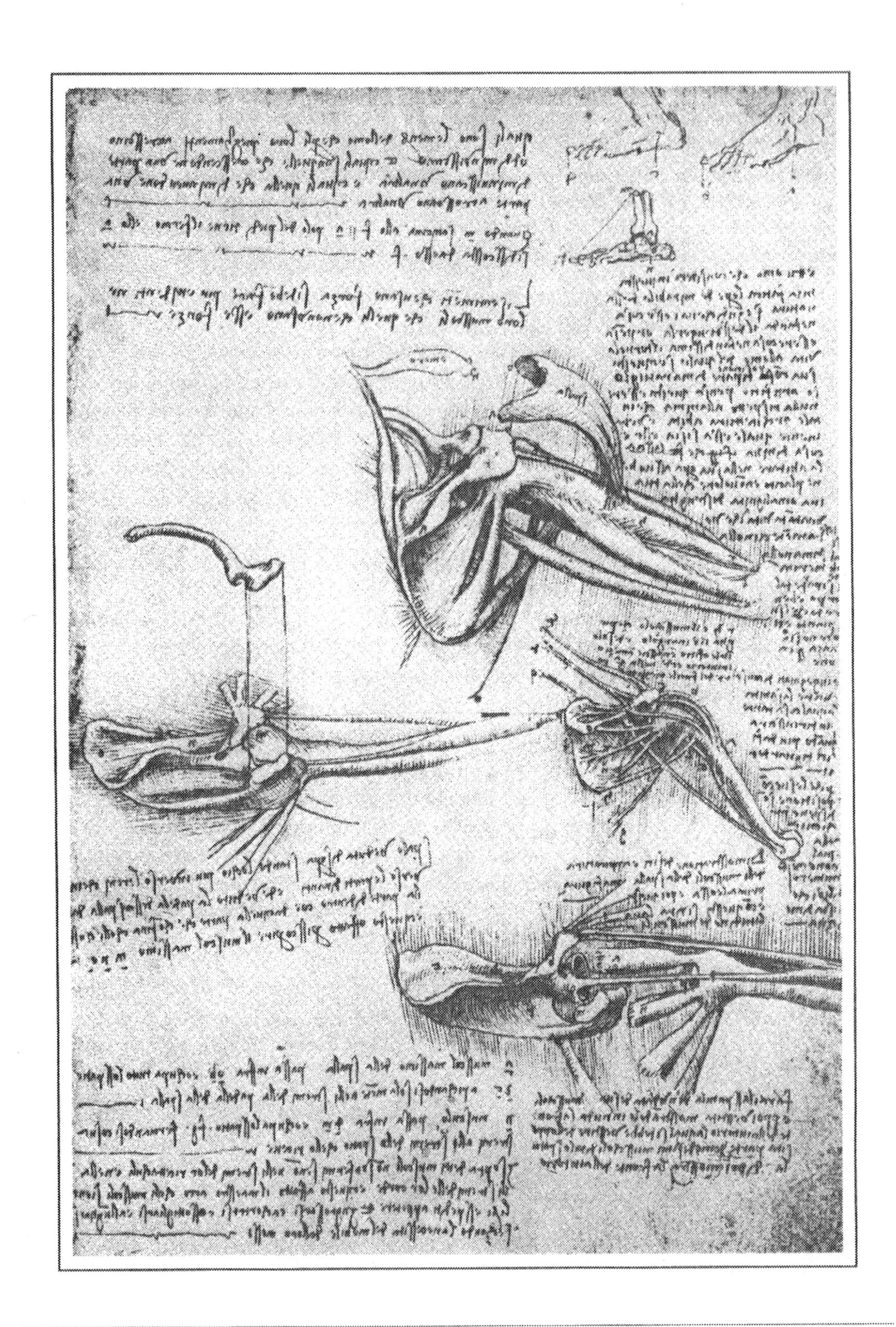

The Evidence: Empirical Data

In the huge variety of technical documents—whether they are field reports, R&D reports, maintenance reports, procedures, proposals, environmental impact statements, or any of dozens of other—there is a common thread. It is always apparent that the documents designed by technicians, engineers, and scientists are very evidence oriented. They are data based. People often say that technical writing is "dry." It is bone dry if we don't see the music in it, but the music is in the eye of the beholder. In a sense technical writing is *transparent;* that is why it is dry. It lacks color because color may confuse the truth of the evidence we present. In our world of scientific understanding, engineers and scientists pursue truths external to us. The poet is just as likely to concern himself or herself with what is internal to us.

This difference in perceptions is quite significant and very much influences the methods of technical writing. Do not think, however, that it is simply a matter of style. Evidence is tantamount to truth for the scientific disciplines. It is interesting to note that "truths"—mathematical truths, for example—suggest a larger issue. The larger question concerns knowledge itself and whether truths of any kind are true of the world or whether they are only our internal fictions that help us maintain our sanity by creating order out of chaos. Two historical figures are well known for their thoughts concerning this issue: René Descartes (1596–1650) and David Hume (1711–1776). Descartes essentially argued that what is true is in the eye of the beholder; truth is in us. This idea makes for great poetry, but it also makes life difficult for the scientist.

Descartes' concept was called *idealism,* a term unrelated to "high ideals," the meaning we assign to the term today. The world is our *idea* of it, he explained. We do not really know the world; we only perceive it. This matters because the problem of our limited understanding is real enough even if it is an impractical world view for scientific inquiry. We really do not see much more than what our humble resources allow us. A typical satellite infrared photograph reveals a great deal about what we see and cannot see. What we see is such a small aspect of our universe that it is indeed a mere idea of the world. Our vision, never mind our intelligence, is the merest wink on the electromagnetic spectrum, as we well know.

Shoulder dissections by Leonardo da Vinci. Over 1200 of his anatomical drawings are known to exist. Da Vinci was an artist and not an anatomist, but he was a master in both disciplines.

512 Le vingt-vniéme Liure,

Figure du Camphur, animal amphibie.

B

Or il y a pluſieurs autres animaux marins qui n'ont qu'vne ſeule corne, & beaucoup d'autres animaux terreſtres : car on a veu des cheuaux, chévres, & daims, pareillement des taureaux, vaches & aſnes, auoir vneieu le corne. Parquoy Monoceros ou Vnicorne, eſt vn nom qui conuient à tout animal qui n'a qu'vne ſeule corne. Or conſiderant la varieté des Eſcriuains, & des cornes qui ſont toutes differentes les vnes des autres, l'ont peut croire veritablement qu'elles ſont de diuerſes beſtes engendrées en la mer, & en diuerſes contrées de la terre. Et pour la renommée des vertus qu'on attribuë à la Licorne, chacune nation ſe plaiſt a luy donner le nom de Licorne. C

Chapitre LI.

Idatz Aga, Orateur de Soliman. philoſtrate, liu. 3. chap. 1.

DATZ AGA, Orateur de Soliman, atteſte auoir veu en l'Arrabie deſerte, des Licornes courantes ça & là à grands troupeaux Quant à moy, ie croy que c'eſtoient pluſtoſt des Daims, ou Chévres de ce païs là, & non des Licornes. Philoſtrate en la vie d'Apollonius Tyaneus, chap. 1. liu. 3. dit, qu'aux mareſts voiſins du fleuue Phaſis ſe trouuent des Aſnes ſauuages, portant vne corne au front, auec laquelle ils combattent furieuſement comme taureaux: de laquelle corne les Indiens font des taſſes qui guarentiſſent l'homme de toutes ſortes de maladies le iour qu'il y a beu, & s'il eſt bleſsé ce iour là, il ne ſent aucune douleur. Dauantage, il peut paſſer par le trauers d'vn feu ſans ſe bruſler nullement. Meſme il n'y a venin ny poiſon beu, ou autrement pris, qui luy puiſſe nuire : & que pour cette cauſe il n'y a que les Rois qui boiuent dans leſdites taſſes : de faict que la chaſſe deſdits Aſnes n'eſt permiſe qu'aux Roys du pais ; de ſorte que l'on dit, qu'Appollonius Philoſophe graue, regarda curieuſement ceſte beſte ſauuage ; & auec grande admiration conſidera ſa nature. Quoy voyant Damis, luy demanda s'il croyoit ce qu'on diſoit de la vertu deſdites taſſes : Ie le croiray, dit il, quand i'entenderay que le Roy de ce pais ſera immortel. Reſponſe que ie delibere d'oreſnauant faire à tous ceux qui me demanderont, ſi ie croy ce que l'on dit des vertus de la corne de Licorne. D

Apollonius. Tyaneus.

Reſponce fort ſubtile.

Descartes' ideas were not to go unchallenged, of course. The age of science was beginning, and the business of learning the truths of our universe was underway. It was not a scientist, however, but another philosopher who challenged the Descartian (called *Cartesian*) reasoning. The Englishman David Hume argued that truth *is* in the world and we can know it. Let's not talk nonsense, he argued. If you pick up two of Newton's apples and add two more you will have four. Furthermore, according to Hume, you will *always* have four if you have two and add two. The truth is absolute and we can know it by looking at the concrete facts (data or evidence). Like Hume, scientists today give much importance to evidence. We accept the evidence. The data are everything. This is an accepted basis for our reasoning in science. We begin with evidence, not with ghosts or devils. We begin with data, not with UFOs or Sasquatch. Scientists want the facts—and the numbers. This belief that there *are* realities and truths in the world is called *empiricism*. Hence the term *empirical data*.

The influence of this brand of reasoning has been profound for several hundred years. We have become empirical. We expect evidence; we expect demonstrations of truths. The empirical shift has been slow in some ways, but it is often dramatic. Notice the illustration of the unicorn. It comes from an early zoology book—called a Bestiary—written for Louis XIV, the Sun King (1638–1715). The textbook goes into great detail about this fabled character. Of course, the unicorn never existed, but Louis, who sent many a rake off to the dungeons, probably never doubted the knowledge of his court surgeon who assembled the book. Why? Because this text predates the empirical mentality, the prove-it spirit of our age. A hundred years later the court surgeon would have been sent to the guillotine as a charlatan. The least Louis could have done was ask about the webbed feet.

Fortunately for science, within a hundred years, legends were displaced with evidence, along with the obligation to demonstrate truth in a credible and scientific way. The most adventurous aspect of this development in scientific strategy can be found among the log books of the circumnavigators and the scientists who later traveled with them. These daring figures serve as one example of the growing significance of evidence (and graphic substitutes for evidence) in demonstrating scientific thinking.

After Columbus made his first trip to the New World, for almost three hundred years the explorations were conducted without any particularly strong scientific purpose. Great strides were made in cartography, navigation, geography, and related technical disciplines, but these scientific pursuits served the purposes of the explorers more than the explorers served the purposes of science.

An illustration of the fabled unicorn. The text, from the late seventeenth century, assumes the animal's existence and explains the creature's habits at length without any question concerning its whereabouts.

Slowly the explorations took on a more scientific character. In 1768 the first of the famous voyages of Captain James Cook was conducted on the *Endeavour*, a ship that carried ninety-seven men on a vessel hardly one foot longer from stem to stern. Given the tight quarters, it is a reflection of the growing importance of research-oriented exploration that a scientific team of five men was assigned to the ship. The British Royal Admiralty had appointed the team of three scientists and two illustrators. There were equipment stores on board that included telescopes and other gear for the astronomer, the botanist, and the naturalist. The two artists supported the mission with their scientific illustrations. One of the illustrators, Sydney Parkinson, completed over fifteen hundred drawings and paintings during the voyages. His subjects ranged from Maori tribesmen to penguins—from anthropology to zoology. Obviously, the recovery of specimens was greatly limited by the preciously little space of the ships of the day. Written explanations and illustrations often had to function as the empirical data by proxy. The purpose of scientific illustration gained new vigor.

Johannes Gutenberg's invention of movable type preceded the age of global exploration and the emergence of modern science. The forces of change converged. Colorful and vividly illustrated books fulfilled people's curiosity and allowed everyone to explore with the explorers. Books could bring evidence to life in language, in pictures, and in the formulations of math and the sciences. As one example among many, Charles Darwin's five-year voyage on the HMS *Beagle* began in 1831. He published a popular history of his voyage, and the text was illustrated.

In the tech writing business the historical development of different concepts of evidence has shaped what we write and the way we write. The visuals or graphics we import into our documents are used to provide evidence, to quantify ideas. The numerical calculations we develop in the many math systems we use provide more evidence. And we multiply our evidence if we gather ten thousand statistical instances of a bit of data or model a million simulations on a computer. We amplify evidence if we use a field telescope or an electro micrograph. The evidence is everything, and so technical documentation has the responsibility or obligation to represent empirical data in whatever ways seem appropriate. We use any number of media to create or render the data. The evidence is the way we try to universalize the truth of a presumed bit of knowledge we think we see. We use statistical probability just in case Descartes was right. If we can demonstrate our truth a thousand times in the lab, it greatly diminishes our lurking concern that Descartes is having the last laugh and is whispering to us, "It's all in your head."

Banksia serrata *(red honeysuckle) by Sidney Parkinson. Parkinson completed fifteen hundred illustrations for the research team on the voyage of the* Endeavour *in the late 1760's.*

more surprising when we consider the size of the planet on which they are found. For Mars is only 4220 miles through, while the earth is 7919. So that

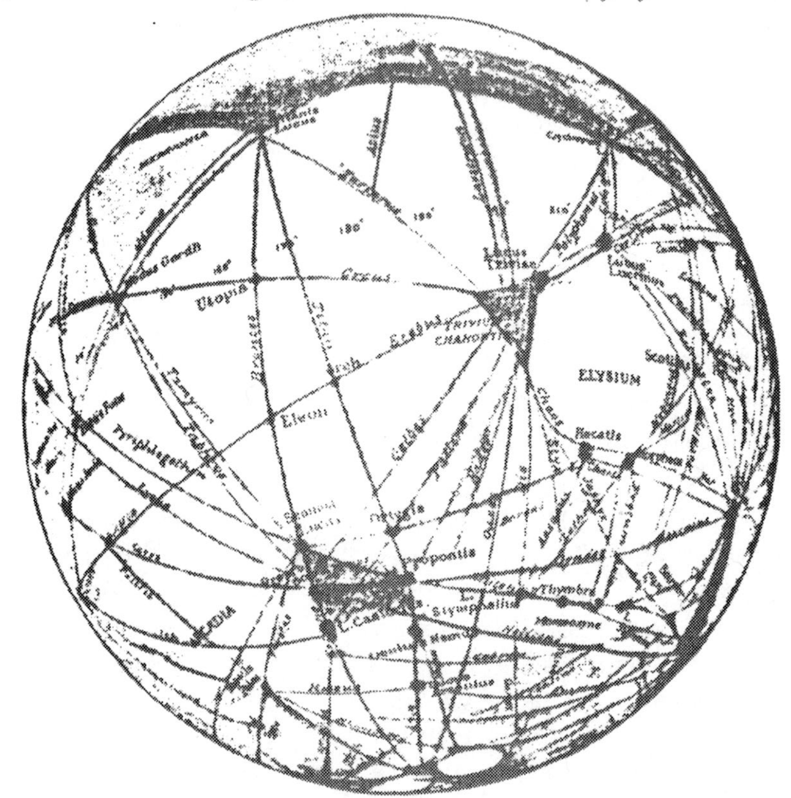

A SECTION OF THE CANAL EUMENIDES-ORCUS TERMINATING IN THE JUNCTION TRIVIUM CHARONTIS

The length of this canal is 3500 miles. The remainder of the canal may be seen on the hemisphere shown on p. 156, where it starts from Phœnix Lake (Lucus Phœnicis).

a canal 3450 miles long, for all its unswervingness to right or left, actually curves in its own plane through an arc of some 90° round the planet. It is much as if a straight line joined London to Denver, or Boston to Bering Strait.

The Writer: Objectivity as a Style

There is another aspect of the empirical issue that we must consider. Amassing data is our responsibility as we investigate or research a project we intend to write. We must also consider the writer. There is a style to technical writing—the one we bluntly called dry. For us, style is a tactical consideration, because of the distinction between objectivity and subjectivity. To be objective is to think without bias. To be subjective is to think with the bias of emotions and values. The objective observer is open to the truth perceived in all experience. The subjective observer brings perceptions to the way in which experience is understood. The objective observer experiences conditions; the subjective observer conditions experiences.

Objectivism and subjectivism go hand in hand with empiricism and idealism. If we accept the empirical data around us, we can understand it correctly only if we do not blunder into the data with misconceptions, false assumptions, or any other distortions of the evidence. Of course, by implication, we can create or at least influence what we construct—but we assume that we can *correct* our all-too-human ways. We assume we can objectify; we *can* study the "object" without bias.

Notice the drawing of the canals of Mars. The first drawing of Mars in Percival Lowell's journals dates from 1904. An Italian astronomer, Giovanni Schiaparelli (1835–1910), had mapped the canals as early as the 1880s. Even with far better telescopes (18- and 24-inch diameters), Lowell persuaded himself that he was seeing canals too. The photographs he gathered do not suggest canals to me; I do not "see" what he saw. I am more objective and more empirical, and I am under no persuasions to see anything in particular. The Mariner and Viking missions, moreover, did not reveal any canals either.

Technical writing curiously accepts both of the traditional perceptions of human knowledge that Descartes and Hume proposed. Evidence is everything. Truth is out there and it is available to us. But, alas, we must guard against distorting it, so we must keep ourselves out of the analysis or else we do run the risk of creating a world of our own ideas. As a result, tech writing has a noticeable neutral style that helps maintain the objectivity. Furthermore, it is seldom written in the first-person style.

Mars as the Abode of Life *was published in 1910. Percival Lowell (1855–1916) founded the famous observatory in Flagstaff, Arizona. His canal theory captured the imagination of the world. Though perhaps less than rigorous science, there was a charm in his speculations.*

If we write as though our writing is nothing more than a transparent lens for observing the data, we are using the tech writing style. In the end, our premise is both empirical and Cartesian. Why? Because the two truths coexist. We are operationally empirical. It is a practical way for us to be. But when Descartes argues that the truth of our ideas is internal or in our minds, we do not blindly dismiss him. The science of psychology, although it did not develop for well over two hundred years after the time of Descartes, is true to Cartesian reality. The psychoanalyst's couch is the ongoing demonstration of how real our psychological world is for us. It is real enough to control what we think we know to be true.

In our technical writing, then, we must balance this subjective force with objectivity. We must be strict in vocabulary and neutral in tone, we must seldom refer to ourselves, and we must control every bias possible. The evidence and whatever truth we can ascertain from it are the only concern.

The Researcher: Objectivity as Truth

We can look at the matter of objectivity in several exemplary cases so that you can see the extraordinary significance the objective method has as a feature of scientific reasoning. First, consider Galileo (1564–1642). Galileo gathered evidence for the heliocentric (sun-centered) concept of our solar system. Prior to his time, all Europe believed in the truth of the geocentric idea, that is, that the Earth was the center of the universe. This notion fit conveniently into religious perceptions upheld by the Roman Catholic church and was thought to be more or less consistent with the Book of Genesis insofar as humankind appears to be at the center of God's Universe. Everyone in Europe essentially assumed that geocentric astronomy was true.

Galileo was prompted to study the night sky because of the invention of the telescope. He observed that the stars traverse the sky in a fixed pattern but that the paths of the planets are eccentric and far more complicated. In time he realized also that the apparent orbits did not function in the geocentric, Earth-centered concept that everyone accepted. His *evidence* demonstrated the heliocentric theory that the sun was the center of the solar system. Rather like Lowell, he was under the influence of an earlier observer. Copernicus had already proposed the heliocentric model. Gallileo's telescope allowed him to gather the evidence—and to be the first human being to see the rings of Saturn as well.

Earlier we spoke of the issue of bias and a scientist's preference for evidence, and for the objective viewpoint. Beyond mere personal perceptions, objectivity helps us overcome not only our individual prejudices but also entire systems of logic that our culture constructs that may be in error! Objectivity, then, is a path to perception that may help us see beyond even the vast cultural belief structures in which we live. The discovery of the heliocentric solar system is a demonstration of the value of evidence.

Consider also the case of Louis Pasteur (1822–1892). Until the mid-nineteenth century the "science" of medicine was in a fairly sorry state. Disease was little understood, and popular theories of the day were wildly off the mark. Not too many years before Pasteur was born, many people, including George Washington, were more or less bled to death in the name of science. The doctors of the day were not seen as quacks, however, because everyone *believed* what they were taught or told about yellow bile and black bile and many other nonempirical ideas of the time. Pasteur inherited this grim state of affairs just as Galileo had inherited geocentric astronomy. Indeed, in a letter to his father, Pasteur's wife

290 Dialogo terzo

risoluo in tal modo. Ticone, come si vede nella nota, osseruò la stella nell'altezza Polare di gr. 55. 58. mi.pri. E l'altezza Polare del Landgrauio fù 51. 18. mi.pri. L'altezza della stella nel meridiano, presa da Ticone fu gr. 27.45. mi.pri. Il Landgrauio la trouò alta gr. 23. 3. mi.pri. Le quali altezze son queste notate quì appresso, come vedete

*Ticone Po.55.58.m.p. * 27.45.m.p.*
*Land. Po.51.18.m.p. * 23. 3.m.p.*

Fatto questo sottraggo le minori dalle maggiori, e restano — 4.40.m.p. 4.42.m.p.
queste differenze qui sotto. | Parall. 2.m.p.

*Doue la differēza dell'altezze Polari 4.40.mi.pri. è minore della differēza dell'altezze della * 4.42.m.pri. e però c'è differenza di parallasse gr. 0. 2.mi.pri. Trouate queste cose, piglio l'istessa figura dell'autore, cioè questa nella quale il punto B. è il luogo del Lādgrauio. D. il luogo di Ticone. C. luogo della * A. centro della terra. ABE. linea verticale del Landgrau. ADF. di Ticone, e l'an-*

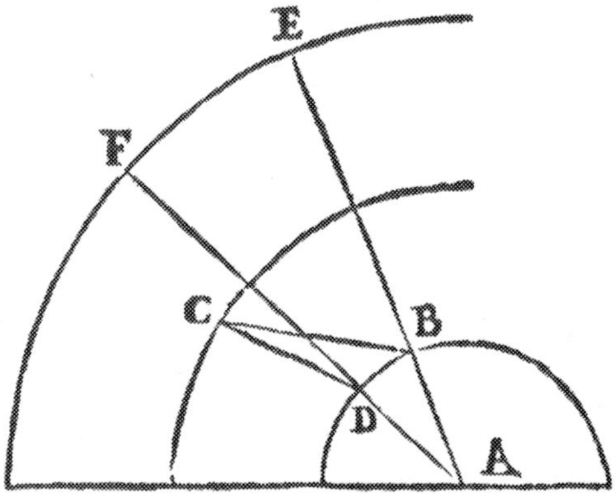

E
F
C
B
D
A

Ang.BAD. 4.40.m.p.corda sua 8142.parti di
BDF. 92.20.m.p.(quali il s. AB.è100000
BDC.154.45.m.p. } *sini* 42657.
BCD. 0. 2.m.p. } 58.

58. 42657. 8142.
8142.

85314.
170628.
42657.
341256.

58 | 59. 3473. 571. 5. | 13294.

wrote, "Louis . . . is always preoccupied with his experiments. You know that the ones he is undertaking this year will give us, should they succeed, a new Newton or Galileo." She was, indeed, correct.

Pasteur was an early example of a funded researcher. He was paid by the European brewing industry to study the causes of deterioration and souring in fermented beverages. His work led to the process of pasteurization of course, but that was incidental to the real discovery of the astounding truth of microbial theory. Until Pasteur's day, cleanliness was not a consideration socially, scientifically, or otherwise. Pity the soul who had cause to be in a hospital of the time. There was little likelihood that doctors would wash their hands between visits to patients. The notion that microscopic organisms were the causes of disease simply did not exist—until Pasteur presented his first paper on microbial behavior in 1857 before the Paris Academy of Sciences. This theory was challenged by the old school, and Pasteur was denied membership in the academy of sciences twice and almost did not receive acceptance the third time around! In fact, he was barely voted in—and only because of the merits of his earlier work on crystals.

The microscope contributed to Pasteur's ability to gather evidence. He adopted the microscope early on in his career and used it to demonstrate that air contains microbes, which was one of the many major firsts in his research. Optics—the telescope and the microscope—certainly changed our understanding of our sciences as dramatically as electronics altered our understanding in the twentieth century. In Pasteur's case he unseated hundreds of years of hopelessly misguided medical *science*. Objectivity and empirical evidence were his tools. He did not impose any truths on his research; the research revealed the truths to him.

Galileo (1564–1642) published his famous Dialogue Concerning the Two Chief World Systems *in 1632. The book was ordered burned by the holy Roman church, Galileo was placed under house arrest, and the new heliocentric world view was off and running.*

MOLECULAR STRUCTURE OF NUCLEIC ACIDS

A Structure for Deoxyribose Nucleic Acid

WE wish to suggest a structure for the salt of deoxyribose nucleic acid (D.N.A.). This structure has novel features which are of considerable biological interest.

A structure for nucleic acid has already been proposed by Pauling and Corey[1]. They kindly made their manuscript available to us in advance of publication. Their model consists of three intertwined chains, with the phosphates near the fibre axis, and the bases on the outside. In our opinion, this structure is unsatisfactory for two reasons: (1) We believe that the material which gives the X-ray diagrams is the salt, not the free acid. Without the acidic hydrogen atoms it is not clear what forces would hold the structure together, especially as the negatively charged phosphates near the axis will repel each other. (2) Some of the van der Waals distances appear to be too small.

Another three-chain structure has also been suggested by Fraser (in the press). In his model the phosphates are on the outside and the bases on the inside, linked together by hydrogen bonds. This structure as described is rather ill-defined, and for this reason we shall not comment on it.

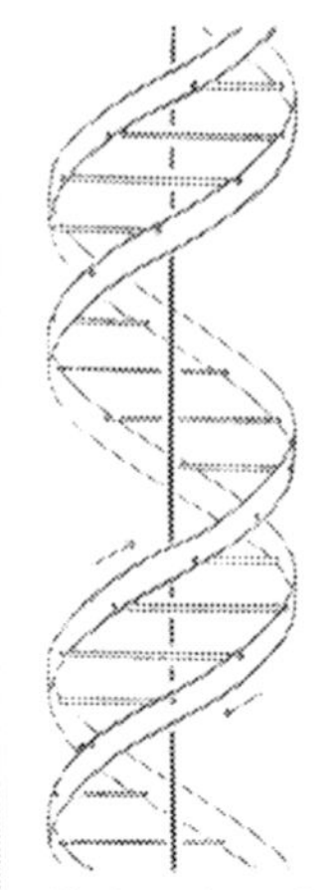

This figure is purely diagrammatic. The two ribbons symbolize the two phosphate—sugar chains, and the horizontal rods the pairs of bases holding the chains together. The vertical line marks the fibre axis

We wish to put forward a radically different structure for the salt of deoxyribose nucleic acid. This structure has two helical chains each coiled round the same axis (see diagram). We have made the usual chemical assumptions, namely, that each chain consists of phosphate diester groups joining β-D-deoxyribofuranose residues with 3',5' linkages. The two chains (but not their bases) are related by a dyad perpendicular to the fibre axis. Both chains follow right-handed helices, but owing to the dyad the sequences of the atoms in the two chains run in opposite directions. Each chain loosely resembles Furberg's[2] model No. 1; that is, the bases are on the inside of the helix and the phosphates on the outside. The configuration of the sugar and the atoms near it is close to Furberg's 'standard configuration', the sugar being roughly perpendicular to the attached base. There is a residue on each chain every 3·4 A. in the z-direction. We have assumed an angle of 36° between adjacent residues in the same chain, so that the structure repeats after 10 residues on each chain, that is, after 34 A. The distance of a phosphorus atom from the fibre axis is 10 A. As the phosphates are on the outside, cations have easy access to them.

The structure is an open one, and its water content is rather high. At lower water contents we would expect the bases to tilt so that the structure could become more compact.

The novel feature of the structure is the manner in which the two chains are held together by the purine and pyrimidine bases. The planes of the bases are perpendicular to the fibre axis. They are joined together in pairs, a single base from one chain being hydrogen-bonded to a single base from the other chain, so that the two lie side by side with identical z-co-ordinates. One of the pair must be a purine and the other a pyrimidine for bonding to occur. The hydrogen bonds are made as follows: purine position 1 to pyrimidine position 1; purine position 6 to pyrimidine position 6.

If it is assumed that the bases only occur in the structure in the most plausible tautomeric forms (that is, with the keto rather than the enol configurations) it is found that only specific pairs of bases can bond together. These pairs are: adenine (purine) with thymine (pyrimidine), and guanine (purine) with cytosine (pyrimidine).

In other words, if an adenine forms one member of a pair, on either chain, then on these assumptions the other member must be thymine; similarly for guanine and cytosine. The sequence of bases on a single chain does not appear to be restricted in any way. However, if only specific pairs of bases can be formed, it follows that if the sequence of bases on one chain is given, then the sequence on the other chain is automatically determined.

It has been found experimentally[3,4] that the ratio of the amounts of adenine to thymine, and the ratio of guanine to cytosine, are always very close to unity for deoxyribose nucleic acid.

It is probably impossible to build this structure with a ribose sugar in place of the deoxyribose, as the extra oxygen atom would make too close a van der Waals contact.

The previously published X-ray data[5,6] on deoxyribose nucleic acid are insufficient for a rigorous test of our structure. So far as we can tell, it is roughly compatible with the experimental data, but it must be regarded as unproved until it has been checked against more exact results. Some of these are given in the following communications. We were not aware of the details of the results presented there when we devised our structure, which rests mainly though not entirely on published experimental data and stereochemical arguments.

It has not escaped our notice that the specific pairing we have postulated immediately suggests a possible copying mechanism for the genetic material.

Full details of the structure, including the conditions assumed in building it, together with a set of co-ordinates for the atoms, will be published elsewhere.

We are much indebted to Dr. Jerry Donohue for constant advice and criticism, especially on interatomic distances. We have also been stimulated by a knowledge of the general nature of the unpublished experimental results and ideas of Dr. M. H. F. Wilkins, Dr. R. E. Franklin and their co-workers at King's College, London. One of us (J. D. W.) has been aided by a fellowship from the National Foundation for Infantile Paralysis.

J. D. WATSON
F. H. C. CRICK

Medical Research Council Unit for the Study of the Molecular Structure of Biological Systems,
Cavendish Laboratory, Cambridge.
April 2.

The Proof: Scientific Induction

We have looked at the uniqueness of technical writing from several angles. First we observed that technical writing speaks many languages and that it is a multifaceted medium. Scientific methodology depends on many languages to examine the world, and technical documentation unites these languages. We also noted that engineering documentation is a bridge between ideas and the finished constructs, so there is a need for graphics. Then we observed that the empiricism upon which science depends has led to a very specific writing style in recent times. Documentation must be true to the evidence. Data are the heart of the scientific process. Evidence can topple widely held misconceptions.

There is a final element in this discussion: the document itself. The scientific "paper" as today's research documentation is often commonly called, is now the stock-in-trade method of communication in the scientific and engineering communities—particularly for researchers.

The presentation of papers is not exactly a result of any scientific trend. Papers were presented at the Royal Academy of Sciences in London and at other major European societies well before the twentieth century. These presentations have always represented, at least among academics, an accountability of sorts. The papers present the outcomes of research or work completed and are now a very well established modern tool that has evolved as a highly analytical device that can appropriately combine the data, the languages of the sciences, and empirical reasoning. The paper has proved to be the most effective device both for reporting the findings and for explaining whatever logical reasoning was involved in the research.

For our purposes, we want to see the logical process of the scientific paper in clear and fundamental terms. The empirical procedure is based on the assumption that the world around us is analyzable, as we have seen, but we have not discussed how the process of empirical research is conducted. What actually occurs once we have accumulated the data? What is the point of all the evidence? The answer is that evidence demonstrates outcomes of research so that we take the outcomes to be truths or probable truths. If we use placebo studies with control groups numbering 150 students rather than 10, we increase

Another revolution began with this modest page and the simple graphic, which introduced to the world the double helix of DNA and created a leap into the future for the science of molecular biology. (Adapted from the original and reprinted by permission from **Nature** *Vol. 171 pp. 737–738 Copyright © 1953 Macmillan Magazines Ltd.)*

the evidence to increase statistical probabilities. A sample of 150 is certainly superior to a sample of 10. Studies may involve decades and thousands of people, as in the famous Framingham Heart Study research that suggested the correlation between cholesterol and heart disease. Computers can simulate millions of instances of a given circumstance. *Demonstration* is a logical procedure that utilizes inductive reasoning, which is the fundamental process controlling research.

Induction is a logical process that derives observations or outcomes from data. It is understood to operate on the basis of a shift from specific evidence to general truths, based on the evidence. In simple terms, the process is a movement from specific to general. As you can surmise, the inductive method is a primary tool for scientific reasoning, since the scientific system is based on evidence as a central force in the premise that we can know the truth of our world. Most scientific research employs this process. A technical or scientific paper is then developed to report the process. The text in a technical paper is used to discuss the logic and to present the evidence and outcomes. The scientific paper is the model of choice for reporting this inductive activity of the researcher.

One of the most practical applications of scientific induction more or less emerged in 1812, in Paris with the creation of the Sûreté, one of the world's first criminal investigation bureaus. The application was the new science of criminology. Edgar Allen Poe wrote the first detective stories, and they were about a French investigator who was imagined to be a student of the new science of criminology. These stories appeared in the 1840s, decades before Holmes and Watson set to work with their magnifying glasses. Mystery stories reflect the scientific process reduced (or raised) to the level of an entertainment. Magnifying glass in hand, we solve the mystery. The clues are the evidence. The process? It is inductive, my dear Watson.

In order for Dr. Watson to relate a Sherlock Holmes story, he had to explain the scientific process of the criminal investigation. Similarly, the scientific paper is designed to report the procedure of a scientific investigation. The technical writer often performs a similar task because inductive reasoning is, in great measure, a fundamental process in scientific reasoning.

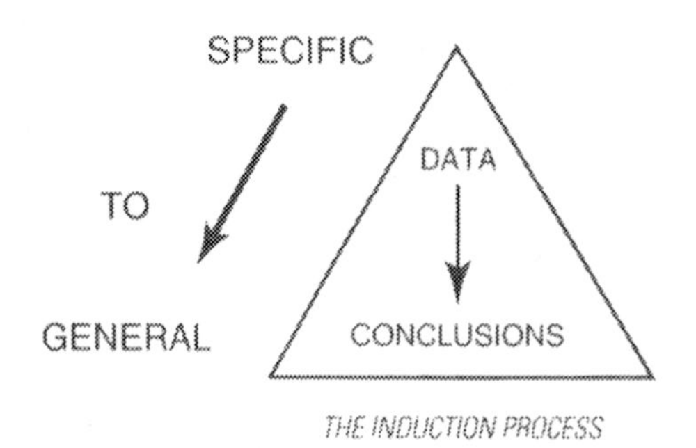

THE INDUCTION PROCESS

A great many variant forms of technical documentation report the stages of induction in that they present *evidence, analysis, discussion,* and *outcomes.* Lab reports of every sort reflect this pattern in its clearest form, but many scientific papers follow a similar design. The pattern is not only one that moves from specific (data) to general (outcomes) but one that is pegged to predictable stages of activity such as we see in lab reports. The components simply reflect the inductive path or procedure: evidence → analysis → outcomes. It is for this reason that the languages of science are critical to tech writing. They are elements of or interpretations of the evidence that is gathered by the researcher for analysis. All the graphics, all the charts, and all the printouts are features of evidence, without which the writing lacks the ability to persuasively show the logic. The text proceeds as a logic structure, organized as though it were a calculation. The process is quite mathematical in some sense. The conclusions of a scientific paper embody the knowledge we gather from the evidence under discussion.

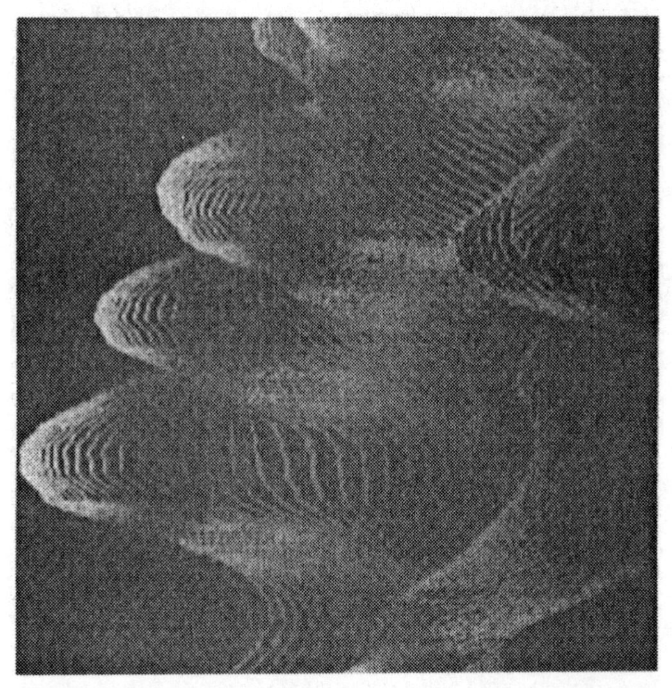

Photographic evidence of the double-helix structure is now available as a result of the one million power magnification of STM (scanning tunneling microscopy). (Courtesy of Peter Raven and George Johnson, **Understanding Biology**, *1993, with permission of the McGraw-Hill Companies.)*

As we noted earlier, the fact that English is used less in technical writing than in other written media does not decrease its importance. The inductive reasoning process of a technical or a scientific document is largely handled in the spoken language. The spoken language also must bind the numerous systems we may be using. Readers depend on the basic language, English, for much of the conceptual structure of a document, for most of the vocabulary control of the discipline under discussion, and nearly all the transitions that propel a reader forward through the maze. The mixture of logic systems creates the multilingual effect; these systems are also part of a multimedia effect that is usually obvious in technical documents. The languages and the media embody the evidence. The challenge is to connect the systems with the English text in order to unite the languages and the media. The challenge is to construct the inductive proof.

The following chapters are a practical discussion of how technical documents usually evolve in terms of layout, formats, illustration, documentation, and so on. Each feature is a reflection of practices that evolved, sometimes slowly, and sometimes in less than a century, into the writing methods we use today. Notice, for example, that the method of referencing graphics by figure numbers was very much in place by the mid-nineteenth century, as we observe in Darwin's commentary about the finches.

346 GALAPAGOS ARCHIPELAGO

Geospiza is shown in Fig. 1, and the smallest in Fig. 3; but instead of there being only one intermediate species, with a beak of the size shown in Fig. 2, there are no less than six species with insensibly graduated beaks. The beak of the sub-group Certhidea, is shown in Fig. 4. The beak of Cactornis is somewhat like that of a starling; and that of the fourth sub-group, Camarhynchus, is slightly parrot-shaped. Seeing this gradation and diversity of structure in one small, intimately related group of birds, one might really fancy that from an original paucity of birds in this archipelago, one species had been taken and modified for different ends. In a like manner it might be fancied that a bird, originally a buzzard, had been induced here to undertake the office of the carrion-feeding Polybori of the American continent.

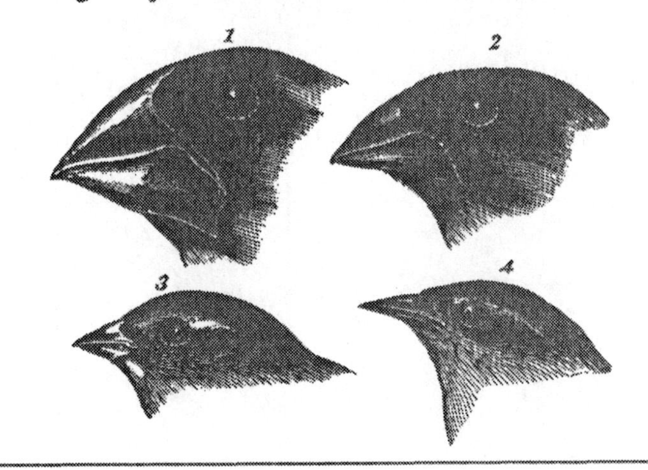

Certainly these are the most famous finches in history. This illustration was part of the evidence Charles Darwin surveyed in his published journal The Voyage of HMS Beagle *(1839). The journal was published many years before his controversial* Origin of Species.

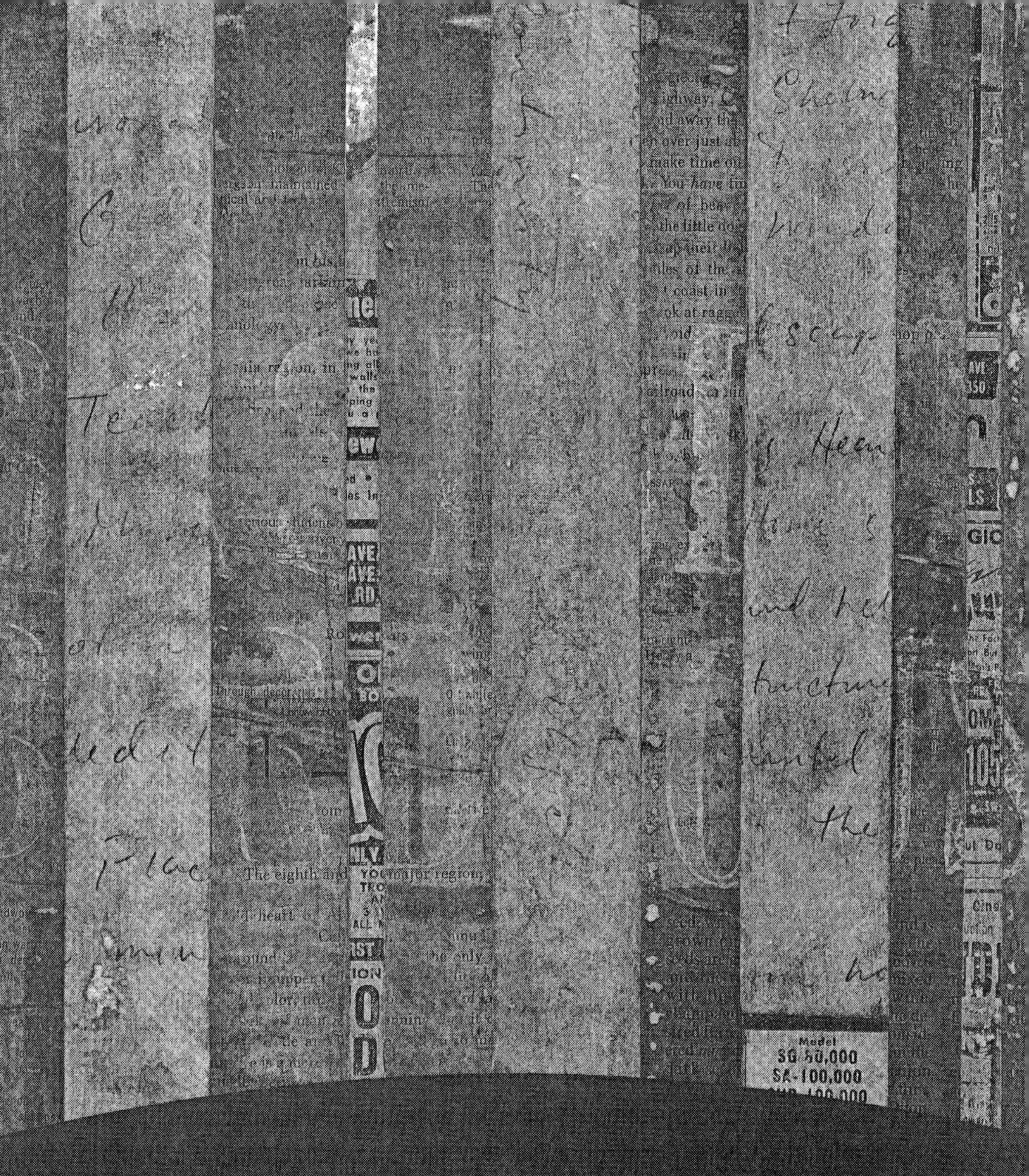

PART I

Document Structure Fundamentals

Work in Progress

2. Task Definition

The majority of documentation assignments I have been assigned have been initiated as the result of a problem, a clarification, or a change in the actual way something is handled. The ISO 9001 standards require companies to be "current," which means that if there is change in a manufacturing process, the associated documentation must reflect the change and be current.

I had received an e-mail from my supervisor requesting that I review the process by which meter stock (or inventory of meters) was received, inspected, routed into stock, and issued to departments. There was evidently some sort of routing confusion that I was to study and resolve. The project had to be done within three weeks. The e-mail contained references to the departments that were dealing with the problem, so at least I knew whom I would contact. However, the e-mail was not specific enough for me to know what was really going on.

Prior to making my first contact, I had to understand how the distribution process is currently documented. This would tell me who (departments), what (handling of meter stock), where (building areas), when (incoming shipments), and how (on-line transactions and document formats) the process was documented.

I called Mitch, my supervisor, and we reviewed the distribution process and found that more departments were affected by it than he had indicated in the e-mail. He told me that any change I was going to make would have an impact on more than the initial groups because departments that were probably not having a problem with the existing process could have a problem in the future. A new change could affect them adversely.

Genus, like many companies, has been converting to a "paperless" documentation method, so I knew that my project would be handled in various computer applications we use. The process documents are primarily produced on Microsoft-type software such as Word, Excel, Team Flow, and Adobe Acrobat. The Material Resource Planning (MRP) tracking system is Reflections (MANMAN) version 5.10. To complete this project I used all these applications because any new changes must appear in the existing formats.

Even though the medium is digital, all the conventions of paperwork remain. The project documents usually contain title pages, text, tables, graphics, a record of revisions, and a "change approval sign-off" for approval verification. Whenever possible I try to keep the documentation simple; Genus is multicultural, and anyone in any position should be able to understand it with little training.

S.B.

The Basic Practices

For most of the text you are reading, you will be concerned with a kind of management process that any writer must be involved in as part of the task of writing. In the earlier volume, *Basic Composition Skills* I discussed at length the process of writing as an activity. There the task was *production* management. Now you are ready for *product* management. In this text I assume that you are prepared for a more advanced perception of the writing craft. Here you should focus attention on the shaping or designing of the text based on the application of the document in a specific environment. The document you write is, after all, a tool. It is shaped and designed, or at least should be, to serve certain applications or purposes. At the most basic level these purposes require the obvious: orderliness, clarity, and precision, for example, as opposed to disorganization, confusion, or vagueness.

First, consider the document at the simplest level of organization: layout. There is an established approach that most writing layouts share that is more or less standard operating procedure (SOP). Although it is unlikely that a company will have an unusual writing policy when it comes to the basics, in the samples in subsequent chapters you may note frequent distinctions in document management that vary from company to company or from application to application. Among the many samples and models, expect to see some variations on standard practices. Consistency *within* a document is very important, and using the usual conventions a reader expects to encounter will make a document more accessible. There is, however, plenty of room for flexibility, as you will notice in the samples. They need not conform to each and every suggestion in this or any other text.

"The basics" include the parts of the documents that are quite commonplace:

Title Page

Table of Contents

Table of Illustrations

Text

List of Terms

Bibliography

Appendices

You will often see these elements, usually in the order listed, followed like clockwork in an endless variety of technical documentation. Consistency is the point. When I speak of standard operating procedures I am referring to convenience. If a reader has a pattern of expectation that is relatively fixed, then a writer is wise to follow the expectations. The less confusion the better. In France the table of contents is not in the front; it is in the back. But if a writer puts it in the back, the result will simply annoy an American reader. Authors want to write in order to communicate, and conformance to standards is one step toward that goal.

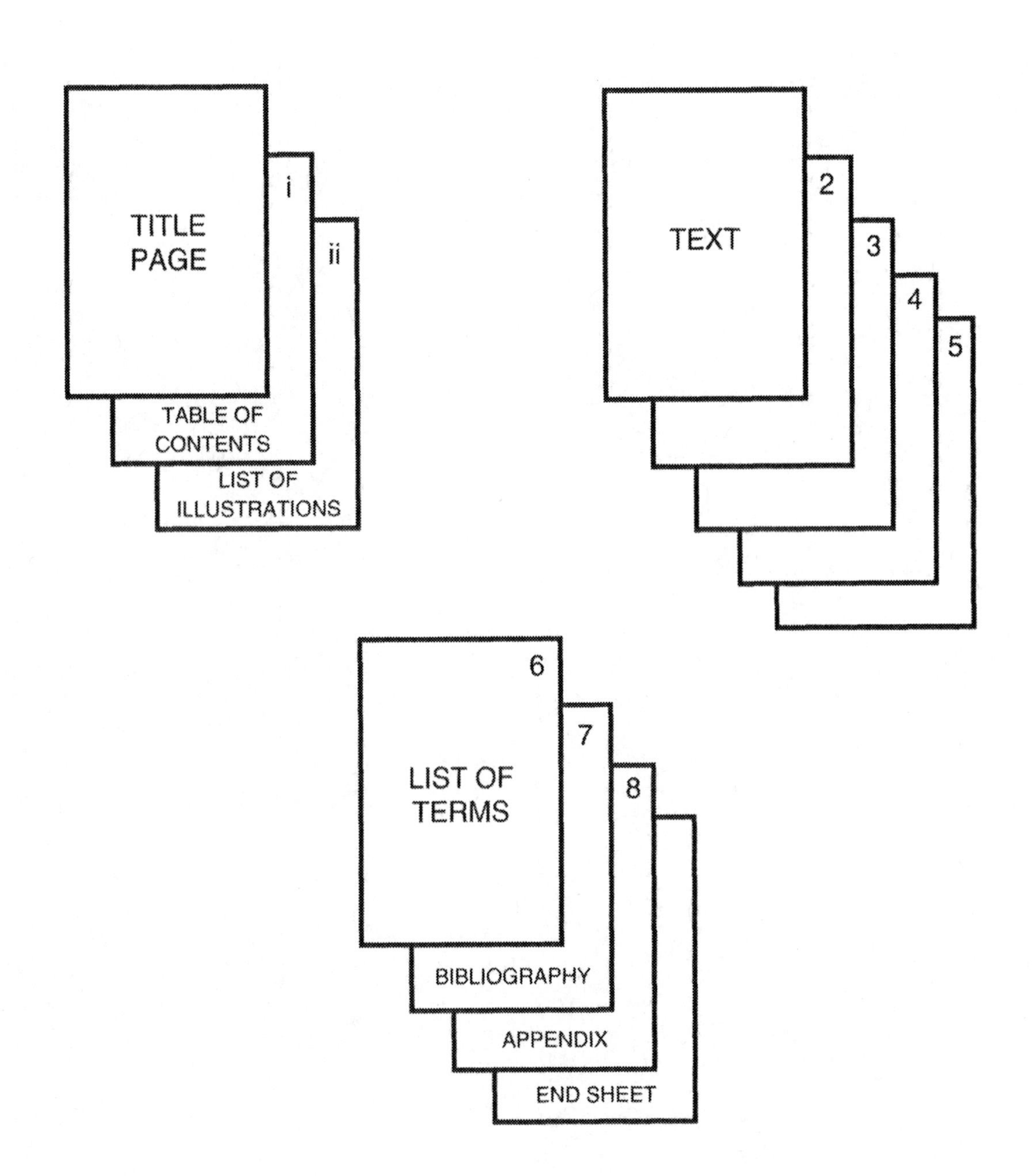
TITLE PAGE
i
ii
TABLE OF CONTENTS
LIST OF ILLUSTRATIONS
TEXT
2
3
4
5
6
7
8
LIST OF TERMS
BIBLIOGRAPHY
APPENDIX
END SHEET

The Physical Product

There are a few very basic layout considerations that can serve as general guidelines. Use 8-½ 11-inch paper and generate copy on only one side. Two-sided copy is a complicated issue, and unless a document require a facing-page format, you do not have to deal with it. Each of the texts in the *Wordworks* series presented frequent facing-page dilemmas, and this text, in particular, was very difficult to organize so that samples of the topic of discussion would face the page containing the discussion. Your technical support staff in a company setting will often handle the job of finishing such a project; what you need to do is design the text at the manuscript level, which calls for one-sided copy.

By "manuscript" I mean that it is not the final draft and it will undergo some editing. But it is not a rough draft, so you want to do your best. It should be viewed as a presentation, and you want it to be as presentable as you can make it. For this reason, if you double space the text it can be conveniently edited by you or your supervisor or any committee to which you may be reporting.

Use standard 12-point type for whichever font you select, although a few of the fonts run small. Do not write the text in all caps. If the product is going to follow the specifications of an employer, then all the suggestions I make are negotiable. A company may use a specific font as a standard practice, or perhaps a company will change the size to 11 or 13. The use of 12-point is simply a historical carryover from the Elite type found on most typewriters, and it is easy to read.

Indent paragraphs, usually with a five-space indentation format. The indent easily cues a reader that a new subject or topic is about to begin. The drawback of using the contemporary method of double spacing to the next paragraph is that, if you are already double spacing the text, you now must double the double space to cue a reader. The gaps are not always appropriate. Besides, in tech writing authors use a fair amount of space anyway, and so the paragraph cue can easily be lost. The space above and below a graphic can be particularly vexing, and readers might not notice a new paragraph. The old convention of indenting probably disappeared as a matter of convenience once the IBM Selectric typewriter became popular about twenty years prior to word processors. Automatic carriage returns and other mechanical assists seemed to contribute to the new business letter formats at that time. Obviously if you *need* the indentation as a tool, it is easy enough to deploy once again.

The next section of this chapter will discuss each of the major features of a technical document or paper. I must first explain the facing-page format you will frequently see. This book is designed to be a series of discussions about the writer's craft, and the writing practices are reflected in samples you will see on your left. Throughout the book I have incorporated the work of a number of men and women who are experienced in various engineering technologies or engineering fields. The samples are often discussed directly, and they will always coincide with the text on the facing page regardless of whether I identify details of the documents. As you read, keep the samples in mind as examples of the discussions in each chapter. In this chapter I have chosen bits and pieces of a number of documents that were written as coursework. These samples illustrate major parts of typical technical documents as engineering tech students and engineering transfer students developed them.

You will notice that there are two types of samples. The full-page "Models" are on your left, run from one to five pages in length, and are selected to illustrate the discussion. I include smaller "Samples" as part of particular discussions that emerge in the text. They are usually on the left also, and they, too, are student products. The samples are single-page illustrations. As I explained earlier, many of the models and samples are "co-op projects" that found their way into corporate applications in a variety of industries. Several others are very typical "term papers" in the college and university tradition; they will also help you see the benchmarks of good writing practices. All the models should provide you with a perspective on peer-group standards. There is nothing here that you could not do as well—or better.

NOTE:
You will occasionally see a disk icon and a text box containing basic tips. The directions concern Microsoft applications that usually work on both IBM and MAC platforms.

The Title Page

Include the following data on your title page simply as a matter of routine.

1. **TITLE**
2. **Audience: (Statement)**
3. **Objective: (Statement)**
4. **Name**
5. **Date**
6. **Project Number**
7. **Course Number or Company Department**
8. **Instructor or Supervisor**
9. **Institution**

Usually you will need a title page, if for no other reason than that it takes a lot of space to present all the relevant information a title page of a document should usually contain.

1. The title Of course, every document that is presented as a "paper" needs a title, usually a precise and matter-of-fact one for technical work. You might capitalize the entire title because titles should run a little larger than the other information on the cover. You could also use 14-point or 18-point bold type to dress it up but nothing much more. Avoid extravagance. Too much of the capability of our processors is misunderstood. There is simply too much horsepower when no one is required to have a driver's license. As a result, you will now see title pages in inch-high letters and other such devices. The results look corny on manuscripts or on unpublished material. Keep the document conservative and as dull as granite unless there is a genuine purpose to some other approach.

2. Audience If you profile your readership, it will help you when you write the document, and it is also quite necessary for the editing of technical documents. You will see a number of audience profiles in the text, and some are quite thorough. Identifying your audience will help you, and those who are working with you on your project, to control the success of the finished project. If you develop a reader profile, and if you clearly understand your writing objectives, you have two very important "controls" in position before you begin to write.

3. Objective It is often helpful to state the objective in a brief statement—a few sentences or a paragraph. It is also helpful to develop a one-page outline of the project structure before drafting the actual document. You will find that this tactic is an effective way to shape concepts early on, and the outline helps clarify the objective statement.

Sample 2.A

AUTOLISP:

AN INTRODUCTION

Audience: Individuals with a comfortable grasp of AutoCAD's menu positions and the command hierarchy and syntax. Some programming familiarity in any language will be helpful. Basic knowledge of the DOS text editor is necessary to aid in constructing and troubleshooting programs.

Objective: This paper attempts to instill in frequent AutoCAD users a rudimentary confidence in LISP. Manipulations of the AutoCAD database are one of the most powerful tools in this software program. Often, the first exposure to AutoLisp bewilders users. It is hoped that the simple program presented here will encourage them.

Michael H.
EGR 231 Technical Writing
May 10, 199X

NORTH SEATTLE COMMUNITY COLLEGE

What is an objective? Simply ask the question: Why am I writing this project? (Your professor's request, by the way, might not be the reason. Think in real terms and not in terms of classroom environments, unless you are doing an upper-division term paper).

Always remember that tech writing is very content driven and usually linear. You usually want to agree on content and organization before you begin to write. You will hear quacks used as the fault signal on some computers, and they are good for a laugh, but "organizational faults" have to be done away with before an author sets to the real work of writing. No computer can signal a fault in organization.

4. Name Credit coauthors if there are any. In the end, be prepared for the rude reality that salaried employees seldom own their writing, and your name will vanish from company documents if they are anything other than business letters. Be sure to keep copies of the rough drafts and finished products so that you can authenticate the authorship and build a portfolio of your work. If it is all in computer memory, archive your work, keep your own disks, or otherwise arrange for safe retrieval for your use.

5. Date You want to date technical material because of the speed with which it becomes obsolete. Some electronics and computer corporations look at eighteen months as a turnover calendar for new generations of equipment. Even conservative companies in heavier industries have to respond to competition in twenty-four to thirty-six months. In such environments, the dishes may as well go out with the dishwater: the support documentation becomes as obsolete as the product lines it assists, so you always want to know the age of a document.

6. Project number Some businesses and some college instructors organize their file controls around numbers. My system, for example, is as simple as one, two, three.

7. Course Number or Company Department Be sure to identify your location. Projects can and do get lost. A campus environment is deceptive. I may have a class of twenty-five, but the members of the group do not realize that they are lost among the real numbers, which are upwards of 27,000 here in my college district. The same holds true for corporate work. If you write a document for the Boeing corporation here in Seattle you have to make your identity stand out sufficiently to be recognized among 100,000 or more employees stretched between here and Wichita, Kansas. If you identify your team or department or division, readers can see the source of the material.

8. Supervisor Similarly, if you identify your college instructor or your project manager, the point of origin is easier to identify.

9. Institution It is appropriate to identify the college or business for which you created the project. Besides, this touch makes the title page look professional, which is always one of a writer's goals. Come up from the bottom of the title page about 1½ inches and

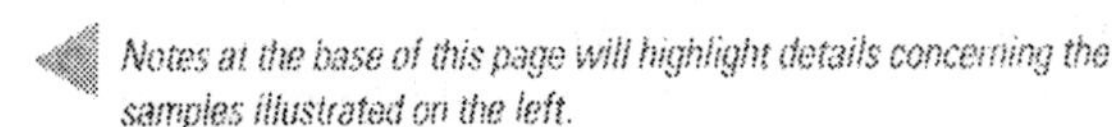
Notes at the base of this page will highlight details concerning the samples illustrated on the left.

identify the full name of the college or business. Consider adding the logo here also. On business stationery the logo is usually on top, but on support documents it may look more appropriate at the bottom.

There are two popular layouts for the title page, variations of which you will soon notice in the project samples. It is a commonplace tool in advertising to imagine a cross on a piece of paper, which more or less identifies the maximum point of interest or impact on a page. Emphasis, as any newspaper editor knows, always moves from upper left to bottom right. It is for this reason that the widespread college student practice of single-spacing title page information at the bottom right-hand corner is ineffective. The page is otherwise blank, and the data sit at the point of least interest, in fine print. Use the cross as the approximate point of impact and either develop a diamond or triangle of information, all on center, or else move the imaginary cross to about a 1 ½-inch margin on the left and place all your information on the margin line. The layouts will look something like the following figures. First, here are the crosses for each of the styles:

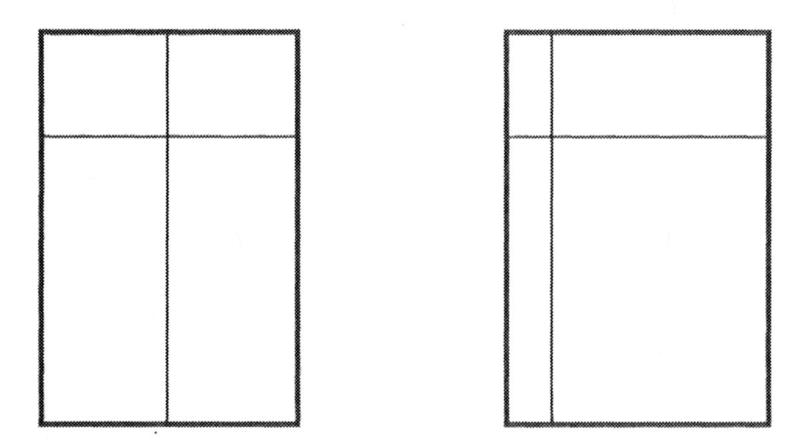

Use the center or left axis and align your information accordingly. You would double-space, or more, to spread the material down the page. The sample title pages you will notice throughout *Technical Document Basics* have several single-spaced sections on the cover page. These are descriptions of audiences and objectives, which you do not need to mention on the final drafts of your projects.

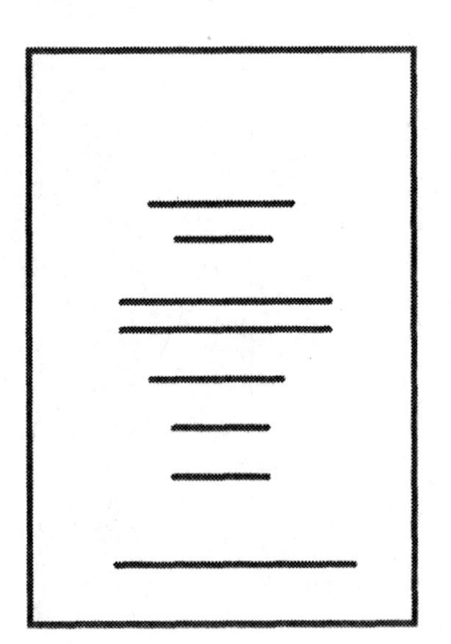

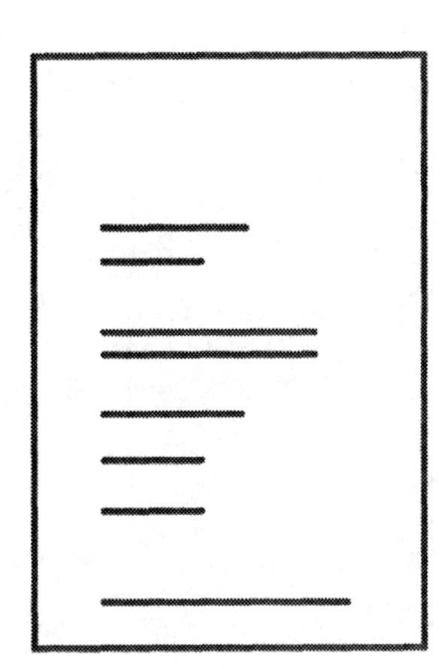

Sample 2.B

i

Contents

TOPIC PAGE

The Tables for the Text

Table of Contents The use of a table to identify the location of the material inside your document is a matter you must consider. Magazine articles and short documents often omit the table of contents. If the reading time of the actual document is quite short—perhaps five minutes or even half an hour—there may be little point in outlining the contents. An outline is very similar to a table of contents, and so it may be redundant to use a table if you briefly outline your intentions early in the text. When the project is long or complex, the table becomes more of a tool. When the project serves as a point of reference for people who might consult it, then the table becomes critical.

Double-space the table at the manuscript stage. If room allows, leave the table in double-space or one-and-a-half-space format even after final approval, at which time the text is often reduced to single-spaced text. The wider space between the lines allows a table to be more easily read or referenced. As a convenience for the reader, type periods across the page to the page number on the right. The eye is helped by the dots, and page numbers are easier to identify properly. A stream of periods usually looks less attractive than a dot-space-dot format. Solid lines seem effective also. Hyphens do not look very attractive. You will seldom see any other devices. Check to see whether your software already has a table of contents formatted for your use.

NOTE:
Many of the basic features discussed here are available in template format on your computer. See Appendix B, Templates and Tips. The appendix will explain how to locate the templates for the following software:

Word 98 for Macs
Office 98 for Macs
Word 7 (Office 95) for Windows
Office 97 for Windows
Office 2000

This table of contents (TOC) is more elaborate than you will usually find necessary. Note that in this case, only indentation is used to designate the headings, subheadings, and subsubheadings.

Sample 2.C

- ii -

List of Illustrations

Figures 1, 3, 4, 5, 6, 7, 8, and 9 courtesy of the Hewlett-Packard Company

Figures 10 and 11 courtesy of the Xerox Corporation

As you create the table, indent subsets of headings just as you would for an outline. Readers will recognize the subset ordering procedure immediately and will understand the order you present. The sections may or may not be numbered in some way, depending on whether the sections in the text are numbered. Needless to say, it is very difficult to add a table of contents to a text that lacks headings. Headings in the text must coincide exactly with those that appear in the table of contents. Headings simply identify sections of the text by content.

To Build a Table of Contents
You can use Word to build a table of contents automatically. Go to **Insert** Menu on the menu bar and go down to **Index and Tables.** If you have used headings in your document, the table of contents will build itself automatically.
To use headings in your text, select the type you need from the **Style** menu found on the **Formatting** toolbar. Rest the mouse on the icons to locate the **Style** menu. (Each pause calls up the screen tip that identifies the icon.)

Begin the table with the page number for the list of illustrations. Do not mention the table of contents; if you are looking at the table, it does not need to tell you where it is. By the way, short sections in a document may not always begin on a new page. Several sections in your table of contents may then repeat the same page numbers.

The table of contents page and the page for the list of illustrations, unlike the pages of the document itself, are identified with lowercase Roman numbers, usually in italics: *i, ii*. Any lower case *i* on a typewriter will do. A preface is also numbered this way. The practice of using Roman numbers in the front matter, and Arabic numbers for the text helps avoid repaging if you later make minor changes in the opening pages.

List of Illustrations If there is going to be a list of illustrations, you must establish some location guide to the graphics by numbering them all—including graphics that may be in an appendix.

Number the illustrations consecutively throughout the text and the appendix. Do not introduce new numbering at any point because the reader will be confused.

A table of illustrations is a formal tool and it is often omitted. For college work, if the document is long, the table adds a little dash. In corporate environments the issue may be one of policy. The first chapter discussed the multimedia effects of technical writing and explained the importance that graphics have in technical documentation. The heavy emphasis on graphic components is unique to the tech writing medium. At times, in fact,

In the case of the list of illustrations presented here, there are two simple credits for all the plates and they are placed below the list. If there are a number of sources for the graphics, this location is not convenient.

Sample 2.D

3

Safety First

Before you get all excited and get yourself hurt, here are a few precautions to keep in mind:

1. Always work with the power cord of your computer disconnected from the main when working on the internal components. This keeps the components from getting damaged by electricity and, more importantly, keeps you from electrocuting yourself.
2. Wear an antistatic wrist strap and ground it while working with the components. Static wrist straps can be bought from your neighborhood value-added reseller (VAR) or from one of the large chain computer stores. If you do not have an antistatic wrist strap, at least ground yourself by touching the metal case or a metal window frame.
3. Always leave the sensitive electronic components in their silver antistatic bags until ready for assembly. . Handle them by the edges and avoid touching the circuitry with your fingers.
4. Never use excessive force when fitting components like memory, cables, and CPUs into their sockets.
5. Study all the manuals that come with your equipment, especially the motherboard manual, before starting.

there is little written text and the document is primarily visual. Obviously, a table of illustrations can be just as useful as a table of contents. Remember, these tables are optional and are often omitted. Decide on or discuss the need for the tables based on the content or on the needs or expectations of your readers.

The tables of contents and illustrations usually appear on separate pages, but short tables can be placed on the same page. You also must identify the source of each graphic you are borrowing. If you borrow all your graphics from the same source, a practical location for identifying the source is just below the table of illustrations. This practice saves you the task of identifying each graphic by using separate credits throughout the text.

You will notice that most publishers accept the common practice of adding a third table when it is appropriate: a table of tables. Table numbers usually identify tables because they are not considered illustrations or figures. You may follow the convention of adding this third table if you wish. I encourage writers to rethink this practice. Tables are visual products. A table is visually unique, like a chart or a graph. In my work I decided not to distinguish among such features, and everything that is not text gets the same moniker: "Figure." Discuss the utility of the issue with your supervisor (or your instructor), and let company practices be your guideline.

Cautions, Tools, and Parts A number of features are frequently placed after the tables of contents and illustrations. Typically, three considerations might be important to a reader:

Cautions and warnings

Tool lists

Parts lists

These three features are particularly common if the document is a set of instructions. The warnings and lists can be structured as separate pages or as part of the text commentary. Warnings address the safety of the user, the safety of the equipment, or both. The safety warnings on the left were a separate page in a text addressed to computer enthusiasts. Notice the breezy style the author uses to address his fellow computer club members. The following list of warnings concerning bus speeds could appear as a separate page or as part of the text commentary. Notice the formal style of the sample. Sample 2.E explains appropriate cautions in a chemistry laboratory.

A common concern in business and industry, safety is often handled with some elaboration in documentation. These warnings are usually presented early on in the text.

Sample 2.E

SAFETY REQUIREMENTS FOR SAMPLE PREPARATION

- The digestion of water and soil samples for metals determination requires heating the sample with strong acids. The fumes given off by the acids, especially hot acids, are very irritating to the mucous membranes and lungs. Prolonged exposure poses a risk of permanent tissue damage, even death. It is therefore necessary that acids and hot digestion solutions be kept in a fume hood at all times. The fume hood allows the chemist to manipulate the digesting samples without exposure to the acid fumes, which are exhausted out of the laboratory.
- When working with concentrated acids, you must have eye protection. The minimum requirement is safety glasses with side shields to protect against splashing. Goggles or a full face shield are preferable to safety glasses because the goggles and face shield offer considerably more protection.
- When working with concentrated acids, you must wear a laboratory coat and long cuffed gloves that protect the forearm. It is easy to become unconcerned about handling acids if your only exposure is to the thick and callous parts of the hand such as the fingertips. A drop of acid on the face, wrist, forearm or neck will be quite a different story. A rule of thumb is that a severe chemical burn over 15% of the body can be fatal. A lab coat is the best defense against such a massive accidental exposure.

Special Precautions for 75 and 83 MHz Bus Speeds

There are some important restrictions for running at external bus speeds of 75 and 83 MHz:

1. At these speeds, the PCI bus runs at 37.5 or even 41.6 MHz. This can lead to several problems with PCI devices. Typical troublesome devices are SCSI controllers, some video cards, and network cards.
2. The speed of the EIDE interface not only is determined by the PIO or DMA modes but also is highly dependent on the PCI clock. This is why the EIDE interface is always slower on systems with 60 MHz or lower bus speeds. This also works in reverse—at higher bus speeds, the interface runs faster. This sounds good at first, but not all hard drives are up for the faster rate. The PIO may need to be lowered to the value of "2" at the higher bus speeds.
3. We have mentioned that the type of RAM is critical in terms of quality. With the higher bus speeds of 75 and 83 MHz, you will need to have high-end EDO or SDRAM. Most 60 ns EDO will work fine if your board supports it.

Of course, brief warnings are often inserted in the running text when a particular point of safety or concern is appropriate. The following sample and the one on page 90 used clip art to signal the specific concern. The text of the sample below was printed in red.

When you are working inside the case, it is a good idea to unplug the power supply from the wall. Some manuals recommend leaving the system plugged in for ground protection, but it is too easy to hit the power switch while working on the system which could result in disastrous consequences. Instead, use a ground strap or ground yourself by touching part of the case before handling the electronic components inside.

In addition to concerns for the welfare of the user-reader, it is also important to identify safety measures that will protect apparatus, particularly since harmless considerations such as static electricity can be a serious problem in certain environments.

Sample 2.F

3. PLANNING AND PREPARATION

Planning:

- Set a day aside for installation; allow for a full day so you are not hurried.
- Before the day of installation, open the computer and review the cable placements.
- Locate the Old Drive and determine where you will locate the New Drive.

Note: The DriveCopy software copied the Old to New Drive in about 20 minutes, but opening the computer, installing the New Drive, documenting, labeling, setting jumpers, breaks, and interruptions can add hours.

Hardware and Software Requirements:

- Computer with original configurations
- New Hard Drive (IBM 38451 8.4 GB)
- DriveCopy Boot Diskette
- Record of Primary IDE Master configuration settings for Old Drive (Maxtor 85120 A8) from previous section

Prerequisites

Diagram of Slave jumper settings for OLD Drive Maxtor 5.1 GB

Diagram of Master jumper settings for NEW Drive IBM 8.4 GB

Recommended preventive maintenance:

- Create a tape backup (optional)
- Scan and Defragment Old Drive using ScanDisk and Windows 98 Defragmenter.

If you have Norton Utilities:

- Run Speed Disk to defragment drive.
- Run Rescue Disk. This can help if startup files get corrupted.
- Run WinDoctor or SystemDoctor to check and correct existing problems.

Recommended References:

- Diagram of how to remove the outer case from your computer.
- Diagram of how interior components are connected.
- Diagram and label any cables and ports to be disconnected.

Lists of tools and parts are common features of a wide variety of technical documents, though perhaps most commonly used for assembly and troubleshooting documents. The following list of equipment involved over a thousand dollars worth of materials for the construction of a computer:

Parts List

PART	BRAND	COST
Processor	Intel Pentium III 450 MHz	$XXX.XX
Motherboard	ASUS Model P2B	$XXX.XX
Hard Drive	IBM 10.1 GB 7200 RPM	$XXX.XX
Memory	64 MB RAM	$XXX.XX
CD-ROM	ASUS 40X	$XXX.XX
Floppy Drive	3.5" TEAC Drive	$XXX.XX
Video Card	Matrox G400 16 MB	$XXX.XX
Sound Card	Sound Blaster Live	$XXX.XX
Modem	3Com Fax Modem	$XXX.XX
Case	Enlight ATX Mid-Tower	$XXX.XX
Monitor	Optiquest 17" V73	$XXX.XX
Keyboard	Micro Innovation Ergo	$XXX.XX
Mouse	Microsoft Basic Mouse	$XXX.XX
		$XXX.XX

Occasionally a manual is constructed by using a very rigorous and repetitive technique of organization that incorporates such features as a parts list into a larger organizational structure.

Purpose:
Create your DriveCopy Boot Diskette.
Requirements:
- One IBM formatted 3 1/2" Floppy HD diskette.
- DriveCopy CD-ROM.
- Serial number from CD-ROM sleeve.

Reference:
- DriveCopy User Guide, pages 1-3.

Estimated Time:
15 minutes.
Planning Tip:
Complete this on a day before copying drives.

The sample on the left includes a parts list, and is unusually elaborate in its suggestions.

Sample 2.G

Iteration Commands

All programs, no matter what programming language is being used, use one of three types of instructions or commands to accomplish the desired task. These types are sequence, selection, and iteration commands.

Sequence commands are the simplest of the three to understand. They are just lists of actions, called "statements" in computerese. The following would make up a sequence:

Step 1 **Get data from the keyboard.**

Step 2 **Display that data on the monitor screen.**

Step 3 **Write that data on a disk . . . .**

Sequences follow a progression, from start to finish, and are usually not difficult to follow when trying to understand programming code.

Selection commands can be a little more tricky for the novice programmer; however, there is really not much difference between them and sequence commands. The difference is that you add the element of multiple choices to the sequence. The IF statement (and the CASE statement once one gains a little experience with the applicable language) allows the flow of logic in the program to jump around. Using selection commands, the programmer can check specified conditions and, based on what these conditions are, can take different actions. An example of this would be to have the program check a variable called X_Test.

The Text

The text may be a running text with no page breaks, or it may be heavily subdivided, or there may be headings but without new page starts. All manner of practices are appropriate, as you will see in the next chapter.

Double-space the text initially for editing. Use 1¼-inch margins on all sides. Do not either pinch or exaggerate margins. You will not save paper, and you will make a dense topic look even more threatening with ½-inch margins. The printing facility of your company will request standard margins so that there is room for error when a document has to be reproduced in quantity. If you plan to bind your document, increase your left margins to 1½ inches so that there is room for staples. If you do not allow for staples, the margins will appear to be unequal.

By all means justify the right margin. Twenty years ago there were few people who had ever heard of this printer's term. Justification was a process of spacing type so that words aligned at the right margin to make perfect columns of type. Now any processor will justify at the press of a command. Be wary of justifying anything less than a full page in width. Computers have no aesthetic values and a two-, three- or four-column format will often look awkward because of the way a computer will space the words.

To Develop a Two-Column Text

1. Go to **Format** on the top toolbar.
2. Click on **Columns.**
3. Select the number of columns you need.
4. Increase or decrease the width of columns or the spacing between columns as necessary.

Indent paragraphs, as suggested earlier. It is now popular to signal a new paragraph with spacing rather than with indentation, but spacing is a poor signal in technical projects that use graphics or math displays. Under the graphic or display, readers tend not to see space as a paragraph cue. This can be a frustration to a reader if there are many such situations. There can be dozens of these problems in one technical writing manuscript.

The first-page sample presented here has a wide top margin to signal that it is a lead page. A text may or may not start with a heading.

Sample 2.H

Users experienced speeds slower than a 14 kbps modem during peak hours. Overall, these considerations mean average download speeds varying between 1 mbps and 2 mbps.

That is only the average speed for downloading, but how is your uploading speed? Upstream information, which goes out every time you click a hyperlink or send an e-mail, degrades the system even further. @Home has already limited the uploading speed to 128 kbps in some areas and is considering this limitation nationwide. Sharing cable lines also leads to a security issue. A neighbor who also has a cable connection can accidentally receive your web pages.

Additionally, the lack of competition could eventually lead to poor customer service and higher prices. @Home and Road Runner almost never compete against each other, and there are even rumors of a merger between the two. Therefore, you're stuck with whichever one your cable-TV company has aligned with. Currently TCI is the main investor of @Home, and Time Warner, MediaOne, Microsoft, and Compaq are the main investors in Road Runner. High infrastructure costs, including fiber-optic upgrades to regular coaxial cable systems have kept out new competitors.

Cable Summary

Consult Table 1 below for a list of the main advantages and disadvantages of getting a cable connection. Then read on to see how a Digital Subscriber Line measures up to a cable line.

Advantages
, Fastest possible speed for home users.
, Constant connection to the Net—no dialing in, permanent IP address
, Service provider installs it
, Cheapest setup price
, Convergence with cable TV

Disadvantages
, Available only in limited areas
, Bandwidth slows as local users increase
, Fast downstream, slow upstream
, No choice of service providers
, Must buy equipment and service from same company

Table 1. Advantages and Disadvantages of Cable

Digital Subscriber Line

What is DSL?

DSL are phone lines just like the ordinary telephone line to your house, but

3

Number the pages with Arabic numerals: 1, 2, 3. Do not use any other conventions, especially clumsy ones such as "page 3 of 7." This tool is used on business letters at times and in other circumstances where pages are loose. The convention is ideal for manual updates. You toss out dated material and update the hard-copy file with the new materials. The pages are technically loose, even though they are kept in three-ring binders. As a result they must be numbered per page and in small units. You cannot put a ten-page replacement in a five-page space if the text has been numbered sequentially in the conventional fashion.

After concluding the text—which *never* says "The End" by the way— you can add a blank page as a dust sheet.

Note: For technical reasons, the publisher of this textbook could not fully double-space most of the samples and still respect the original content of the samples. Either the samples had to contain less material, or else some had to be single-spaced. Because the content is important, I elected to vary the spacing to preserve enough material for you to review. In your projects, use the double-space format at the manuscript stage.

The two-column style has gained in popularity in recent years because of the ease with which the format can now be constructed on a computer. This style is most appropriate for published documents.

Sample 2.I

11

Glossary

Analog Sound Recording. An audio recording system in which the electrical sound signal fluctuates over its entire range exactly like the original sound stimulus.

Audiophile. One who is enthusiastic about sound reproduction, especially music from high-fidelity recordings.

Coding. A set of instructions that cause a computer or digital recorder to perform specified operations.

Compatibility. The ability of the software of one system to work with that of another system.

Data Compression. Reduction of digital information by elimination of unneeded data.

Digital Sound Recording. An audio recording system that translates original sound stimuli into many computer-type, on-off pulses. Compared with analog sound recording, this system has a better signal-to-noise ratio (less noise and interference in the signal).

Encoding. The process of converting a body of information, such as a signal, from one system of information into another.

Optical-Based System. A digital storage system that uses laser beams to create holes (pits) in a disc. The disc is read by a low-power laser that measures the energy that is reflected off the disc between the pits and the unburned areas between the pits, called "lands."

Psychoaccoustics. The branch of science that deals with hearing and sensations produced by sounds.

Random-Access. Permitting access to stored data in any order the user desires.

Sampling Rate. The number of times per second a digital recorder samples (records) a sound. The higher the sampling rate, the more accurate the reproduction of the sound.

Sonic. Sound waves.

Concluding Sections of the Text

List of Terms Although I will discuss the issue of vocabulary at considerable length, for now simply observe that it is fairly commonplace to define and explain words or concepts in scientific and engineering writing of all types. What to define and how extensively to do so is the issue. If you decide to include a separate page that lists and defines terms, the usual location for that list is at the end of the text. The glossary is a practical tool everyone depends on at one point or another. This is usually a new page or pages with a heading, "List of Terms" or "Terminology," for example, in the top center. This is a numbered page. Any page with writing on it is numbered, except the title page. The page will read easily if you double space between entries that are single-spaced. In other words, separate the definitions. You might boldface the terms or otherwise design the page so that the words you are defining are easy to consult and locate.

Appendices When your document is finished, you may find a group of documents you hesitate to omit, and yet you did not "use" them. If there is value in including *secondary* material, you may do so in an appendix. The appendix also serves as a location for oversized material such as foldouts or specialized material such as a cost analysis that has to be included in an engineering report.

Number all the appendix material as additional pages of text. You may design this section in a number of ways. Each document may be a new appendix (A, B, C, and so on), or all the documents may be part of one appendix (Figures 8, 9, and 10, for example, might be one appendix). The appendix title can go on a separate page or at the top of the appropriate page. You will see samples at several points in this text; for instance, notice the samples on p. 67. For the project concerning windows, the appendix names are at the top of pages that serve as the title pages for the appendices. The pages are otherwise blank, and the contents of the appendices begin on the following pages.

The terms presented in a glossary are a courtesy to readers. The terms are most efficiently defined by the author because the specific context of the document or the readership might call for tailored definitions rather than dictionary definitions.

Sample 2.J

APPENDIX 2

SEARCH STRATEGIES

Searching For . . .	**Best Bets:**
A broad overview of a subject.	Looksmart is a topical directory of the Web, with more than 24,000 categories to choose from.
A subject using a short phrase of two or three words.	Infoseek recognizes phrases, and won't ignore the most common words that other search engines discard.
The answer to a plain English question.	Ask Jeeves has the answers to more than 7 million common questions, and understands plain English queries.
A topic using rare or uncommon words.	Because AltaVista is the largest Web index, it will find the most documents about rare or uncommon subjects.
The most popular Web sites on a subject.	Search using HotBot, then click on the "Get the Top 10 Most Visited Sites For [Your Query]" link, at the top of your re-sults list.
Web pages that were published on a spe-cific date.	AltaVista's Advanced Search and Hot-Bot's Super Search find documents cre-ated within a specific date range.
The most comprehensive results for a topic.	Dogpile and Metacrawler are meta-search engines, meaning they simultane-ously scan the indexes of multiple search engines and present you with all of the results.
The ability to search within the results of a search.	Infoseek lets you refine your search using only the results from a previous query.
Sites that have paid to get top listings in search results.	GoTo.com allows Web masters to bid for position on search results for specific terms
A browsable directory of quality links that have been evaluated by experts.	All links in every About.com guide site have been carefully chosen by topic experts.

Replacing Windows *Page 8*

Appendix I

Component U-Values

If the appendices contain a number of pages, the full-page appendix headings help divide the material. In the case of our samples here, there was a second appendix with substantially more material, and so the author used another full-page appendix heading to separate the two groups of documents.

Replacing Windows *Page 11*

Appendix II

Glazing Material Chart

The sample on the left is one of several appendices that appeared in the same document. Because they were brief, they could be identified at the top of each page.

Sample 2.K

16

APPENDIX B. SPECIFIC DVD-ROM DRIVE FEATURES

Following is a list of features to expect from a DVD-ROM drive:

- Most 1× DVD-ROM drives have a seek time of 90–200 ms and access time of 100–250 ms.
- 1× DVD-ROM drives provide a data transfer rate of 1.321 Megabytes/second with burst transfer rates of up to 12 MB/s or higher.
- The data transfer rate from a DVD-ROM disc at 1× speed is roughly equivalent to a 9× CD-ROM drive (1× CD-ROM data transfer rate is 150 KB/s, or 0.146 MB/s).
- DVD spin rate is about 3 times faster than that of a CD (that is, 1× DVD ~ 3× CD).
- Almost all DVD-ROM drives increase motor speed when reading CD-ROMs and achieve 12× or faster performance.
- 2× DVD-ROM drives are available (providing a transfer rate of 22.2 Mbps or 2.6 MB/s from DVDs, equivalent to an 18× CD-ROM rate).
- Most 2× DVD-ROM drives read CD-ROMs at 20× (max) speeds and higher. Currently, 4×, 4.8×, 5×, and 6× drives are also appearing, although they usually do not achieve a sustained transfer rate at their full rating.
- A 5× drive can theoretically transfer data at 55.4 Mbps or 6.4 MB/s, equivalent to a 45× CD-ROM data rate.
- Most 4× and faster DVD-ROM drives read CD-ROMs at 32× (max).
- Connectivity is similar to that of CD-ROM drives: EIDE (ATAPI), SCSI-2, and others.
- All DVD-ROM drives have audio connections for playing audio CDs.
- No DVD-ROM drives have been announced with DVD audio or video outputs (which would require internal audio/video decoding hardware).
- In order to hook a DVD-ROM PC to a television and a stereo receiver, the decoder card or the video card must have a TV video output and an audio output.
- Some cards have SP/DIF outputs to connect to digital audio receivers.

A word of caution: in the tech writing industry the phrase "data dumping" refers to the efforts of some authors to unload round after round of data on their readers, especially in an appendix. I myself have dumped so much data into appendices that one project—which ran only 150 pages—had a 500-page second volume to hold the appendices! Do not pad the appendix. It usually fools no one. There are a few guidelines that will encourage you to use the appendix with discretion.

A sloppy appendix is a very unconvincing document. Make all elements of the appendices, photocopies in particular, conform to standards. Do not use crooked copies with shadows on the sides with your fingers showing on the edge. Cut out and remount each page on 8½ × 11-inch paper. Remove the original page number from each page, and then add the current page number of your text. Give each document an appendix number or a figure number or both. You may need a caption (title), and you must identify your sources at some point. Appendix material often looks particularly impressive if bordered and scaled down to 80 or 90% of the size of the original. If you go through all these tasks, you will certainly be selective, and, besides, the document will look more professional.

Notice the following sample appendix (Sample 2.L). It describes the construction of a student project, but it is obviously secondary to the text. This appendix is a latter day variation of what is known as a "colophon," a description of the text (although the focus here is more concerned with constructing the product).

You will see an occasional sketch here and there in this text in which authors explain their word processing procedures and desktop practices. I have writers attach this explanatory document as an appendix to the technical projects they are constructing. It has become a popular part of each project for curious classmates who want to see how the projects were generated—but it remains secondary to the purpose of the project. The appendix, in this case, is a part of the learning experience of the classroom and has no purpose in terms of the content of the project. A description of the project development is simply an excuse to build an appendix for practice.

This appendix sample indicates that it is one of several. It was used to identify technical details concerning one point of discussion in a paper that examined CD and DVD technologies.

Sample 2.L

7

Appendix

About This Document:

This document was created on a Dell XPS D Pentium2-233 system running MS-Windows95 OSR 2.1.

A whole array of software was used:

- MS Word for Windows 98—Word processing and layout
- Adobe Photoshop version 4.0—Graphics layering, annotation, and layout
- Metatools's Bryce2—3D renderings
- MS Paint—Freehand line art
- Fractal Design's Painter—Miscellaneous freehand art

Don't let the long list fool you—I used this many tools only because I'm not very proficient in any of them.

Appendix

PRODUCTION NOTES

This paper was created using the following:

- Microsoft Word 2000
- Panasonic Copier
- Epson Stylus Color Printer

The main body of the paper was created on Microsoft Word 2000. I left spaces for the figures and then printed the pages on the Epson Stylus Color Printer. I then photocopied the pictures from the manuals using the Panasonic Copier and enlarged or shrank the pictures to the size that gave the best detail. If necessary, I added more space where the figures were to be placed. I glued the pictures in the proper spaces using rubber cement.

As you read the descriptions of the procedures that were followed to develop student projects, notice also that everything from glue sticks to *Photoshop* can be part of the job. The craft can be as mechanical or as electronic as you make it. As the expression goes, "Whatever works." Here is how another author developed an appendix to explain the technology that was used to create the paper (each smaller sample originally appeared on a full page).

Appendix

SOFTWARE AND HARDWARE

This project was created on my IBM PC at home and a DeskJet Hewlett-Packard Color Printer. I downloaded the keyboard image shown in Figure 3 from Netscape and modified it. Besides that, I took a computer picture for Figure 1 from Corel *Clip Arts*. I used Paint in Windows 98 Accessories to draw the other four figures of my project. Finally, I used Word 6.0 to type my text document and inserted all the pictures in the project.

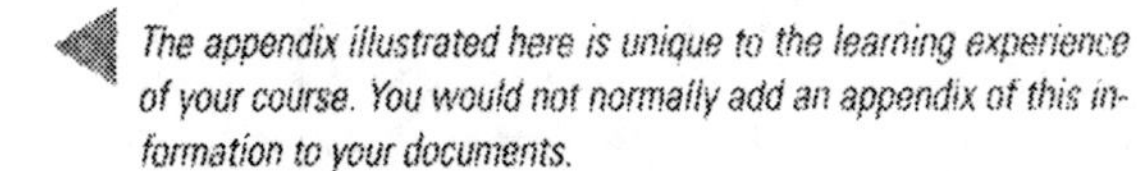
The appendix illustrated here is unique to the learning experience of your course. You would not normally add an appendix of this information to your documents.

Sample 2.M

Bibliography

Embedded Controller Handbook. Santa Clara: Intel Literature Sales, 1988.

Gaonkar, Ramesh S. *Microprocessor Architecture, Programming, and Applications with the 8085/8080A.* Columbus, Ohio: Merrill Publishing Company, 1989.

Maloney, Timothy J. *Industrial Solid-State Electronics: Devices and Systems.* Englewood Cliffs, N.J.: Prentice Hall, Inc. 1986.

Microprocessors. Benton Harbor, Mich.: Heathkit-Zenith Educational Systems, 1981.

Single-chip Microcomputer Data. Austin, Texas. Motorola Inc. MOS Integrated Circuits Group, 1984.

Streitmatter, Gene A., and Fiore, Vito. *Microprocessors: Theory and Application.* Reston, Va.: Reston Publishing Company, Inc. (Prentice Hall Company), 1982.

Bibliography Adding a bibliography is a popular practice in college papers, although you may seldom need to create this feature unless you are in an academic setting. In general, the bibliography is a page or pages by itself. It is numbered and identified as "References" or "List of Resources" or "Bibliography." This tool is simply an alphabetical list of resources used to create a document. Each entry conforms to very precise standards. For a practical discussion of the procedures for building a bibliography see Chapter 9 of the *Writer's Handbook,* the second volume of the *Wordworks* series. As a layout practice, if you revert to single-spaced text at a later date, use double spaces between entries in the bibliography so that they can easily be read. The unique indentation that is used in a properly designed bibliography, in which the second and successive lines of each entry are indented, makes the alphabetical listing of the authors' names stand out clearly on the page.

As a final point, most authorities place the bibliography before the appendices. The bibliography is the last component identified in this discussion simply because it is often omitted in corporate documentation. It is, however, a popular tool in academic presentations.

In this chapter you have looked at the basic architecture of the technical document, particularly the common practices involved in a technical article or essay. Most of the layout conventions are, of course, standard procedure. In the next chapter you will see particular formats that demonstrate a variety of practices and conventions, depending on the content of the document or the use it is intended to serve.

You should feel confident in following the practices you have examined here. However, corporate policies will set different standards that can vary from company to company. Consider the text sample (Model 2.A) by Mark, an engineering student, that concludes this chapter. This project was first drafted to standard format and then converted to the corporate format after his supervisor approved it. Everything from the page header to the font size to the text format is unique to the company. Each of nine pages in the text features a centered graph with very brief commentary. All the graphics are in color in the original document. The entire document conforms to a specific set of guidelines for technical promotions released by the company.

The bibliography or list of references is a common feature of college projects. It is somewhat less common in corporate documentation. The layout you see here is very readable.

Model 2.A (1)

PRODUCT INFORMATION BULLETIN: DIESEL TOP OIL

Written for international distributors of Bardahl Manufacturing Corporation

Mark E.
November 9, 200X
Project 2
Supervisor
EGR 231

BARDAHL MANUFACTURING CORPORATION
1410 N.W. 52ND ST.
P.O. BOX 70107
SEATTLE, WA 99107-9607

Model 2.A (2)

TOP DIESEL REPORT

Diesel fuel quality is decreasing on a worldwide basis. Surveys of diesel fuels by EXXON suggest a gradual decline in cetane and cold flow properties and an increase in boiling end points. The literature also notes reduced fuel stability, reduced lubricity, and increasing particulate emissions from current generation diesel fuels.

Diesel engine performance is related to the quality of fuel and the efficiency of the fuel injection system. Knight and Weiser note in their SAE Paper #76721, *"Lubrication Requirement of European Automotive Diesel Engine"* that the importance of proper combustion of diesel fuels cannot be underestimated. Engine performance and fuel economy suffer directly from poor combustion. Poor combustion also increases soot levels. Increased soot results in excessive exhaust and smoke emissions as well as higher levels of varnish, lacquer, and gum deposits and increased engine wear. Other authors have reported that soot will adversely effect the antiwear performance of motor oils.

More recent studies reported in SAE Papers 861524 *"Diesel Fuel Quality and Performance Additives,"* 861179 *"Gasoline and Diesel Fuel Additives for Performance/Distribution Quality—II,"* 872146 *"Benefits of Multifunctional Diesel Fuel Additives Demonstrated in a Fleet Test,"* 892527 *"Can Lubricants and Fuel Enhancers Improve Diesel Fuel Economy?"* and 921199 *"Relationships Among Oil Composition Combustion-Generated Soot, and Diesel Engine Valve Train Wear"* have confirmed a relationship between fuel economy, emissions, engine wear, and diesel fuel system deposits. Reduction of diesel fuel system deposits, particularly injector deposits, improves fuel economy, reduces smoke and other exhaust emissions, and helps control engine wear by reducing soot formation.

Bardahl Top Diesel, also known as Bardahl Diesel Top Oil, is a multifunctional diesel fuel additive. Bardahl Top Diesel will boost the performance of diesel fuel in critical areas, helping to restore and maintain engine performance as well as controlling soot formation and exhaust emissions.

Top Diesel Will:

- Increase Fuel Stability
- Improve Injector Cleanliness
- Enhance Corrosion Protection of Fuel System Parts
- Raise Fuel Economy
- Reduce Exhaust Smoke and Emissions
- Enhance Lubrication of Fuel Pumps and Other Components
- Increase Injector and Fuel System Component Life

Model 2.A (3)

TOP DIESEL REPORT **Page 2**

TOP DIESEL IMPROVES FUEL STABILITY

Bardahl Top Diesel chemistry has been evaluated using standard stability tests which measure the ability of a diesel fuel additive to control oxidation and disperse existing deposits in diesel fuel.

ASTM D 2274 Stability Test

ASTM 2274 measures the amount of sediment and gum formed during the heating of a small fuel sample at 95∞C in the presence of oxygen introduced at the rate of 3 liters per hour for 16 hours. Increased levels of sediment and gum indicate reduced fuel stability.

Test Fuel	Total Insolubles mg per 100 ml	ASTM color, before aging	ASTM color, after aging
Diesel Fuel 2A	0.61	<1.0	<2.5
Diesel Fuel 2A with Top Deisel	0.17	<1.0	<1.5
Diesel Fuel 2B	0.68	<2.0	3.5
Diesel Fuel 2B with Top Diesel	0.17	<2.0	<2.0

Bardahl Top Diesel improves fuel stability, which reduces deposit formation on fuel system components. This improves fuel economy and reduces emissions. ASTM D 2274 fuel stability testing showed significant reductions in deposits with use of Top Oil chemistry.

Samples and Models

Comments

- The samples for this chapter are illustrations of standard practices. Each page is drawn from a different project. The table of contents and the list of illustrations are original designs, although you will find templates for these pages in several popular computer programs.
- The sample concerning iteration commands is a typical first page that uses a drop-start at the top. The wide margin indicates that it is the first page (see p. 60).
- The appendix can be identified at the top of the page containing the appendix material (see p. 66), or if the appendix material consists of several pages, an appendix banner page can be developed, samples of which are on p. 67.
- The three-page model (Model 2.A) reflects the design standards and layout policy of a specific company. Notice variations of standard procedures. All the practices discussed in this chapter are commonly modified to suit the specifications of a corporation.

NOTE

Any multipage sample in this text is referred to as a "model," and the sequence will be indicated in parentheses, for example, 2.A(1,2,3).

Summary

- The components of a document are as fixed and as traditional as the parts of a business letter, although features of the components can be modified to meet specific needs.
- A number of elements may precede the body of the text:
  - A title page
  - Tables of contents and illustrations if desired
  - Additional pages itemizing tools needed, materials needed, and the like (see p. 88).
  - A list of terms, a symbology, or a graphic sometimes appears in the front of a text (see p. 357).
- Format details are standard:
  - The body of the text is produced in a 12-point font.
  - The text is double-spaced.
  - Paragraphs are indented.
  - Graphics are integrated into the text unless otherwise instructed.
- At the conclusion of the text, additional sections are added as needed:
  - A glossary, a symbology
  - A bibliography
  - Appendices

Activities Chapter 2

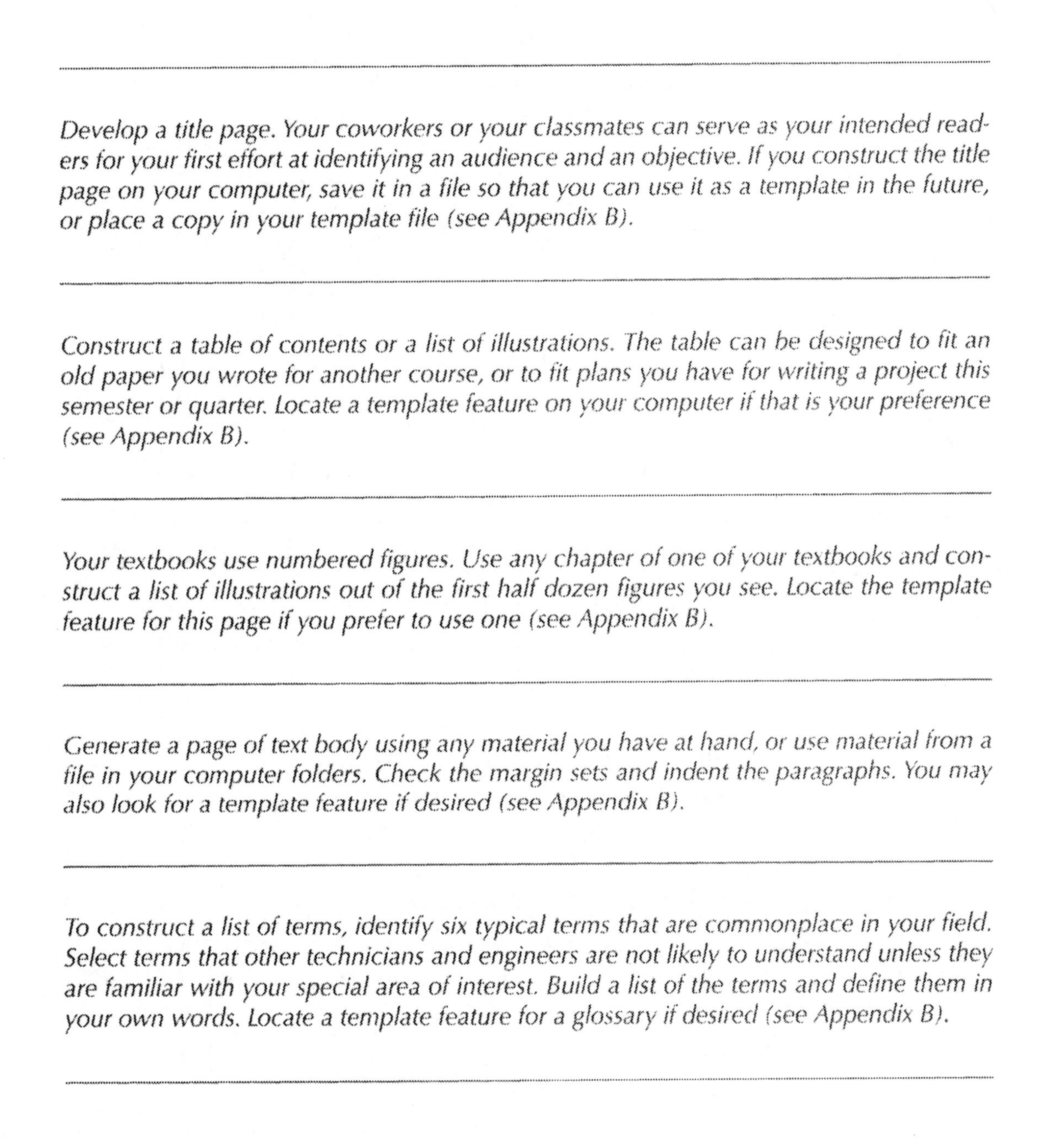

Develop a title page. Your coworkers or your classmates can serve as your intended readers for your first effort at identifying an audience and an objective. If you construct the title page on your computer, save it in a file so that you can use it as a template in the future, or place a copy in your template file (see Appendix B).

Construct a table of contents or a list of illustrations. The table can be designed to fit an old paper you wrote for another course, or to fit plans you have for writing a project this semester or quarter. Locate a template feature on your computer if that is your preference (see Appendix B).

Your textbooks use numbered figures. Use any chapter of one of your textbooks and construct a list of illustrations out of the first half dozen figures you see. Locate the template feature for this page if you prefer to use one (see Appendix B).

Generate a page of text body using any material you have at hand, or use material from a file in your computer folders. Check the margin sets and indent the paragraphs. You may also look for a template feature if desired (see Appendix B).

To construct a list of terms, identify six typical terms that are commonplace in your field. Select terms that other technicians and engineers are not likely to understand unless they are familiar with your special area of interest. Build a list of the terms and define them in your own words. Locate a template feature for a glossary if desired (see Appendix B).

Copy an important page or table from one of your textbooks. Construct an appendix out of it as though you intended to put it in a project. This is an opportunity to use a scanner but it is not necessary. A copier and a glue stick will do.

If you are also using the Writer's Handbook, *you may be interested in the proper construction of a bibliography. The usefulness of this exercise will depend on your program. See your instructor for suggestions.*

You may use the page sample that you develop for each of the exercises as a template for any future projects that call for the same page and structure—if you built the exercises on a computer. Put this timesaver to work by saving all the samples and by placing the samples in your existing template file (see Appendix B).

Share a Project

Collaborate with two other members of the class to build the parts of a technical project. Your group will be most efficient if all the members are from the same engineering technology program.

- *Decide on a subject that can be constructed into a technical document that will include the following parts:*

Title page	**First page of text**
Table of contents	**List of terms**
List of illustrations	**Appendix**

- *Decide on two pages that are to be developed by each team member.*
- *Assign a due date and bring four copies of each page to the meeting. Staple sets of the six components for each member and discuss the procedures and outcomes of the activity.*
- *The fourth copy is for the instructor.*

Work in Progress

3. An Agenda

Once I understood the general situation, I next had to understand what the problem was. It was time for a little fieldwork. I took a hard copy of the existing process with me. I always do this so I can write notes on working documents to remind me of pertinent items. Because Genus is in four very large buildings, it is easy to get confused by other issues when going to and from the work areas.

I contacted the employees noted on the e-mail and went to their areas to understand their role and discuss the issue. It became apparent that some departments did not know and understand what the other departments were doing with respect to the issue I was to examine. Everyone understands things a little differently and has their own opinions.

To clarify the issues, I used e-mail to set up a meeting to get the applicable employees together to discuss the problem (which was not yet clear to me). Since I was calling the meeting, I had to get a meeting packet together and an agenda for each person. I got copies of the existing distribution process documents together and assembled them in a sequence to show how meter stock is currently ordered, received, inspected, routed to stock, and issued to departments.

The meeting was held, and another problem surfaced. The production planner came to clarify changes in the process. The senior inspector provided additional specification requirements, and the stockroom issuer provided transaction information. They could not agree on the issues and requested that their supervisors also attend due to the potential impact of any proposed changes. As is frequently the case, most employees know a great deal about their part in a process and are not aware of how they have an impact on others. I scheduled the follow-up meeting. I kept meeting minutes and forwarded these to the supervisors who would be attending the follow-up meeting.

The follow-up meeting was held, and many little problems were brought out in the open that aided all the departments involved. It turned out that a failure to track certain system transactions between departments was the root cause of the problem. Basically, meter inventory was documented as being in one location when in reality it was already physically issued to departments. Everyone left the meeting with a promise to "red-line" the agreed upon changes in the documents they use and to give these to me so that I could assemble the formal production change.

The agenda was now clear to me. Each department would analyze the changes needed and send me the documentation. I would develop the overall system changes based on the documents and forward the project for approval. I still had two weeks to go.

S. B.

The Outline Format

The next matter of concern is the internal structure of a document of any substantial length. The last chapter explored the overall structure in terms of layout conventions. Most variations of the shape of the product are adaptations of the basic layout practices. The variations are usually alterations to the *body* of the text that are made to allow the text of a document to more readily meet the needs of either the subject of the project or the readers. To the extent to which the production is going to be content driven, writers change the basic model by using three primary techniques of presentation: the outline method, the running-text method, and the heading-and-paragraph method. You should look at the utility of each of these formats so that you understand when and how to use them. As you move along, study the models. They will coincide with the formats under discussion.

Model 3.A (1)

THE PRODUCTION OF BLOOD COMPONENTS

FROM WHOLE BLOOD

Audience: This document is intended to instruct new technicians at the Pacific Sound Blood Center (PSBC) in the details of a specific procedure. New technicians are assumed to have at least two years of college education with a science background.

Objective: To explain the separation of a unit of whole blood into its component parts, how to properly document component creation in the PSBC computer system, and how to properly store the new components.

Matthew W.

02/02/2000 Project 1

EGR231.01 D. R.

The first option is the outline format. Outlining is something of a lost craft but it is critical to success in tech writing. Outlines are usually composed of numerical and topographical controls that are used to structure writing. A variety of very important technical support documents are vital in business and industry: maintenance and repair manuals, troubleshooting manuals, procedural documentation, regulations, specifications, and a wide variety of contracted agreements. These groups of documents usually function best if the content is organized in the manner of an outline. Conventional outlines may serve the purpose quite adequately. Equally popular is the outline style called "MilSpec." MilSpec is simply a numerical outline technique standardized by the U.S. Government for uniformity and practicality. Also, many variant systems have been designed by many corporations and state and local government agencies to serve their own purposes.

Outlines place a heavy reliance on two organizing tools that are easy to use and that, you should note, take *no* writing skills whatsoever. The outline depends on alphanumeric controls for organization and topography for layout logic. The outline is an architectonic tool in ways: it is a descending staircase in design, and it is alphanumerically and spatially ordered, but there is a task involved in organizing the material. Usually a writer will outline when the content more or less demands an outline. A complex set of instructions calls for the attention to detail that an outline can provide. For training a young surgeon in bypass surgery, a little hard copy reviewing is always available in the office. (And the references *are* often used although the public is not encouraged to think surgeons have to read up on procedures. They do.) Accomplished professionals use their resources for consultation. I have spent many hours in a surgeon's library, and it is obvious enough that, at least in general surgery, a surgeon cannot remember all the countless procedural matters, detail by detail, without review. The outline method is the most thoroughly organized tool any professional can refer to, whether the issue concerns a heart problem or a hard drive problem in the office computer system.

Is an outline format the way to go? In order to make that decision, you must understand the *advantages* and the *drawbacks* of the various formats. Because outlines can line-item the content of a document more readily and more thoroughly than any other format, the strengths and weaknesses of outlines will be fairly apparent.

On the Plus Side

As noted already, instructions, manuals, regulations, specifications, and similar products work effectively in the outline style. This is because they are most often seen as *reference* documents. Outlines are particularly easy to consult because readers can get into and out of the text rather easily since outlines serve as their own index. Outlines are also *obvious,* which is why they can be referenced quickly. There is very little mystery to the designs and the ordering systems of outlines. An outline follows a topographical ladder that is hard to overlook. Instructions, for example, may be easily developed with indentations and numerical controls for ease of presentation.

This project was an in-house set of instructions that was designed to meet the specifications of a medical laboratory.

Model 3.A (2)

I. Principal

Whole blood is made up of several constituent components. Each blood component has a different therapeutic value when transfused to a patient. Not every patient needs to benefit from every component; hence, the utility of each donated unit can be maximized by separating it into it's component parts. The process of separating components is called *fractionation*. Fractionation consists mainly of spinning a unit in a centrifuge and expressing the plasma into an attached satellite bag. By collecting blood in collection sets having three attached bags, it is possible to convert a unit of whole blood into Red Blood Cell (RBC), Platelet (PLT) and Plasma (FFP or RP) components.

II. Objective

The procedure illustrates the process of converting a Whole Blood unit into an RBC, PLT, and FFP component. Additionally, the procedure covers the necessary steps for documenting the component creations, and the proper storage conditions they should be placed in.

III. Scope and Responsibilities

A. The production process described in this SOP will be performed by trained technicians in the Inventory Production Department.

B. Storage equipment and areas designated as "controlled storage" will be monitored and maintained by the Facilities and Engineering Department.

Also, although the alphabetical and numerical systems can be used for *any* purpose in an outline, they are frequently chronological or prioritized. Again, these are content-driven matters. If the subject matter is ticking like a clock, then sequential reasoning dictates the order of the events— *and* the writing format the author will select.

A final and very important point on the plus side has to do with you as a writer. Most engineering tech students and engineering students quickly realize that writing in an outline form is easy. In other words, of the several formats identified here, the outline technique is clearly the best available option for a person for whom writing is not a strength. If you know that writing is a limitation for you, you may find the outline to be a convenient tool that will allow you to meet the demands on your own terms.

On the Minus Side

Imagine a James Bond story in outline format. What a bore. Outlines do not entertain; they *inform*. They are dry. They are mechanical. They are lockstep. They are also much too convincing. They are so well organized that people trust them to be right. Outlines have warm-bodied sources that can lead to errors. This is one reason why software comes with an 800 number.

After I made this comment in a lecture a few years ago, a person in our industrial power program said that the local utility company had to use highlighters to indicate sections of their manuals that, for one reason or another, were *not* to be followed to the letter. There were frequent glitches for a host of reasons, from typos to technical inaccuracies to localized adaptations or variations on the manuals. The class was quite surprised when he brought in one of the manuals and explained a few of the highlighted sections. The point was that a rookie had to be told *not* to trust the procedures blindly if they were coded with highlight coloring. For the writer this risk should be a concern. Authors have quite a liability, since readers take writing on faith—particularly in the case of an outline.

Two other negative points involve the lack of depth in many outlines and, a similar matter, the possible lack of alternatives. Outlines by their nature may not contain much explanation, so the result may be a little bare-boned. There are few frills. Outlines get to the point and are done with the job. Also, unlike flowcharts, which usually have decision boxes to lead a reader into options, the outline is very often prefigured in a fairly static way, and options are limited. Other problems include the fact that readers scan outlines and may miss details. They may also plug in at any point of interest and miss sequential processes.

In this sample notice that the text begins with paragraphs. Paragraphs are used for the preliminary discussion to set the stage for the outline to follow.

Model 3.A (3)

IV. Quality Controls

A. Prior to processing into platelets, all units must be kept at a temperature of 20–24°C.

B. All blood components must be processed and handled in a manner that prevents the introduction of external air to the component bags.

V. Materials

A. Pacific Sound Blood Center—Inventory Production—Standard Operating Procedure Manual.

B. Donated Whole Blood (WBD)

C. Product Labels
- Platelets
- RBC
- FFP
- RP-Frozen

D. Centrifuge cups

E. Two-pan balance

F. Sorvall RC-3BP refrigerated centrifuge

G. Plasma expressor

H. Tubing clips

I. Dielectric sealer

J. Electronic scale

K. Biohazardous-waste container

L. Computer terminal with attached bar-code scanner

M. 20–24°C monitored platelet incubator

N. 1–6°C monitored refrigerator

O. -18°C or less monitored freezer

Usage

Regardless of how much of the text is outlined, open the presentation with a few paragraphs or pages to explain your document. In other words, if the entire text is an outline provide the reader with a little steerage in an introduction just in case there are questions. Also, remember that the success of topography depends on vividness. If subcategories are not apparent, you will defeat your purpose. To that end be *sure* you indent successive subgroupings five spaces (or more) so that the reader's eye perceives the categories. As you know, there is a limitation to this sublogic process. After about four subcategories are developed, your subsubsubsub category is about two inches wide on the right side of an empty sheet of paper. Avoid too many divisions.

The outline style called *military specification,* or MilSpec format, is a very important technique for any corporation that deals with government contracting. If your company demands MilSpec, you will need a procedures manual. Of course, millions of Americans learned a variety of MilSpec practices as a routine part of their duties in the armed forces. Whether or not they wrote documents in MilSpec, most servicemen and -women had to read military specifications daily.

MilSpec is a structure that uses only numerical elements. That may sound complicated but it is as simple as the following illustration:

Outline	**MilSpec**
I. ______	1. ______
A. ______	1-1. ______
1. ______	1-1-1. ______
a. ______	1-1-1-1. ______
b. ______	1-1-1-2. ______

In other words, it takes one number to indicate a primary category in MilSpec, but it takes four numbers to indicate a subsubsub category. It seems complicated, but it is actually a convenience. Notice that no line of the traditional outline indicates where you are in relation to other sections of the outline. In MilSpec you always know precisely what the category is as well as the step number.

MilSpec often follows a convention for the first set of major categories so that the documents always conform to a basic architecture. Here is the order of the parts for a typical manual.

In this sample page, notice that the text does not rigorously hold to the outline style suggested. There are many variations. This sample uses headings for the lists that are developed beneath, and bullets were used for the subsets.

Model 3.A (4)

6

VI. Method

[Sections A through C are not shown here. DR]

D. Perform Platelet Rich Plasma (PRP) spin.

1. Units must be loaded into the centrifuge.
2. Set centrifuge to the Platelet Rich Plasma (PRP) settings.
3. Start the centrifuge by pressing the RUN button.
4. When the run cycle is complete gently remove the cupped units from the centrifuge and place them by the plasma expressor.

   • **Important Note**: Do not to disturb the RBC/Plasma interface after centrifugation. Handle the units very gently to prevent the RBC from re-mixing with the plasma.

5. Express Platelet Rich Plasma (PRP)

   a. Place the unit in a plasma expressor. Then apply a tubing clip to the tubing leading into the satellite bag without a Platelets product label. Note the following setup of apparatus (Figure 4).

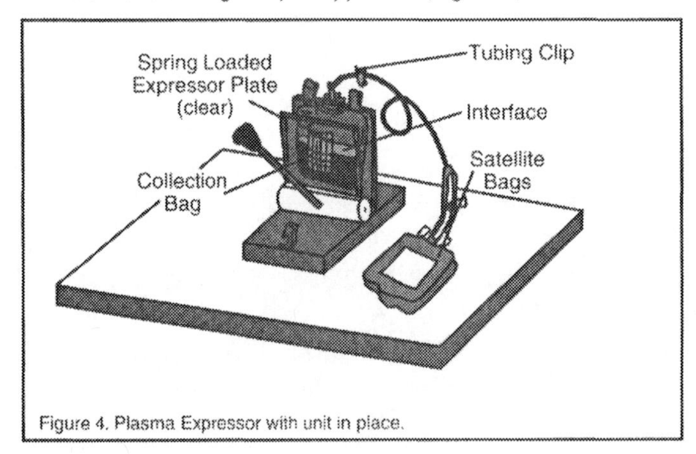

Figure 4. Plasma Expressor with unit in place.

Severe injury may result from improper centrifuge loading.

1) **Receipt and Installation**
2) **Theory of Operation**
3) **Operating Procedures**
4) **Maintenance Procedures**
5) **Troubleshooting Procedures**
6) **Parts List**

This configuration allows *all* documents to be uniform yet still adaptable to their particular needs. The subcategories are dependent on the content. The subdivisioning proceeds in logical and orderly fashion, so that oiling an instrument and cleaning it would be 4-1 and 4-2, respectively.

Manual updates are an enormous expense to government agencies, so MilSpec documents are page-numbered for replacement in a manner similar to one noted earlier. Sequential numbering stops at the end of each of the six major categories. The page number is prefixed with one of the section numbers, so that 5-392, for example, is p. 392 of the troubleshooting section, which can be updated as a separate document. If two new pages are installed, they simply become 5-392a and 5-392b. The system is efficient. The same logic applies to appendices, which essentially become category seven. Here the pages are numbered with an alphabetical prefix because appendices are alpha listed. So if Appendix B has three pages, they will be B-1, B-2, and B-3.

This system also extends to graphics, which are subdivided into figures and tables. Because graphics can be updated just like the text, a new schematic, for example, has to have a two-part numbered tag. The first number is the number of one of the basic six divisions, and the second number is the position in sequence. Thus, the sixth schematic in the maintenance procedures is Figure 4-6.

To give you a sense of the MilSpec style, a former federal employee, Troy, constructed the following directions for baking a cake.

Once the instructions begin you see that the conventional outline is used. Notice that the indention is very much a part of the logic of the outline structure. Distinct indention is as important as the enumeration.

4-5 Desserts

4-5-1 Cake, Chocolate layer 9"

Recipe for cake, chocolate layer 9", feeds 6–8 people. Approximate preparation time: 1 1/2 hours.

4-5-1-1 Preparation

Assemble ingredients and utensils listed in tables 4-27, 4-28, and 4-29.

NOTE:

Keep milk and mayonnaise refrigerated until ready to use.

Preheat oven to 350 degrees.

Table 4-27 Ingredients for Cake, Chocolate Layer

Ingredient	*Unit of Measure*	*Quantity*
cups unsifted flour	cup	2
cup Hershey's cocoa	cup	2/3
teaspoon baking soda	teaspoon	1 1/4
teaspoon baking powder	teaspoon	1/4
cups sugar	cup	1 2/3
eggs	each	1
teaspoon vanilla	teaspoon	1
cup Best Foods Real Mayonnaise	cup	1
cups water	cup	1 1/2

4-5-1-2 Procedure for Cake

Makes two 9" layers.

4-5-1-2-1 Step 1

Grease bottoms and sides of 2 9" layer cake pans.

4-5-1-2-2 Step 2

Dust greased sides and bottoms of cans with cocoa (see figure 4.36).

4-5-1-2-3 Step 3

Mix 2 cups unsifted flour, 2/3 cup Hershey's cocoa. 1 1/4 teaspoons baking soda, and 1/4 teaspoon baking powder.

4-5-1-2-4 Step 4

4-35

You may have occasion to use MilSpec, but it goes beyond the simple matter of a subject that calls for outlines. The determining factor in this case is more a matter of uniformity—*and* conformity to government or company practices. Conventional alphanumerical outline methods should serve your purposes.

The project for the production of blood components (Model 3.A) is a simple and vivid set of instructions using the outline method. The format can be made more complex, as can be seen in the cake instructions. Given the opportunity, simplify complexity.

Model 3.B (1)

BUILDING DESIGN FOR SEISMIC FORCES

Audience:
First-Year Drafting Students

5/27/200X
INSTRUCTOR: DAVID R.
PROJECT 3
EGR. 231.01

South Seattle Community College

The Running-Text Format

You are not likely to see any academic articles handled in the outline style. Small outlines can be inserted here and there in any document, but academic articles are seldom based on outlines from start to finish. The technique of running paragraphs one after another for the duration of the document is quite popular for scholarly work, and it is probably the oldest of the methods. That does not make it venerable—except at colleges and universities, where the method remains the most likely entrant in most course work that involves major investigations and term papers. The method is also notably used as the chapter-by-chapter format for graduate school productions and for very large documents at the dissertation level. In the corporate environment the running-text approach is not as prevalent a practice. Perhaps the learning communities are more traditional. Perhaps, also, the density of the method is of little concern to knowledgeable academic readers who are at ease with it.

To some extent, of course, paragraphs will show up in any format. However, extensive paragraphs, with few if any demarcations to divide a text into subsections, are the trademark of this style. The usual approach to teaching English 101 and English 102—the national benchmarks of essay-writing basics at any college—is a strict focus on the running-text method to the exclusion of outlines or heading methods. There is probably an unspoken enthusiasm for the skill it takes to construct a document in a running text. This means that from the viewpoint of many engineers and engineering techs, the first strike against the concept is that it is a difficult method. There is, of course, a positive utility in the format.

On the Plus Side

In academic environments the frequent preference for some variation of the running-text style is a reason to master it for transfer students who plan to complete a four-year degree. More to the point, however, is the *reason* for this preference. The issue has less to do with the run of paragraphs than the paragraphs themselves. Paragraph logic allows for a great deal of elaboration of thinking and provides discussion of such details as evidence. The author has virtually no constraints in a running-paragraph text. The paragraphs can be as numerous and as detailed as the author needs. In other words, for long discussions there is *no* replacement for the running-paragraph format, and so, as a rule, it is the choice for all lengthy academic projects. In addition, the method is conversational, though formal, and reveals the evolution of the author's thinking. Also, since there is no restriction on the topics or how they are addressed or how long the paragraphs are to be, these documents can be argumentative. They can also be subtle, which outlines were never intended to be.

This is a conventional title page for an academic project handled in the running text style.

Model 3.B (2)

Earthquake Risk

Designing buildings to be resistant to earthquake hazards requires the analysis of many factors. Every new building design presents a new combination of factors to be considered. New problems and solutions come to light in the aftermath of large earthquakes when the damage to structures is inspected.

Minimum building requirements for seismic resistivity have been standardized and written into local building codes in most areas. Good construction, especially for large or critical structures, will often go beyond the minimum. These building codes are based on complex engineering formulas and calculations, but an overview of some of the general requirements is possible.

Structural design for earthquakes involves two special considerations. One is that the principal application of forces is in a horizontal direction, which makes it perpendicular (or lateral) to the direction of the force of gravity.* The second consideration for earthquake forces is that they are dynamic.

---

*Earthquake forces are not limited to horizontal forces. Shock waves cause motion in all directions, including up and down, but because of the safety factors used for gravity design, the vertical effects are not usually critical.

On the Minus Side

The running-paragraph style is somewhat demanding of a writer. One problem is that the organization of the running-paragraph method lacks obvious mechanical tools; that is, transitions cannot simply be enumerated, for example. Instead, the writer must know how to construct sentence transitions between paragraphs and how to build entire paragraphs that sum up major divisions of the composition or introduce new divisions. The internal continuity is also difficult to create because the running-paragraph style depends on paragraph logic and the skill to write paragraph logic structures.

There can also be drawbacks for the reader. The text is less obvious than other formats, so a reader can miss central points. The writing can also be hard to follow if the text is lengthy and windy. These are usually *not* quick reference documents, even with an index.

There is a final subtle issue in the running-text style. In the same sense that outlines are trusted for curious reasons, readers may find the running-paragraph style talky and not entirely objective.

Usage

The running-text format depends on various internal transitions for continuity. Even if you are familiar with developing transitions, you need to be very sensitive about transitions as tools in tech writing. Transitions are assists for the reader. For example, a sentence that begins with or includes the word *then* or the word *however* creates an immediate logical continuity for a reader. Authors use dozens of these simple transitions. Phrase transitions are just as common, such as "On the other hand . . . ." Equally prominent are entire sentences: "We must also consider the possibility of interference in the circuit performance." These short transitions may be linking sentences or they may lead a reader from one paragraph to the next. Larger sections of text can be held together with entire paragraphs that sum up and guide the reader onward. These paragraphs also act as transitions.

Unique to tech writing is the additional obligation of moving readers into and out of the other "language" systems—into and out of chemistry, math, physics, tables, charts, and graphics. Be alert to this special need, particularly in the running-text format. Technical documentation needs a special type of transition—a *display transition*—to move the reader from one language to another (see pp. 458–460). Many documents are choppy because of the abrupt shifts between math, graphics, and English. With the convergence of visuals

Because a great many documents open with paragraphs, the actual format is not always immediately apparent. Note that this author favors brief paragraphs.

Model 3.B (3)

-2-

Forces on buildings can be analyzed in terms of either static loads or dynamic loads. Static loads are concerned with forces in a state of rest or equilibrium. Consideration of dynamic loads on structures must include the effect of time.

In discussions of the effects of dynamic forces on structures, two ideas should be understood: energy capacity and harmonic motion (vibration).

Energy can be defined as the capacity to do work. In structural analysis the concern is mechanical energy, which occurs in one of two forms. *Potential* energy is stored energy, as in a compressed spring or an elevated weight. Work is done when the spring is released or the weight is dropped. *Kinetic* energy is possessed by both bodies in motion; work is required to change their state of motion, to slow them down or speed them up.

Harmonic motion is a type of kinetic energy, which is of major concern in structural analysis. A commonly used illustration of this type of motion is a swinging pendulum. In neutral position the pendulum will remain at rest. If the pendulum is moved from this position by being pulled sideways, it will tend to move back to the neutral position and will be carried past it by momentum to a position of displacement in the opposite direction. See the motion illustrated in Figure 1.

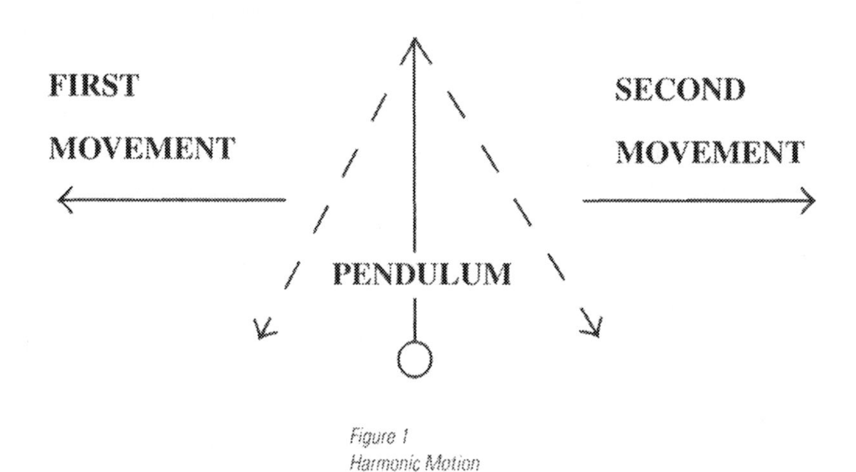

Figure 1
Harmonic Motion

and other symbol systems, the English text easily becomes a halting trail of starts and stops, so you must take the reader by the hand:

> The calculation was easy to determine: . . . .
>
> Notice the time curve on the following chart.
>
> The formula is easily expressed as . . . .
>
> This procedure is visible in the lower left corner of the preceding illustration.

Essentially, you have to *write* the readers *into* and *out of* every major shift in medium from writing to graphic to calculation and back again. Use conventional transitions to bridge the logic of the paragraphs, and use display transitions to bridge the logic of the languages or mediums. Each model in this discussion of formats has a graphic display and a display transition.

In the running-paragraph style, be aware of the need to hold the document together and the need to hold on to your reader. Every shift from language to another medium such as math has the potential to create two disjunctions—one on either side of the math display. You must write your way into the display and you must write your way back out. The same holds true for graphics. The preceding brief expressions show you the two most practical strategies for linking the language to the technical elements. One option is to address the reader with a command: "see . . .," "notice . . .," "observe . . . ." The other option is to explain the logic of the link between the text and the display.

> This reaction is complex:
>
> The belt (2) is held by the pulley (3) in Figure 3.
>
> This calculation demonstrates . . . .

In sum, *connect*. The task of holding all the material together is one of the major challenges you will face. In truth, with a little writing practice, transitions—display transitions in particular—become an easy tool for document organization, and you will place a premium on this element in your writing. (For more discussion of displays see Chapter 8 of the *Writer's Handbook*).

Once the text body is underway the format is more apparent. This sample is a running text that is broken only by occasional use of graphics that support the paragraph logic.

Model 3.C (1)

PROCEDURES FOR CASH APPLICATIONS

A SECTION OF CUSTOMER AND BUSINESS SERVICES

PART A

AUDIENCE PROFILE: Anyone requiring a brief description of the role of Cash Applications within the department of Customer and Business Services.

OBJECTIVE: To provide an overview of the activities of Cash Applications and its relationship to other departments. An overview of the procedures for each activity is included.

April 25, 200X

Project 1

Dave R., Instructor

The Heading-and-Paragraph Format

There is a hybrid format that combines the best of both worlds of the different approaches of outlines and running texts. Magazines are particularly notable for using the heading-and-paragraph method, which basically adapts the outline style to a text of paragraphs but without numbers or letters. Each section of paragraphs is given a major heading, and, of course, there can be layers of subheadings. The paragraphs are not numbered, and the headings frequently are not numbered either.

The package is organized by the headings, which act as cues for the upcoming subject matter. As you realize, I am using this method throughout this text, and it is used in the other volumes of the *Wordworks* series. This chapter has a banner page (see p. 83), which is the major heading. Then I constructed the page formats for the key subsections. The heading styles identify the levels of subsections. There was little need for the outline numbering system and certainly no need for MilSpec complexities. This has been the preferred technique of magazine publishing for over a century.

On the Plus Side

As a hybrid, the heading-and-paragraph format preserves the best of both worlds. Headings guide the reader directly and boldly. The text style is reasonably rapid for referencing. The author is saved much (but not all) of the difficulty of carrying the reader forward with transitions.

The method leaves paragraph logic intact, and there is no noticeable choppiness, since the subdivisions are seldom divided as extensively as you see in an outlined text. This does mean, however, that you still need to know the craft of constructing paragraphs, which is the core skill of writing once the overall organization is under control.

Usage

The headings are the key. They are usually kept short and precise. Their placement and size and consistency matter. Readers are quite conditioned by a lifetime of reading, and they always understand the logic of heading design—which runs from a shout to a whisper by changing from larger to progressively smaller letters. I suggest that you do not change the font of the text. If you do, use the heading font for *all* your headings and rely on size changes for helping the reader see the sets and subsets. I prefer boldface to a

This project was designed for the accounting staff of a hospital and the author chose to use headings throughout the document.

Model 3.C (2)

3

REVIEW OF PROCEDURES

LIVE CHECKS

Live checks (checks not yet deposited to the bank) are processed every day for deposit in the daily deposit (the deposit of checks made each day by Cash Applications). All payments require research to determine where to apply the funds. The bulk of the live checks belong on patient accounts. Following is a description of the different types of payments.

Patient checks are from patients themselves for the balance due on account after the insurance has been paid. Many of these checks are forwarded from the credit department, which works with patients who have trouble paying. Most of these checks are photocopied and are deposited daily without fail. All patient checks are posted in Cash Applications.

Insurance checks are also deposited every day. Some payments require more research than others, depending upon how much information has been included. All insurance checks are posted by Voucher Processing.

Other payments belong to AR. Either the payment is forwarded to Cash Applications, already coded with the GL number, or Cash Applications contacts Accounting for the number. Although Cash Applications deposits these checks, they are not part of the patient payments. They are recorded separately.

The steps for processing the live checks are as follows:

- Mail is sorted by the mail room clerk. Only envelopes containing checks are delivered to Cash Applications.
- Cash Applications then opens the mail.
- The checks together with their EOBs are reviewed and sorted into batches.
- Any EOBs without a patient account number are researched at this time.
- Any nonpatient payments are identified and assigned a GL number, and the check is photocopied.
- The checks are separated from the EOBs and placed in a pile.
- A tape (adding machine ticker tape) is run on each group of EOBs to create batch totals.
- The batch totals are taped (added on the adding machine).
- The checks are taped twice.
- All three tapes must balance (match each other).

change in font style. If you use a typewriter, you can use caps and underlining for your emphasis, or perhaps italics. Let's structure a typical group of possibilities:

HEADING	**18-pt bold**
SUBHEADING	**14-pt bold**
SubSubHeading	**14-pt bold, upper- and lowercase**
SECTIONAL DIVISIONS	**11-pt bold**

One of these headings, whichever is most important, could also be underlined. There are many variations to this system—including color—but two elements are critical. First, *always* use larger letters to indicate a more important division, and smaller letters to indicate a lesser division. Second, you must be doggedly consistent because readers quickly understand and depend on your system of choice. It seems simple enough, but I correct heading errors all the time. The errors are easy to find, as any confused reader knows.

Notice that headings are displayed, that is, they are set off from the text on their own. Two other popular practices involve the use of headings *on* or *in* the line of text. The first device is the *on-line heading.* Follow it with a period (usually) for maximum clarity. For example,

> **Calorimetry.** Hess's law can be demonstrated by measuring the heat of neutralization of aqueous HCl and aqueous NaOH.

Boldface allows you to construct *running headings* also. The heading is simply part of or in the sentence.

> **Ionization Constants** can be derived from acid-base titrations without difficulty.

Any number of headings look sharp with ticks, bullets, dingbats, or wingdings (avoid the corny ones). The modest black dot is probably the most popular.

- **Phase Annulus.** A precentered phase annulus is attached to the substage mount of the microscope.

The content of a project using the heading-and-paragraph style will still need to be oiled with transitions. Headings are a great help to you as the writer and to the reader, but you cannot depend on them exclusively. You must also continue to develop *written* transitions—particularly the large ones that explain what a section has just accomplished and what you will do next. Map the document. You construct a lot of hills and mountains for

Partly because of the advent of word processors, the heading-and-paragraph format is now very popular for manuscript work and self-publishing. This sample uses a centered heading, a left margin subheading, and boldfaced running headings.

Model 3.C (3)

LOCKBOXES 10B AND 20H

The bank prepares the lockboxes (individual departments within the bank process payments on behalf of businesses) for the medical center. Certain bills go out with a post office box number belonging to the bank. The bank receives these payments, photocopies them, adds and balances the checks, and then deposits them.

Lockbox 10B is for clinic services, Lockbox 20H is for hospital services. Hospital payments are sometimes mailed to the clinic lockbox and vice versa.

The check copies, together with their EOBs, are bundled in groups of 50 or fewer and taped. A tape of the bundle totals is attached to the deposit slip. The bundles, tapes, deposit slip, and all other paperwork are brought by courier to the medical center each day by 4:30 p.m. The lockboxes are processed the next day. The two lockboxes are not mixed together for processing because to do so would cause confusion. Following is a description of the different types of payments:

Patient checks for the clinic are bundled together by the bank. These bundles can then be posted immediately, since no other sorting or research must be done.

Insurance checks for both the clinic and the hospital must be researched and sorted into batches, as for live checks.

Other payments occasionally come through one of the lockboxes. These payments are assigned a GL number and recorded separately.

The steps for processing a lockbox are as follows:

- The bank prepares the lockbox (which consists of bundles of check copies attached to their EOBs with a tape listing each check amount and the total of the bundle), a tape of the bundle totals, and a deposit slip listing the grand total.
- The lockbox is delivered to Cash Applications by a courier at 4:30 p.m.
- The bundles of patient payments are immediately removed for posting.
- The check copies together with their EOBs are reviewed and sorted into batches.
- Any EOB without a patient account number is researched at this time.
- Any nonpatient payments are identified and assigned a GL number.
- A tape is run on each group of EOBs to create batch totals (no more than 50 payments per batch).
- The batch totals are taped.

your readers, so you must direct them with road signs made up of headings and transitions. Take the reader by the hand. Rather than thinking of yourself as a writer, think of yourself as a leader.

Remember that the reader does not have a clue. A writer always faces a problem referred to as "omniscience," which means that the writer *knows* what he or she plans to say and what the strategy will be. As a writer you *always* know what will be on the next page. Obviously, the reader does not. You know what you want to say. You know the order in which you will say it. You know exactly what is coming next and why. This foreknowledge is a barrier to your ever seeing your work the way readers see it. As a result, it is difficult to anticipate their needs—though you must try. And it is difficult to accept their complaints—though they provide you with the best possible insights for revisions.

When you see this in your document,

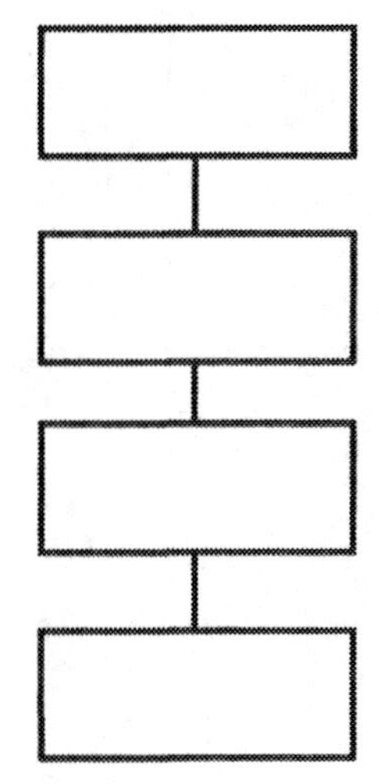

you do not want the reader to come up with this

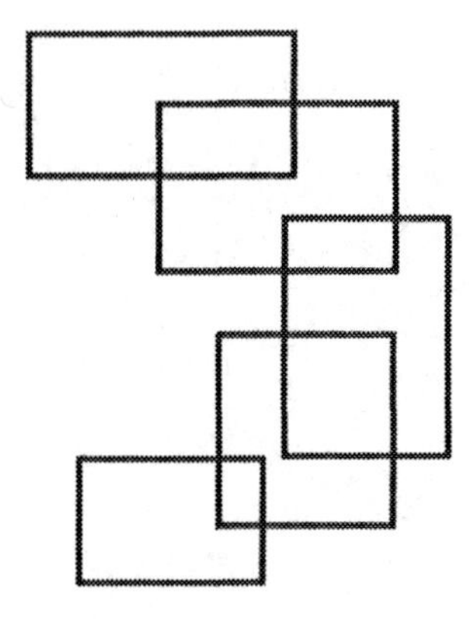

This page maintains the heading continuity, as does the balance of the text. Note that bullet lists add clarity because bullets function as an organizational device for the reader's eye.

Sample 3

Risk Factors and Incidence

Risk factors are:

- Being of 50 years old
- Being female
- Obesity
- Having some sort of thyroid surgery
- X-ray or radiation treatments

The incidence for having hypothyroidism is 2 out of 100 people.

Symptoms

There are many different symptoms of hypothyroidism. Some of them come on early and some late. Early symptoms include

- Pale color
- Thinning of the hair
- Thin, brittle fingernails
- Joint or muscle pain
- Depression
- Unintentional weight gain
- Cold intolerance
- Fatigue

Late symptoms include:

- Hoarseness
- Thinning of the eyebrows
- Decreased taste and smell
- Puffy face, hands, and feet
- Thickening of the skin
- Dry flaky skin
- Slow speech

There are many other, lesser symptoms to hypothyroidism. If three or four of these symptoms occur, consult a physician and get some tests run to see if it really is hypothyroidism, or something else.

Tests

There are a couple of different ways to diagnose hypothyroidism, but the most efficient and effective way is by measuring the blood TSH level.

Other diagnostic tests are

- Checking the thyroid hormone level
- Doing a thyroid biopsy
- Doing a thyroid scan and or a sonogram
- «Checking for thyroid antibodies: antiperoxidases (anti-TPO) or antithyroglobulin (anti-T_g).

Treatment and Management

There is only one way to treat hypothyroidism, and that is by hormone replacement therapy. The drug used in this replacement therapy is *levothyroxine* (Figure 5). Levothyroxine tablets provide

Figure 5
Sodium L-3,3′,5,5′ tetraiodothyronine,
Levothyroxine

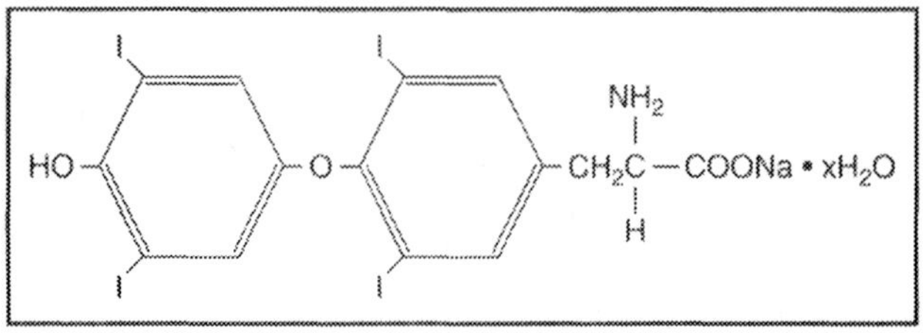

The three varieties of formats identified here are subject to variation. They are quite flexible. Outlined materials can easily be inserted in the running-paragraph model or the heading-and-paragraph model. A particularly popular device is the tick list. A list of any kind can be inserted in any of the models. It is quite popular to use bullets for a list, as illustrated in Model 3.C (3).

A somewhat more complicated issue results from a two-column format. A conventional running text of paragraphs is easily designed for a two-column presentation. If, however, there are headings, the layout calls for more attention to details in the design. If there are subheadings and bullet lists, the layout can become difficult for the reader to follow. The page on the left (Sample 3) is quite well handled, but you can see that the headings and tick lists could pose a problem for both an author and a reader. **(73)**

I will conclude the discussion with a comment on "margin numbering." This chapter explained three basic formats that are the usual paths that are used for technical documentation:

- Outlines (and MilSpec)
- Running texts
- Headings and Paragraphs

All three methods can be *externally* numbered in a variety of ways. You have seen legal paper; every line of text is numbered in the margin. This is a common tool used by attorneys and lawmakers that allows them to isolate any page and line of text immediately during discussion. This tool may prove to be of practical use from time to time. A little less fussy is a technique in which, if paragraphs are the major logic blocks used in the document, the paragraphs are numbered in the margin. This device has the advantage of appearing less legalistic. Thus the number (73) that you see to the right of the preceding paragraph is a handy reference point to the 73rd paragraph of this chapter. Since this is Chapter 3, the numerical reference could also read (3:73).

In this two-column format the author has used subheadings, bullet lists and a primary heading for a discussion that continues on a subsequent page.

The Models

Outline Format: (Model 3.A)

The blood component project uses numerical tools throughout, and page 4 of the sample exhibits the topography of the traditional outline. The sequence is chronological.

Running-Text Format: (Model 3.B)

The discussion of seismic forces was handled using the conventional paragraph method. Written transitions hold the document together. Of particular interest is the second paragraph on the second page of the text. The one-sentence paragraph is designed to introduce the subsequent discussion of energy and harmonic motion.

Heading-and-Paragraph Format: (Model 3.C)

The heading system can be used at any time, with outlines or with paragraphs. The cash applications paper illustrates the use of headings and subheadings and tick lists in the context of a paragraph block presentation.

Summary

Select the format that fits your needs and the needs of the reader. Each system has positive and negative features.

OUTLINE FORMAT

Virtues	Drawbacks
• Abbreviates text	• Boring, tedious
• Quick reference	• Lockstep, too trusted
• Obvious order	• Does not detail well
• No frills	• May lack depth
• Often chronological	• May lack alternatives

PARAGRAPH FORMAT

Virtues	Drawbacks
• Elaborates the logic	• Less obvious
• Highly detailed	• Windy
• Argumentative	• Can be hard to follow
• Subtle	• Can be hard to write
• Great for discussion	• Needs transitions

HEADING FORMAT

Headings build clarity. The use of headings with paragraph blocks is a compromise that conveniently combines the best features of both of the other systems.

Outlines are	Paragraphs are
• What oriented	• Why oriented
• Composed to divide (to construct relationships)	• Composed to combine (to explain relationships)

Activities Chapter 3

Develop an outlined set of instructions for someone at work or at home. You might explain some detail of a work-related technical skill or a hobby or sport or explain how to change an ink cartridge in the office laser printer. How about a leaky faucet repair? Could your neighbor use a set of house and pet care instructions while you are on vacation? Even cooking instructions will do, but remember, a recipe (a topic outline of sorts) is a list of ingredients and you have to write instructions (a sentence outline). Use full sentences and develop at least two pages. Save the outline and place a copy of it in your template file (see Appendix B).

Read an article in a technical magazine and highlight the key points with your highlighter. Develop an outline that shows how the author constructed the document. Focus on the major topics, subtopics, and the sequence of the overall logic structure.

Using another technical article of interest to you, highlight the text transitions and the display transitions. Are there any longer transitions that involve full sentences or groups of sentences?

Share a Project

Work with two other members of the class to construct brief projects of approximately three pages in length (750 words). It will be helpful if all members of the group are from the same engineering technology program.

- *Decide upon a familiar technical topic that would allow you to develop all three papers in three distinct formats: the outline format, the paragraph format, and the heading-and-paragraph format.*

- *Each member should select one of the formats.*
- *Agree to the objectives of the documents.*
- *Assign a due date and bring three copies of each project to the meeting.*
- *Discuss the outcomes of the activity.*
- *Provide a set of the projects for the instructor.*

Work in Progress

4. Project Sketches

If the documentation is wrong, incomplete, or hard to understand, there is always the chance that inaccurate documentation will slip into public distribution. Then the customer is the one who will notify your company that things aren't right, gravity will take hold, and things will roll downhill from there.

As I mentioned earlier, because I have both assembly and inspection experience, I have my own ideas and suggestions about the production processes. One reason production process changes take time is that you frequently have to wait for responses from others who are having to deal with their day-to-day duties in addition to attending to the document you need from them. I couldn't count on a prompt response, so I began with my ideas. On this job, I assembled notes for myself so I would have them as a reminder when the "red-lined" (revised) documents arrived. Otherwise I would forget important issues.

While I was waiting I reviewed the distribution process and the preliminary changes agreed to in the meeting and started to focus on the central points of the project. I try to be precise and accurate from the very start of a project. The existing process documents are electronically available on the main Genus computer drive; they are "read only." I printed a copy because when the time comes for the approval sign-off, the supervisors always do a "from-to" hard-copy review, which is a comparison between the existing document (from) and the new changes (to). For identification I hand write "From" on the existing documents. When the new documents are completed, I will write "To" on the changed documents.

The likely changes will affect several areas, so I wrote a descriptive introduction to the document in order to identify those areas. Even though the project was nowhere near done, I started on a conclusion for each subsection.

I was still waiting. I tried to write as much as I could based on my notes and my own previous experience in several of the departments. What should someone do who is not as familiar with the background for a writing project? If you do not have hands-on experience of the subject, your best bet is to keep notes you can understand later, and of course ask questions. Also, if someone assigns you to do something, ask that they write the task down for you. My supervisor knows I prefer to see tasks in writing. As you can see, even with precise correspondence, I still had to investigate the issue myself, hold meetings, and study the problem in detail.

S. B.

Riders: The Alternative Introductions

The most practical introduction for technical documentation is the outline introduction. State you intentions, and briefly explain the parts of the coming discussion. Although you will usually use an outline introduction for technical and corporate documents of any length over two pages, you can put the outline introduction on hold for a paragraph or two while you develop an attention-getter for an opener. I refer to these paragraphs as "riders" because they sit above the outline introduction. There is seldom a problem of redundancy because the two structures are unrelated in design. One is an outline; the other one is not. The classic openers are these:

- **Statement of importance**

  Demonstrate the value of the subject.

- **Historical background**

  Provide the important history that will set the stage for the project.

- **Statement of a problem**

  Explain the problem in the introduction when the rest of the document is the solution.

- **Preliminary definition**

  Open with a brief definition (a paragraph or two) to initiate a text that will define some device or concept at length.

- **Statistical start**

  Open with a paragraph of relevant statistics.

- **Opening with questions**

  Ask a series of provocative questions to engage the reader.
  (Three will do the job.)

- **The striking incident**

  An old strategy is to start with a vivid example.

- **Opposing viewpoint**

  State the other person's ideas first and then oppose them in the outline introduction.

- **Extended proposal**

  In business and industry it may take entire paragraphs to simply state the thesis or proposal.

- **Narrowing the issue**

  If a document will focus on "samples" of an issue, an introduction can make the shift from general to specific so that readers see the big picture but will focus on the little picture.

There are others, but these are probably the only riders you are likely to use occasionally for introductions to technical documentation. All these introduction methods are effective in engineering and technical fields. The first four are perhaps the most likely tools you will use, but the last two are also quite common. Notice the openings of articles you read in magazines, and you will see examples of all these introductions. They build attention.

As I have mentioned elsewhere in the *Wordworks*™ series, authors usually will not write the introduction first. This is particularly true of the alternative introductions identified above. Since these introductions essentially are used to add color, the document should be composed first so that the author has a strong sense of how the document should be introduced. A conventional outline introduction should be developed in any case since the alternative introductions lack an overview of the document content. Once the outline introduction is constructed, the author is then ready to decide if a rider is appropriate.

Authors who will address a general audience—perhaps in the popular press—have more use for the alternative introductions than a writer in a corporate setting or an author in a scholarly setting. College papers composed by students may fit somewhere between these three environments, depending on who the author is addressing. In a writing course, the author is usually free to select the intended readership and might plan to use the alternative introductions to build interest among general readers. In engineering courses, however, the papers usually address professionals, and the alternative introductions of interest will be limited. For academic use the historical background and the statement of a problem will be useful, whereas opening with questions or incidents may prove to be less appropriate.

Sample 4.A

The mass spectrograph (MS) is a device used primarily by chemists and physicists to aid in the study of isotopes. When a sample of material is placed in the MS, it is broken down into ions of its constituent elements, and through the use of electric and magnetic fields the charge-to-mass ratio (CMR) of each of these elements is measured. The discussion to follow will deal with the general function and operating principles of a typical MS. The reader should bear in mind, though, that due to the complexity of the actual device, the diagrams and explanations used in this discussion are necessarily simplified to present only the essential ideas.

A typical MS is comprised of the following five basic sections:

I. Sample preparation chamber
II. Accelerating and directing section
III. Velocity selector
IV. Momentum selector
V. Detector plate

A description of the function of each section in turn will be followed by a detailed discussion of the basic operating principles of each component. For this initial part of the discussion, all references to parts of the MS will refer to Figure 1.

Section I in the illustration is the point where the sample is broken down into ions of its respective elements. This section is the most variable from one device to another because the sample may be gaseous, liquid, or solid, and all three types require very different procedures and equipment to prepare them for testing. Later, in the discussion of operating principles, a device to test solids will be detailed.

You should examine a few sample introductions to identify the techniques. The outline introduction is an appropriate starting point. For example, see the outline take shape in the following sample. This paper was called "The Trend toward Automobile Efficiency."

> With each new model year, automobiles become more efficient and technologically advanced. The cars of only ten years ago seem vastly different from the cars of today. The basic functions of the individual systems remain the same, but the method in which these systems operate has become much more sophisticated. Electronics and computers have replaced tasks previously handled by mechanical devices. Fuel systems have been improved by replacing carburetors with electronic metering devices such as fuel injection. Suspensions and braking systems are designed to provide a much greater degree of comfort and safety. The design and construction of automobiles has also improved as a result of manufacturing parts to closer tolerances, which results in a more solid car. There are older cars that utilize many of today's features, but such technological advances were found only in the most expensive cars of the day and not in the cars readily available to most everyone, as is the case now.

The two samples, 4.A and 4.B, fill the page. In Sample 4.A the author includes an outline, in the text. In Sample 4.B the plan of action is explained in very thorough paragraphs.

The following sample paragraph, concerning programming languages, uses the statement of importance as an opening. I urge writers to use this method in technical writing because it is very easy to write and very easy for a reader to understand. If you simply state the importance of the *subject*, you indirectly attach importance to the *document* you are about to present to the reader. The psychological transference here is a subtle comment on the relevance of the paper. The introduction thereby rationalizes the text and starts off the project. The statement of importance is particularly easy to write—so long as the project *is* important! This project concerned BASIC, C++, and Java programming languages.

In this example the author opens the text with a brief definition and then identifies the basic parts of the mass spectrograph in an outline. The paper discusses each of these topics in sequence.

Sample 4.B

THE PC CHARACTER SET

The PC character set is an important feature of the IBM PC. In this project we will look at the keyboard and how to generate characters from it, and we will examine. ASCII and scan codes, and the keyboard and the function keys. We will explore everything that has to do with generating characters from a computer keyboard.

There are three major sections to this discussion. The first one, the PC Character Set, covers ASCII proper, control codes, and extended ASCII. Its purpose is to cover what the characters are and how they do what they do. The next section, Keyboard Operation, discusses how the keyboard works and what it does with the character set. This section also talks about how to generate and use the characters while in DOS. The last main section, BASIC's Keyboard Operation, deals strictly with the BASIC language. The section covers character generation at the keyboard and deals heavily with the function keys. This last section also has some short, miscellaneous information on abbreviated commands.

Appendix A contains all the BASIC language error messages and their meanings.

Appendix B has some interesting photographs of a RAM chip that could contain this whole paper in its memory.

Although ASCII is the same for most computers, scan codes and control-key combinations are usually different; therefore, this paper will deal with IBM and compatable computers using the BASIC language under the Microsoft Disk Operating System (or PC-DOS). MS-DOS and PC-DOS are identical.

> One problem facing today's computer users who are interested in learning a programming language is the wide variety of languages available. There are languages that are geared for both specific and general purposes, and there are languages designed with both the beginner and the professional programmer in mind. Visual Basic is easy to understand and teaches the user basic programming skills, whereas C++ or Java are more complicated and are hardly ever used by amateur programmers. When the personal computer user makes the decision to learn a programming language, that is just the first step. Which language out of all the possible choices should he or she learn? There are many important factors that one should consider before making this choice. If you, too, are looking at programming languages, you need to know your goals for using the language, the extent of your knowledge of programming languages, and, last but not least, the cost of the programs.

The statement of importance has proven to be the most popular introduction strategy for the writers I assist. This introduction, like all the others identified, is not a substitute for the outline introduction; it sits above the outline introduction to attract the reader's eye to the document. In fact, the preceding paragraph combines both the outline *and* the statement of importance. Sample 4.C was designed as a set of assembly instructions for a corporation. The emphasis on the subject is quite clearly stated. It is intended to be read by assemblers at an electronics firm. The introduction is a page long and makes company policy quite clear: it is *important* to get the job done properly. The subject is important, both to the company and to the employee.

The next sample is the opening paragraph of a text that analyzed two popular microprocessors. This author used a brief historical introduction.

> In the middle seventies Motorola and Intel designed the first 8-bit microprocessors, the Motorola MC6800 and the Intel 8080. These designs evolved into the second generation of microprocessors, which appeared in the early and middle eighties. The Motorola MC 16805 and the Intel 8085 became popular second-generation microprocessors. Then the Motorola 6800 appeared. Two chips, one from each manufacturer, are examined in this paper. The Motorola MC6800 and the Intel 8086 have similar architectures and instruction sets; however, they also have some major differences that may gave one microprocessor an advantage over the other in a particular design application.

These samples are quite effective, and they will give you an idea of the options and techniques available to you.

This is another use of a pair of introductions. The first paragraph is a statement of importance. The second paragraph is an outline introduction, although it uses no numbers or lists.

Sample 4.C

What You Need to Know about Assembling the GX

Assembly of the GX-Series Drive-Through unit differs from normal assembly procedure in that there is no order in which the subassemblies must be made. When you perform the final assembly of the unit, you must have all the subassemblies that go into the unit in a pile somewhere, but the order in which they are placed on the pile is not important.

It is important to realize that the components are "kitted." Someone somewhere sat down at one time and counted out one each of everything needed to make one GX unit (1231 items). All you have to do when you go to "Stores" to requisition the materials for the number of GX units on your work order is to ask for the right number of GX "Wiring kits," "Chassis kits," or "Printed Circuit Board (PCB) kits."

The Wiring kit contains all the components necessary to complete the wiring subassemblies for one GX unit. The Chassis kit contains the parts necessary to make the box, or housing, of the unit. The PCB kit is actually three prepared PCBs (minus a few minor parts) all ready to install in the completed chassis and connect with the completed wiring subasemblies. Go ahead and check out the right number of kits of the right type, and turn to the appropriate section of this document for specific assembly instructions.

Before beginning any wiring assembly you should make sure each wiring kit corresponds to the Aiphone drawing, number DT-23B, "WIRE SUBASSEMBLY PARTS LIST" included in the appendix at the end of this document. **You are responsible for these parts while they are checked out to you.** If you find you have a problem with the contents of your wiring kit, look at the figures and drawings below.

We have tried to make the GX-Series Drive-Through System as easy to put together as possible. To assemble the entire unit, you won't need much more than the special nutdriver and two crimping tools that you will check out from your line supervisor. Once you have done a few of these units, you will know as much about their assembly as anyone else. If you have suggestion or ideas about how to do it better, please give them to your supervisor; we would also like to hear them.

Introductions can take a number of other approaches, as noted in the list on p. 114. The following samples show the different openers at work.

1. The following introduction is a *historical overview* of the development of the Internet. True to form in the computer industry, "history" is yesterday, and the text opens with developments that are little more than twenty years old.

> What is now the Internet was ARPANET in 1969, funded by the Advanced Research Projects Agency, a branch of the Department of Defense. It was created as a data and communication link that could hold the United States together after a nuclear war. Only the military, government agencies, a handful of defense contractors, and select universities had access to the system. In 1979, Usenet was invented. Shortly after, the Computer Science Network (CSN) was founded. For several more colleges and universities that had previously not been allowed to use ARPANET, this alternative provided similar services while not being part of the military's secured network. As with ARPANET, CSN's most heavily used feature was the electronic mail (e-mail). The largest jump in expansion was seven years later.

2. The ozone hole was the subject of another project. Because the damage to the atmosphere is a global problem that requires global attention, the opening strategy was clearly an easy choice. The text defines *the problem* as its opener.

> Stratospheric ozone is a much-discussed and widely misunderstood topic. Ozone is the trioxygen (O_3) form of oxygen and is an important constituent in the atmosphere of the earth. It blocks out much of the lower-energy ultraviolet radiation (UV), which is damaging to most biological life-forms, before the UV radiation can reach the earth. This protective ozone layer covering the earth is rapidly deteriorating due to many factors. The most significant of these is the release of manmade chlorofluorocarbons into the atmosphere, with volcanic eruptions and high-altitude aircraft flight adding to the depletion, also. As a result of the depletion of atmospheric ozone, areas known as ozone holes are being created that allow unfiltered ultraviolet radiation through to the biosphere, where all of earth's life-forms reside. This ultraviolet radiation is damaging to animals, plants and single-celled creatures.

3. *Definitions* are a practical opening tool. The introduction might define one or more terms for the reader. In a project concerning modern technology, one author chose to open with several paragraphs that define the subject.

This sample is quite elaborate. The second paragraph is a statement of importance. The fourth paragraph emphasizes the importance by issuing a warning of liability. In effect, the entire page is an introduction, which is a common practice.

> Modems are very useful for communication with the outside world. They link your computer with other computers by way of standard phone lines, through which services can be used, information can be obtained, and data can be retrieved. Information from your computer is sent to the modem, which in turn sends it out over the phone lines to another computer, where another modem receives the information. There are various applications that take advantage of this type of computer-to-computer communication. These include Internet applications, terminal and data transfer applications, and fax software.
>
> A modem is a device that translates digital signals from your computer to analog signals used by standard phone lines. Phone lines do not use the same type of communication that computer peripherals use. When a computer sends information to a printer or monitor or other peripheral device, it sends digital signals in the form of voltage variations. A modem must translate these signals for transmission.

4. For the final samples, you can examine what may be the most obvious and the most subtle of the introduction types. Opening with *statistics or factual background* is an obvious option that is popular, as you will see in this text concerning aviation fasteners.

> This document is intended as a general quick-reference guide for commercial aircraft mechanics who routinely fasten sheet metal and composite materials with mechanical fasteners. Aircraft fasteners are generally classified as either permanent or removable fasteners. Permanent fasteners are those that cannot be removed without destroying one or more fastener elements; common types include solid rivets, lockbolts, threaded pins and collars (Hi-Loks), blind fasteners, radius lead-in bolts, taper shank bolts, and sandwich panel inserts. Removable fasteners include external wrenching bolts and nuts, screws, pins, and other special fasteners. A large commercial aircraft will contain from several hundred thousand to over a million of these fasteners. A Boeing 727 has more than 700,000 structural fasteners, and the larger Boeing 747 aircraft has over 2,400,000 bolts and rivets installed during assembly.

5. Perhaps the most subtle structure among the introduction strategies is an introduction that narrows the topic. The actual topic may not be initially apparent. There is a movement from the big picture to the smaller focus of the document. This process gives the reader a larger frame of reference but shifts attention to a narrower concern. The following paragraph moves from the general to the specific.

Computer-aided design and computer-aided modeling (CAD/CAM, respectively) improve visual perception skills and understanding in design by allowing a lateral design process. Several different CAD/CAM processes are available to enhance visual perception. The most beneficial process of CAD/CAM is constructive solid geometry. Constructive solid geometry (CSG), or solid modeling, uses shaded solid models. The shaded solid models are then used as the infrastructure for the rest of the product design cycle. Previously, manual drafting methods limited visual perception in the design process. Solid modeling allows everyone involved in the product cycle to perceive an object in its "real" form.

This particular type of introduction is often used when a sample is going to represent a larger issue. The document focuses on one particular point or element or event that is an example of a larger or more general concern. The contents explore a small area of the broader interest. The text following the preceding paragraph, would be constructed to discuss the solid modeling applications of CAD design systems.

A number of other introductions can be found in the multipage models presented in *Technical Document Basics.* As you read the model pages study the introductions, particularly in Chapters 5, 6, and 7. The models provide more of a context for the introductions than we see in the preceding samples.

Sample 4.D

12

SUMMARY

The physical layout of computers on a network is called a *topology.* There are three primary topologies: bus, star, and ring. There are variations on these basic topologies, including the star bus and a star-wired ring.

The bus is the simplest and most commonly used topology. It is a linear configuration, with all computers connected by a single cable. Only one computer at a time can send data out to the cable. Therefore, the more computers on the bus network, the heavier network traffic will become.

In a star topology each computer is directly connected to a central component called a *hub.* It is the most used topology in many organizations. If the central component fails, the entire network goes down.

A token ring network connects computers in a logical circle. It uses a token and moves it around the ring through each computer in a clockwise direction. The sending computer uses the token to attach to the message it wants to send and releases a free token.

Conclusions

Many technical documents do not conclude. They stop. This chapter is going to do just that. It will simply stop. In other words, a set of regulations or a set of instructions will not usually be wrapped up in some final statement. At the most, these documents will conclude with checklists such as you see in Sample 4.E.

The last page of a construction code or electrical code book is the last page and nothing more. This is a convenient element of important types of tech writing. There is another type of document, including the laboratory report, in which the findings are on the *first* page even though they obviously are "outcomes." In chronological terms, the findings are the *results* and occur near the end of a research project, but these outcomes are likely to be presented at the outset. Here again we have a huge category of tech writing with no tail end to the document. Academic articles and some other varieties of technical work—typically written in running text—will use a conclusion.

I encourage engineers to simply sum up (see Sample 4.D). In other words, if you need a final word, you should draw on the highlights of a document for the conclusion. If you used an outline introduction, you might think that the conclusion is saying the same thing. This is somewhat the case, except that the introduction is positioned *before* the text, and the conclusion is in a position to be much more specific because it can draw on some of the evidence of your document to support the proposed outcomes.

The redundancy is not an issue unless the paper is very short. If it is short, you probably do not need much of a conclusion anyway. When I write articles, I write about one hundred words to sum them up. As the articles get longer the need for longer conclusions is logical, mainly because they summarize the document and reflect the length of the overall project.

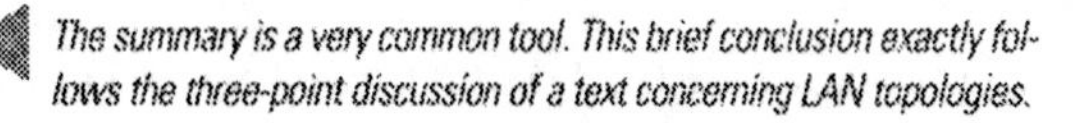
The summary is a very common tool. This brief conclusion exactly follows the three-point discussion of a text concerning LAN topologies.

Sample 4.E

Post Assembly

Your new PC is almost ready to run for the first time. You have the basics installed and connected. All the extras, such as the modem and sound card can be installed *after* the first boot-up, just to make sure everything is connected correctly and working before you add new hardware.

Before jumping right into booting, take a minute and, with a flashlight in hand, check all your work. It is better to "waste" the time than to fry your system after all this work.

Pay attention to the following list:

1. Drives are properly connected to power.
2. The CPU fan is attached to power.
3. The Power switch is off.
4. P8 and P9 are connected properly, with black wires in the middle.
5. The 110/220-volt switch is configured properly for your area.
6. Ribbon cables are attached correctly, with the red edge on pin 1.
7. All connections are tight; no connectors are off by one set of pins.
8. CPU voltage settings are correct.
9. Cards are fully in slots.
10. No wires are protruding into the fans.

Congratulations! You have completed the first stage of building a PC. We will cover the setup of software and drivers, troubleshooting, and optimizing your new PC in the sequel to this guide.

Apart from the summary, two other popular conclusions frequently appear in technical work: checklists and recommendations. Checklists bring many sets of instructions to a close. They are used to verify proper completion of an activity or setup. Recommendations are the logical outcome of a problem-solution analysis in general (see Chapter 7). Recommendations constitute the most important phase of a technical proposal in which the author intends to prepare a proposed bid or estimate for a contract. Depending on the nature of the project, the recommendations might be developed as a brief summary, or they might involve half or more of the length of the presentation. In the latter case, the summary conclusion *and* the recommendations conclusion are combined, since an elaborate presentation of recommended actions must be summed up!

Recommendations are extremely popular but are not always appropriate. Sample 4.F is the concluding page of a proposal for a network installation. The bulk of the proposal is an explanation of recommendations. At the suggestion of one of the major manufacturers of network apparatus, contractors are encouraged to use a summary to highlight the most important features of their contract proposals. In a laboratory setting, concluding observations may be important, as in the following sample. Here there is no concern for marketing the document, and concluding observations simply summarize the laboratory findings.

Conclusion

After the system was breadboarded, it was tested for correct operation. A bit error was simulated by pulling one of the data line leads on the transmitting ribbon-cable. This caused its inputs into the exclusive-or gates in the parity checker section to float to an erroneous high state. Under these conditions, the data acquisition section correctly displayed the hamming code for the line in error. The binary value for the 4 LED incoming data display was also corrected as specified in the project considerations.

The system proved to be very effective. However, to be used for a modem parallel(short range) data transfer system, it would most likely have to be enlarged to handle eight or sixteen data lines.

Checklist instructions are generally used to double-check another set of instructions that have been completed. This procedure is very common in manuals, and functions as a conclusion.

Sample 4.F

-8-

NETWORK INSTALLATION PROCEDURES

- The hours for the installation of cabling shall be from after the office day ends (3:30pm) to 11:00pm each weeknight, and from 8:00 a.m. to 6:00 p.m. on Saturday to avoid any school disruption.
- The installation and configuration of the MDF and IDFs shall take place during the normal workday hours of 8 a.m. to 6 p.m. during the last week of the installation process.
- All cabling shall be run through existing conduit between buildings, and in the plenum area within the buildings.

COST INDEX

Please see Appendix D

SUMMARY OF RECOMMENDATIONS

In order to provide a network able to support the growing needs of Geo. Consulting, several recommendations have been put forth, including the removal and subsequent replacement of the entire network infrastructure with a newer high-speed configuration. The new installation will provide employees with the ability to incorporate and use state-of-the-art computers and applications, access to the Internet for research, and allow the gained knowledge from which to be used in the daily tasks. Finally this installation will allow for future growth both in size and bandwidth and provide the ability to integrate further advances in technology as they arise. The ability to exchange information has become more and more important in today's corporate environment. Networking itself is a necessary part of this exchange, by giving a business the ability to provide access to the latest resources and ideas.

Pronouns and Passives

The use of pronouns—*I, you he, she, me*—in your writing is an important stylistic matter, and there are certain standard practices to use as guidelines. First, you *usually* do not refer to yourself or the reader. Note, however, that there are glaring exceptions. Trying to write a set of instructions without references to the reader is hard going and awkward. Do not hesitate to use the "you style" in the case of instructions, although you do not usually need to do so because directives are commands and need only to imply the presence of the reader: "Find the poser-on indicator light." (Notice the use of the pronoun *you* in Sample 4.E.)

Introductions commonly use self-references such as "I will examine . . ." or "We need to analyze . . . ." There is nothing wrong with these devices, and they can add clarity to the organization. However, most scientific and engineering documentation steers clear of pronoun references. Corporate documents will seldom use the "I" except for correspondence and, possibly, research narratives. (Notice the use of the pronoun *we* in the last paragraph of Sample 4.C.)

As you surely have noticed, the writing style in this text is somewhat uncommon for a textbook. The textbook tradition usually calls for the "objective" style whereby the author "writes" but never "speaks" (of himself or herself). I use the objective method most of the time, but if I choose to shift to a "voice of experience," then the text shifts to the "I format," as you see here. Usually, however, the *I* should be omitted.

It is a long-standing tradition to write in the "objective" style to make the document appear to be without any bias. This style was very much in place by the middle of the nineteenth century as I explained in Chapter 1. Observe the following sentences and notice how I can phrase the same comment in progressively more objective terms.

> I believe there was no evidence that the fever was spread by mosquitos.
>
> I have no evidence that the fever was spread by mosquitos.
>
> We found no evidence that the fever was spread by mosquitos.
>
> There was no evidence that the fever was spread by mosquitos.

A related issue is the passive construction. The passive construction is a popular but nonstandard procedure in much technical writing, and it is also *much* frowned upon by teachers of English. The theory is that a sentence that reads,

> The nut is then attached to the bolt mount.

Contract proposals can be very complex and a summary of recommendation is one way to conclude on a simplified overview.

is somehow less "active" than the sentence,

Attach the nut to the bolt mount.

In the first example the author manipulated the language to avoid pronouns and more or less pushed the performer off stage. This is the passive construction. You should not be too wary of pronouns, and you should feel free to use the pronoun *you*—as well as commands—for instructions.

The issue of the passive construction is probably not pressing if most readers do not recognize a passive construction when they see one. Furthermore, if you have a grammar check application to identify your passives for you on a tech writing project, you may be shocked at the percentage of your sentences that you are supposed to recompose to active status. The percentage may be particularly high in sets of instructions.

On the other hand, you should avoid the *misuse* or *abuse* of the passive. If you are trying to be objective, or if you want to avoid commands and pronouns, the passive style may be helpful. However, notice that the verb in a passive construction has no expressed driver (performer), and this problem can confuse the reader at times. If a reader becomes confused about who is performing the action, there is a misuse of structure in the sentence. Far worse than misuse is the potential for abuse. The passive is a great tool for passing the buck. If you stepped on your spouse's watch this morning, you are probably not going to say,

I stepped on your watch and broke it.

Instead, you meekly say,

Your watch got broken this morning.

The passive construction hides the culprit—or at least attempts to do so. In these situations the passive is certainly not appropriate for clear and honest communications with a reader.

In general, the difficulty most writers have is that everyone speaks with pronouns all day long. In formal, objective writing authors are expected to avoid most pronouns that refer to the author or the reader. Pronouns are allowed very limited use. Every writer then discovers passive constructions, even though he or she may never have heard of the term. Unfortunately, these constructions are considered bad form! I avoid this entire catch-22 by simply writing to achieve clarity. The rules are secondary to the goal of communication. Do not misuse or abuse either pronouns *or* passive constructions, but it is not necessary to pretend that you cannot or do not need both of these language tools. Defer to authority, and follow policy—either your company's policy or the recommendations of your instructors.

Outlines and the Peer Review

Although not an integrated feature of most technical documents, outlines serve an important role in the production management of getting technical projects completed in a well-organized and timely manner. Writing in engineering and science fields can be an organizational mire. Outlines help to establish control over the substance of a project in several ways.

Essentially outlines serve three purposes for the author of technical documents. The first issue is the "designated" task. In business and industry there are assigned writing chores, and in academic settings these tasks are "required." Whether you are a student or an employee, at some point you will be responsible for an engineering report or a term paper. If a task is designated, you must realize that you are working with, or for, someone in an essentially supervisory role who is supporting your work. This means, in turn, that you need to coordinate your work with that person's perceptions of what you should be doing. Use an outline before you start writing. It is the perfect discussion tool and an idealized model of your plan of action. Talk out the outline before the project goes any further.

The second use of the outline is as an organizational tree. The plan of the outline is perhaps speculative; it may even be wishful thinking, but it *is* a start and it does provide a linear progression of activities for you to follow in a methodical and orderly fashion. Writers need discipline of several sorts, and any assist that tells them what to do next is a big help.

The third use of the outline concerns the subject matter. Remember these watchwords: *simplify complexity.* The materials we often have to deal with are very complicated. Complicated documentation simply compounds our problems. Outlines allow us to structure sets and subsets of logical units that have some coherence. Mastering the subject is often the greatest of the three challenges.

Sample 4.G

Outline Second Project:

HOW A THERMOSTATIC EXPANSION VALVE OPERATES

AUDIENCE: This paper is written for the first-year Refrigeration/Air Conditioning student.

OBJECTIVE: This paper is written to give the student an overall view of how a thermostatic expansion valve is used in a system and, more specifically, how it operates.

I. A description of how a Thermostatic Expansion Valve (TXV) functions in the refrigeration system
   A. A refrigeration system moves heat from a location where it is unwanted to a location where its presence is less objectionable.
   B. Refrigerant flows through the refrigeration system and "boils" to absorb the heat to be moved.
   C. The flow is controlled by a metering device; one of these is the TXV.
   D. The advantage of the TXV is that it controls the refrigeration and increases the flow as the load increases.

II. The construction of the TXV
   A. There are many types of valves constructed in different ways.
   B. The metering orifice varies on different types of valves.
      1. Some types of TXVs have a needle-and-seat orifice to control the refrigerant flow.
      2. Other types have a push rod–and-ball-type orifice for flow control.
   C. There are internally equalized TXVs.
   D. There are also externally equalized TXVs.
   E. The power element senses the temperature of the suction line to maintain the flow of the valve.

III. The flow through a TXV
   A. Through the inlet port
   B. Through the needle and seat
   C. Through the outlet and into the evaporator

IV. The pressures that operate the TXV
   A. The bulb pressure is the opening force on the valve.
   B. The suction pressure affects the closing of the valve.
   C. The superheat spring is a closing force on the valve.

If you are familiar with *Basic Communication Skills* (another volume of the *Wordworks*™ series) you realize that there are three types of outlines:

The sentence outline
The command outline
The topic outline.

The topic outline has very limited use because of the absence of complete sentences. Without the logic of sentence structure, it is very difficult for a reader to follow the outline. In a workplace setting, the outline is usually a tool that serves to preview a plan for development and so it must be designed in complete sentences, or as a set of commands, which are also sentences. The samples you see here (Samples 4.G and Sample 4.H) are very thorough. The task does not have to be daunting, however, and an outline half the length of the samples might serve your purposes quite properly. Also note that the two samples are very traditional in use of the alphanumerical system of outlines, but outlines can be simplified and headings and bullet lists can be conveniently integrated into the product. Since the outline is a helpmate, feel free to design it in any way that serves your purposes.

Use full sentences for the outline if you expect to share it with someone. If you use only a topic outline, it is more difficult for anyone other than yourself to follow. Notice the thoroughness of the samples.

Use outlines to encourage cooperation, to encourage the author (you), and to encourage control. Entropy lurks all around you. It is more than a law of physics. It is the law of disorganization that pesters every writer.

Note that a sentence outline can incorporate subsections that are phrases as long as the content is clear.

Sample 4.H

Visual Basic

AUDIENCE: This paper is for people who have programmed in other languages besides event-driven languages. Experience using event-driven applications, such as spreadsheets or word processors, is expected.

OBJECTIVE: To discuss aspects of the environment of the Visual Basic programming language. To teach the reader how to build a simple application in Visual Basic. The reader should be able to expand on this knowledge by him- or herself to produce more complicated applications.

OUTLINE:

I. An introduction on how to get around the Visual Basic environment.
   A. How to get into VB.
      1. This is done from Windows.
      2. From **Start** go to **Programs** then **Visual Basic** then **Visual Basic** again.
      3. Open a project from the **File** menu as you would from an **Office** application.
   B. There are various aspects of the screen seen by the programmer.
      1. The first window seen is the Project window.
      2. This leads to the Form 1 or the Code window.
      3. On the top is an icon bar like most applications.
         a. The Help menu is extremely useful.
      4. Double-clicking on the Form 1 reveals the Properties window for the Form 1.
      5. On the left is the toolbar. This contains the objects that will be placed on the forms.

II. This section expands on the toolbar.
   A. *Labels* are the most basic object. They display information without allowing input from the user.
   B. *Command buttons* allow the user to execute code within the framework of the button based on the "event" the user chooses.
   C. *Text boxes* allow the user to enter data for manipulation by the VB application.
   D. A *frame* contains one or more choices, like "options" or "check boxes."
   E. *Option buttons* are placed inside frames and give the user choices. The user may pick only one option.

III. This section covers building a simple application.
   A. Placing objects from the toolbar on the screen.
   B. Setting the properties of objects.
      1. This is done for every object.
      2. There are programmer naming conventions for objects.
   C. Writing the code for the application.
      1. Showing how to enter code in the Code Window for a specific event.
   D. Running the project and saving it.

Putting It All Together

To conclude this first group of chapters we will look at the human side of the technical writing issue as an appropriate transition to the next group of chapters, in which you will have the opportunity to try your hand at tech writing projects. Our discussions somewhat overlook the mechanical matters of the employee in the workplace, and realistically speaking, we cannot ignore the obvious. All the chapters of *Technical Document Basics* focus on writing a document, but the issue is not just the document. The writer is the creator of the project and diligently goes through the paces to create the end product.

In a preceding textbook in this series—*Basic Composition Skills*—I dedicated two entire chapters to the work schedule involved in document development. You are probably familiar with the process if you used the text. To briefly recall the chapter, I will reproduce the document work schedule chart and touch on the twelve stages that are often involved in our writing activities in engineering and engineering technologies. Because the commentary of engineers who have writing experience is helpful, I will introduce you to Todd and Blair, two engineering technicians who write frequently as part of their workloads. Each technician will briefly explain a project and show you the stages from rough draft to finished product, and they will explain what goes on in between.

The case study you have been reading in the sketch that opens each chapter is a similar example produced by a working author. Shirley's commentary points out the complexity of real-world profit-and-loss writing in a complex setting of people and departments and contract specifications. Todd's commentary and Blair's discussion will show you actual pages of text as part of their stories so that you can see the basic page-by-page writing issues that must be addressed.

Although some writing projects are completed in one day, many projects take days, weeks, or months. The turnaround time depends on the complexity of the project. A rather elaborate project will involve a dozen or so distinct stages in the writing process. The following table is designed to reflect the various stages of a writing project. The lines in the third column represent a *critical path schedule* and indicate the possible length of various phases.

These two outlines are examples of thorough detailing but, depending on the subject, an outline half as long can be of equal value.

DOCUMENTATION WORK SCHEDULE

	Phase	*Project Input*	*Path Duration (accrued)*	*Project Output*
1	Project is Initiated	task memo		perception of project
2	Critical Path Analysis	agenda for the activities		calendar of the events
3	Project Sketches	author's notes		key positions and supportive logic
4	Research	scientific data, developmental data, cost data		statistical findings for support
5	Coworker Responses	critical dialog		recommendations, emendations, negations
6	Form and Content Analysis	logic architecture for the document		document outline
7	Development One	creation of initial document parts		rough copy of components
8	Development Two	creation of organized whole		rough draft copy of full document
9	Revision One	copy modifications		second draft
10	Revision Two (option)	second copy modifications		third draft
11	Final generation edit one	"clean" copy detailing		model document
12	Peer Review	engineering/ technical edits		ready document

1. ***Project is initiated*** On receiving a task memo, an engineer will examine an assigned project to survey the situation. A written assignment is superior to a spoken one because the paper trail should be initiated with the request for document preparation. That request should outline, as precisely as possible, the nature of the task.

2. ***Critical path analysis*** After examining the task, the engineer must determine a list of activities and plan a calendar from start to finish, even if events are likely to change. This is a good time to talk to the supervisor who requested documentation.

3. ***Project sketches*** The work begins. The author develops notes of ideas, perhaps structuring key concepts, or organizing subject matter or other initial ideas.

4. ***Research*** Then it is time for data gathering. In the workplace this could all come from the terminal at an office workstation, or it could involve considerable legwork, conversations, meetings, and phone calls.

5. ***Coworker responses*** Critical dialog is an aspect of research. What people know and what opinions they hold become valuable contributions to company documentation.

6. ***Form and content analysis*** As the perception of the job becomes clear, the author can begin to think about how to shape the document. It is time to outline.

7. ***Rough draft*** The draft of the document is started. Some authors develop pieces and focus on continuity later on. Others follow an outline and build as they go, more or less in order.

8. ***Rough draft organization*** The pieces soon fit together, and the document then exists as a whole. The initial draft is complete.

9. ***Revised draft*** The text is then scrutinized in many ways to correct and perfect the structure, the language, and technological matters.

10. ***Another revised draft*** The process continues. Most authors use several drafts so that progress can be recorded. Making a great many changes in one leap is too complicated.

11. ***Final copy*** The neat and orderly final copy is the cosmetic phase. It is a time for touching up details and enjoying the victory of the challenge of writing. Entropy is once again at bay, and all is calm and under the author's control.

12. ***Peer review*** Everything is under control, at least until the other engineers see the document. A final important ingredient in technical writing for engineering technicians and engineers, the peer review provides the author with a double-check on the technology embodied in the document.

Not all the phases are critical at all times. I used all twelve to write the book you are reading, but the scope of the project clearly was a large one, so much so that the twelve steps were used repeatedly. At the very least, every chapter involved a separate cycle of the twelve steps. On the lucky end of the scale is your e-mail. You can draft a business letter in less than an hour. In between these extremes are the many documents of the workplace. Allow me to introduce you to Todd and Blair who will explain several such projects.

Todd's Workday

I work for Zetron Incorporated, a company located in Redmond, Washington. Zetron is a high-tech company that manufactures a variety of equipment used in many different wireless communications systems throughout the world. Some examples include paging encoders, trunked radio systems, radio dispatch consoles, and alarm monitoring equipment. I have worked for Zetron for almost two years. My job title is Marketing Research Analyst. I work closely with product managers and engineers to ensure that our documentation at various levels stays up to date and is complete and accurate. I also spend much of my day doing company research and reporting on target markets, competition, and market trends.

Purpose

The purpose of this brief explanation is to show you how I accomplished a typical project at work. I chose the work I did on a new product specifications sheet. I chose this because it had very clear-cut tasks and a definite start and finish. Many jobs I work on are ongoing projects that are hard to define, and their outcomes are often just informal reports to my supervisors. Creating the specifications sheet required various writing assignments, and the result was a nice glossy spec sheet that I hoped would be a useful tool for both our customers and our employees.

Project Assignment

Everyone who has ever worked in an engineering environment knows that things often run to the last minute and projects are often finished later than planned, and with more obstacles than were originally anticipated. (Actually, I think this is universal to all companies and not just engineering firms.) The project I am about to describe is no exception, but it has a happy ending.

I received an e-mail one morning from my supervisor, Matt, regarding the project I would put at the top of my priority list for the next several weeks. Zetron had designed a new product called the Site Decoder that was soon to be released. Much of the work needed to bring the product to market had already been done, but one key piece was missing. The product needed a specifications sheet urgently because in six weeks we were headed to a large trade show where there would likely be a great deal of interest in this product.

Typically the engineering group that designs the product creates this document, but in this case they were bogged down, and if it was to be done by the trade show deadline, they needed some help. I started out my career as an electronics technician and have worked closely with many engineers over the last few years, so I was a logical person for the job. I was pretty sure I could accomplish this task in the required time.

Research

The first task I needed to undertake was to find all the technical information that existed on the product so that I could sort through it all and take what I needed. I sent off an e-mail to the head engineer, Jim, to inform him that I was involved and that any information he could give me regarding the Site Decoder would be helpful. Less than a day later, he bestowed upon me a sixteen-page document that thoroughly described how the unit functioned. It was much more than I needed, but better too much information than not enough.

Model 4.A (1)

Introduction

Many hospitals are equipped with a two-way radio system that facilitates communication with ambulances and other urgent care facilities. These are often part of the nationwide system. Most of these systems were installed years ago to provide communications and coordinate resources, especially in times of an emergency or disaster when the telephone system dies. The goal was to create an alternative means of communication should the conventional phone systems become overburdened or disabled. Most of the units were specially built models. Unfortunately, newer base stations being sold today lack the right functionality. Until the development of the Zetron Site Decoder, upgrading the base station meant sacrificing the old system. The Zetron Site Decoder is an external accessory device that facilitates the replacement of obsolete base station radio equipment and provides an upgrade path to newer technology.

Features

Compatible with modern base stations

- Decodes DTMF signaling
- Decodes two addresses of up to 16 digits each
- Provides both individual and group call decoding
- "Transpond" call notification
- Encodes EIA Site Remote Control (SRC) functions
- Compatible with existing Site Remote Controllers
- Front panel keys and display
- Small space requirement (1.5≤ ¥ 19≤ rack mount)

Operation

Most systems are configured for operation on two radio channels. One channel is used for ambulance-to-hospital calls (typically 950.123 MHz), and another channel is used for hospital-to-hospital calls (typically 950.456 MHz). The base station is equipped with two receivers so that users can hear simultaneous conversations on each channel, and calls are not missed.

The system provides selective calling to alert an individual hospital, or group of hospitals. In this way, an ambulance team can place a call to a specific hospital without interrupting all other hospitals within radio range. Hospitals can place calls to other hospitals just the same as the ambulance teams. A "group call" function is also supported, which allows users to alert multiple hospitals with a single radio transmission.

The typical base station is a 100 watt, two-wire remote control, with a second receiver. This configuration allows the hospital to monitor two radio frequencies at the same time.

Next, I knew I could use a copy of the manual, the test procedure, and the bill of materials. These documents would give me direct insight into how the product works and would fill in any gaps that Jim's document left out. However, before I could request these documents from the Document Control Division, I would need their part numbers. I could do this research on my PC. I have a mapped network drive that allows me, in "read-only" mode, to search Document Control's database. I performed a search on "site decoder" and pulled all the existing part numbers (see Table 1). Now I had a list I could use to fill out a document request form and get what I needed. I requested the necessary documents and received them a day later.

Table of Part Numbers

Part #	Rev	RCN	RDate	IDate	Mode	Dept	Title	Notes
009-9951	C	2562	4/23	4/23	SITE	RM	Inverter Software	
011-9951	A	2562	10/27	10/27	SITE	RM	Printed Circuit Board	Same as AX, Verify Plots
026-0285	A	4562	10/27	10/27	SITE	RM	Users Manual	
311-9951	A	5510	10/27	10/27	SITE	RM	Test Procedure	
416-0328	A	5510	10/27	10/27	SITE	RM	Bottom Chassis	Same as AX
416-0329	A	5510	10/27	10/27	SITE	RM	Upper Chassis	Same as AX
703-9950	A	4486	10/27	10/27	SITE	RM	Site Decoder Board	
703-9951	C	2562	4/23	4/23	SITE	RM	Logic Extention Board	
812-7473	A	5510	10/27	10/27	SITE	RM	Logic Cable Assembly	See Model 444

The drawing in the manual reminded me of one other thing I would need for the spec sheet: a picture. This would be an urgent task because Zetron has the glossy pictures needed for spec sheets done out of house by a local photographer, and I was unsure about how long this would take. I went to Heidi in Marketing Services to get this resolved. Fortunately, Zetron was doing a photo shoot in a week and she informed me that if I could get her a unit she would get the picture done.

I called Jim to see if I could get a unit for the photo shoot, and, fortunately, he had a nonfunctional unit that was perfect for the job. With that task accomplished I had the initial legwork done and I could start drafting the spec sheet.

Drafts and Approvals

Using the information I had located, I created a first draft of the spec sheet. (Note the rough draft.)

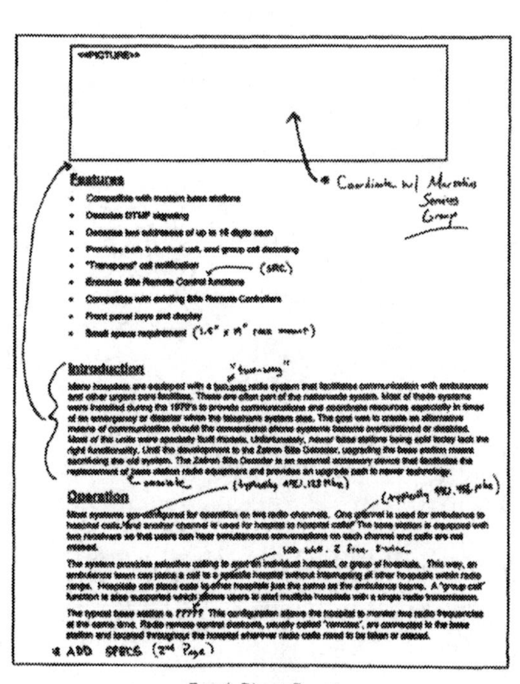
Features

Introduction

Operation

* ADD SPECS (2nd Page)

Rough Direct: Page 1

Model 4.A (2)

Specifications

Environmental	
Power	+0.5 to +10.5 VDC supply
Temperature	−30° to +70° C (except 0 to 50° C for LCD) 90% relative humidity (non-condensing)
Size	5.5" W × 6.25" D × 1.4" H. 1.5" × 19"Rack panel mount included
Weight	1 pound
General	
Adjustments	Rx audio level, line output level, LCD contrast
Jumpers	High/low rx audio gain, carrier detect polarity
Indicators	Backlit 2 × 16 char LCD display
Front panel controls	2 push bottons
Configuration and setup	Using front panel switches and displayprompts
Receiver Interface	
Connector	10 pin × .15≤ DB type connector
Signals required	+10.5 VDC supply, ground, discriminator audio, carrier detect
Input impedance	10k mV
Discriminator audio level	100 mV to 1 V peak to peak
DTMF decoder sensitivity	10 dB sinad
Pulse dial decoder sensitivity	16 dB sinad
Pulse decoder bandwidth	1000 Hz ± 50 Hz minimum
Make / break ratio	50/50
Acceptable dial pulse rate	5 to 20 PPS
Wireless output	
Line type	2-wire 500 site remote control audio circuit
Connector	Dual position screw terminal
Secondary protection	High voltage TJR clamps and fusible resistors
Audio output impedance	> 5000 during Rx, < 500 during alert
Function tones	500-5050 Hz in 100 Hz steps
Function tone accuracy	<1%
Function tone duration	10 ms, -0+5 ms
HLR output level	Ajustable, +10 to −15 dBm
HLR frequency accuracy	1000 Hz ± 5 Hz
HLR tone duration	100 ms, −0 +5 ms
Programmable Settings	
Capital ID	Up to 10 digit individual call address for the hospital
All Call ID	Up to 10 digit group call address
HRTR decode	Enable / disable, allows decoding of "digital dial" and TR if required
Unmute function	500-5050 Hz function tone sent to unmute the receiver
Tx function	500-5050 Hz function tone sent to cause the base station to transmit
Transpond alert tone	None, warble, or up to 4 beeps can be sent for call verification

Zetron's spec sheets for the most part fit into a common format, so looking at a related products spec sheet, I knew what I would need to include. The spec sheet would need an introduction, a feature list, an overview, and, of course, the technical specifications. It is also common to include some diagrams, especially if it is necessary to show how the equipment interfaces with a particular system or other components. In my first draft I included many system diagrams, but they were edited out when it was decided that they simply clutter up the spec sheet. The in-depth diagrams could be used in less expensive mediums like the manual and application notes where space is not a critical issue.

After I created the first draft (minus the diagrams), I sent a copy to engineering so they could give me their input. I also made some notes of my own and gathered some more of the necessary technical data (see figure below). I also took my notes to the Director of Marketing to see if he had any additional corrections or comments that I could use. With a few changes and three weeks to spare I had a copy I was happy with.

Once I had input from the engineers and management, I turned what I had over to the Marketing Services group so they could complete the task and have the glossy spec sheets printed. I filled out the necessary form and gave a copy of my work (in a Microsoft Word file) to Susan in Marketing Services. She would convert the document using PageMaker so that the picture could be added and the layout could be changed to reflect our standard format. We still had two weeks to go, but we would be cutting it close, since it normally takes two weeks to get the job back from the printers.

Once Susan finished her work she sent around copies of the document for approvals. During this stage last-minute changes could be made and implemented if necessary. Once it was signed off it could be sent to the printer to have the final copies printed.

When the copies came back from the printer I took some time to admire the finished product. It felt good to know that I was a necessary ingredient in getting this job done on time. I hope that this explanation has helped to give you an understanding of how a typical task is accomplished at a company like mine.

T. R.

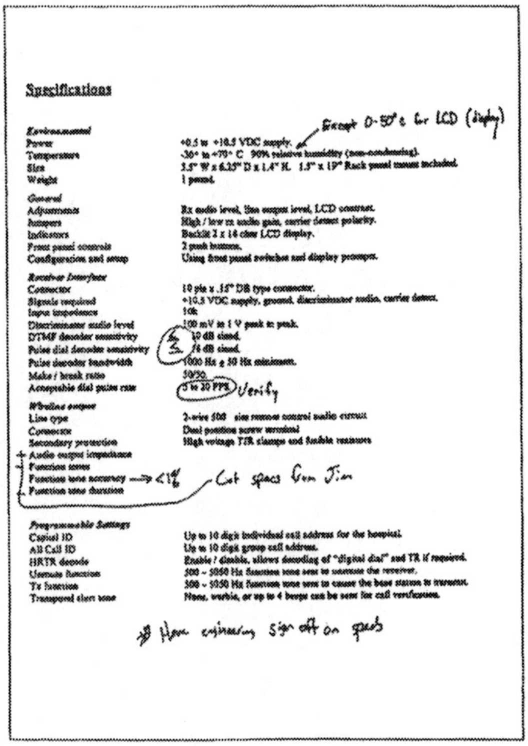

Rough Draft: Page 2

Blair's Workday

Hello. My name is Blair and I've been asked to say a few words about my work. I have a degree in electrical engineering, and I have spent eight years working in the aviation industry. I'm going to explain one aspect of the job I have held as an engineer while working for various manufacturers of commercial airplanes. The part of the job I will describe involves detailed technical writing of the processes or procedures used to manufacture and maintain wire assemblies—also known as wire harnesses or wire bundles—for the many electrical systems on an airplane.

How many miles of wire are in a wide-body jumbo jet? That question never crossed my mind during transistor circuit design theory class. Nor does it cross the minds of people flying to visit grandma for Thanksgiving. More than 100 miles of wire and cable are used to make up the many unique systems in commercial airplanes. Such systems include audio and video entertainment, cabin and external lighting, galley power to heat the food, the autopilot, and many other flight management systems.

When I was in college, English and the electives looked incidental, and I couldn't wait to begin the real fun stuff: crunching numbers, studying science, and getting into engineering. Little did I know that a large portion of my professional future would involve the writing of technical specifications. Specifications became part of my job description. Let me explain the writing process to you.

There are many jobs for engineers at large aerospace companies that manufacture aircraft. The term aircraft *includes wide-body and narrow-body commercial airplanes, military fighter jets, helicopters, missiles, space shuttles, and most any craft that flies through the air. My first job after graduating as an electrical engineer involved working in the "electrical systems organization" for a large aerospace company.*

The electrical systems organization produces the specifications for the design, installation, and testing of all the electrical wire and cable harnesses on the aircraft manufactured by that company. My specific job title was Electrical Components, Materials, and Process Engineer.

An Electrical Components, Materials, and Process Engineer is a specialist in the parts and materials that constitute a finished wire harness, as well as the process and tools to build them. The majority of these parts and materials include wire, cable, connectors, switches, relays, circuit breakers, tapes, sealant, ties, and straps. Electrical design and installation engineers specify many types of parts and materials on their drawings. These engineers often need assistance in selecting the best part to use for their wire harness. In addition, procedures must be available for the assembly and maintenance of these parts and materials into the finished wire assemblies. This is where the Electrical Components, Materials, and Process Engineer fits in.

These parts and materials all require various documentation. This documentation includes specifications for parts, materials, testing, and processes.

Parts or material specifications *define the characteristics of the parts or materials. These characteristics include, for example, the dimensions or maximum temperature of the part.* ***Test specifications*** *define various test procedures that must be performed on the parts or materials. These tests make sure that the parts or materials meet the characteristics that are defined in the part or material specification. The parts or materials then become qualified.* ***Process specifications*** *are the procedures or instructions to assemble, install, or maintain equipment, or in this case, wire assemblies.*

The writing of process specifications is no different than writing other types of technical documentation. First, as in most technical writing, the audience and objective must be defined. For commercial airplanes, the audience is well defined as the personnel in production areas—known as "Wire Shops"—at the airplane manufacturing site or subcontractor site, and the maintenance crews of the airlines throughout the world. These people do not have any special command of the English language.

Writing for these readers involves nuts-and-bolts English. Interestingly enough, English has been defined as the standard language in the world of aviation. For this reason, a writing standard known as "Simplified English" was developed in France by a consortium of European interests known as AECMA (Association Européenne des Constructeurs de Matériel Aérospatial. Basically the suggested vocabulary is the same one I use: nuts-and-bolts English. This AECMA Simplified English is slowly being adopted and is soon to be a requirement for the technical media involved with all commercial airplanes. Its use is a perfect fit for the writing of process specifications used by foreign maintenance crews, as well as production personnel who are often of foreign origin, for whom English is a second language.

The objective of writing these processes is to provide clear step-by-step procedures with easy-to-read information in the form of tables, graphics, and references to part numbers or materials. This must be done with the idea that the assemblers can repeatedly build a durable product, in the shortest amount of time.

Second, in developing the electrical wire processes specifications, I have to completely understand the parts that are to be assembled. This understanding comes from learning the function, science, and composition of the various components or materials. Much of this understanding comes from the suppliers who manufacture the individual components and materials that will be used to assemble the finished wire assembly.

Typically I begin to write by doing a rough draft of the process. This is the third stage. To enhance the process, I add tables and graphics, which make the instructions easier to follow.

After a rough draft of the process is on paper, it is entered into standard documentation formats. Paragraph, table, and figure numbering can be added, generally by company standards.

I usually have a coworker proofread the draft for errors or enhancements. To improve on the process, I can give the draft to technicians to see how well it fits into their production schemes. This depends on the complexity of the process and time availability.

The draft is then submitted to an organization that publishes and stores the corporation's technical data.

For a practical understanding, let's look at an example of a completed document of electrical wire process specifications. My comments will explain the development cycle.

NOTE:
The examples have been modified to protect proprietary rights.

Model 4.B (1)

Telephone 555 Jack Assembly

1. Part Numbers and Description

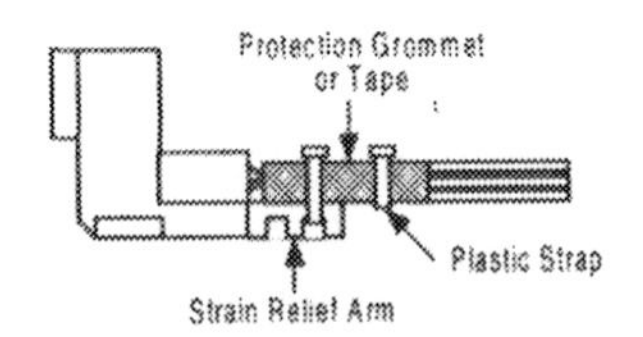

Jack Assembly
Figure 1

A. Connector Part Numbers

Table 1
Connector Part Number

Part Number	Supplier
555	Telephone

B. Contact Part Numbers

Table 2
Contact Part Number

Part Number	Supplier
M39029-555	QPL

C. Necessary Materials

Table 3
Necessary Materials

Description	Part Number	Supplier
Protection Grommet	6705	Grommets R US
Tape, Protection	88	Irish Tape Co.
Strap, Plastic, Adjustable	ST99	Strap Happy

2. Connector Disassembly

A. Strain Relief Removal

(1) Disconnect the telephone plug and receptacle connection.

(2) If necessary, remove the modular jack assembly from the structure or panel.
Make sure to keep the hardware to install it again.

(3) Remove the two plastic straps. Refer to Figure 1.

(4) Remove the protection grommet or tape.

In model 4.B(1) you will find the first half of the finished process for the Telephone 555 Jack Assembly. The figure below labeled "Rough Draft: Page 1," shows an early draft of these process specifications. Model 4.B(2) contains the second half of the process, and the figure labelled "Rough Draft: Page 2" is another early draft.

This process document concerns assembly and disassembly of a basic telephone jack that will be attached to a panel or structure in the airplane. This jack leads to a wire harness that connects to a central information area and then to the flight deck or cockpit of the airplane. The telephone then gets plugged into this jack. The phone is located in the cabin of the airplane, and is usually used by flight attendants.

In writing the process specification, it is my preference to do a basic outline of the process electronically first and then mark it up by hand. Usually the drafts are a combination of both handwritten and computer-generated text, tables, and graphics.

There are often many versions of the draft process as more information is gathered. During the writing of the process, many situations can occur that change the basic steps. These changes may include the completion of tests or development work, or text changes such as specific formatting.

In this case I initially laid out the process in the basic order and format I thought best. Through further development, many details changed.

I relocated the placement of the graphic. It is always a good idea to show a drawing of the final product. This gives the technician an initial quick frame of reference as to what is being built. It is also a good practice to list all the part numbers and materials at the beginning of the process. This identifies the items that are being put together.

In a maintenance situation, the first step that usually occurs is the disassembly of the item, as shown in the rough draft page.

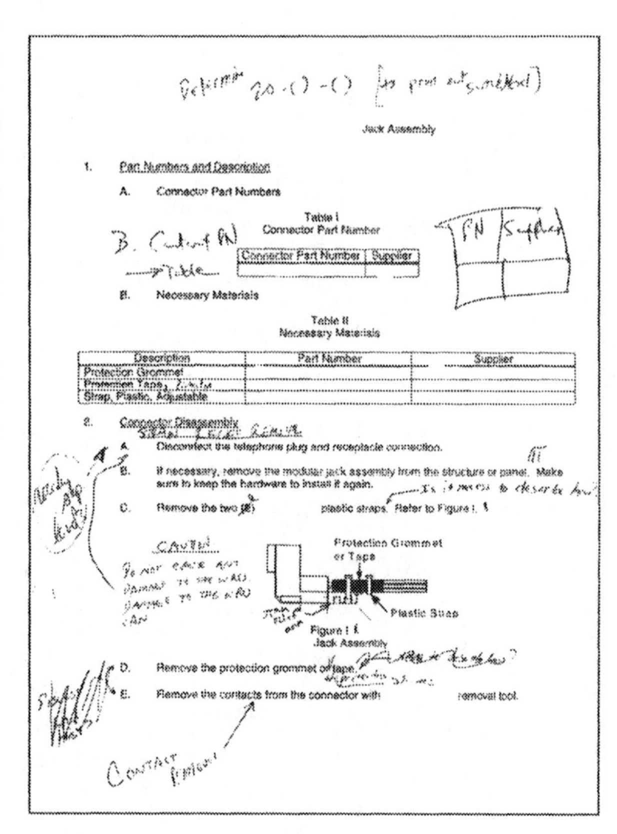

Jack Assembly

1. Part Numbers and Description

A. Connector Part Numbers

Table I
Connector Part Number

Connector Part Number	Supplier

B. Necessary Materials

Table II
Necessary Materials

Description	Part Number	Supplier
Protection Grommet		
Protection Tape,		
Strap, Plastic, Adjustable		

2. Connector Disassembly

A. Disconnect the telephone plug and receptacle connection.

B. If necessary, remove the modular jack assembly from the structure or panel. Make sure to keep the hardware to install it again.

C. Remove the two (2) plastic straps. Refer to Figure I.

Figure I
Jack Assembly

D. Remove the protection grommet or tape.

E. Remove the contacts from the connector with removal tool.

Rough Draft: Page 1

Model 4.B (2)

Telephone 555 Jack Assembly (continued)

B. Contact Removal

Table 4
Contact Insertion and Removal Tool

Part Number	Supplier
M39029-TOOL	QPL

(1) Make a selection of a removal tool from Table 4.

(2) Remove each contact form the modular jack.

3. Connector Assembly

A. Contact Assembly

Table 5
Contact Crimp Tools

Wire Size (AWG)	Crimp Tool				
	Basic Unit			Locator	
	Part Number	Setting	Supplier	Part Number	Supplier
24	M22520-BU	3	QPL	M22520/ZZ	QPL
22	M22520-BU	4	QPL	M22520/ZZ	QPL

(1) Make a selection of a crimp tool from Table 5.

(2) Remove 0.25 inch of insulation from the end of the wire.

(3) Crimp the contact on the end of the wire.

B. Connector Assembly - *Refer to Figure 1*

(1) Make a selection of an insertion tool from Table 4.

(2) Make a selection of a protection grommet or protection tape from Table 3.

(3) Make a selection of a plastic strap from Table 3.
NOTE: 2 straps are necessary for each modular jack assembly.

(4) If the strain relief is the grommet, put the grommet on the wires.

(5) Insert each contact into the correct contact cavity of the modular jack.

(6) If the strain relief is the protective tape, put 2 layers of tape on the wires.

(7) Install a strap around the wires near the front end of the grommet or tape and in the rear slot of the strain relief arm.

(8) Install the other strap around the wires near the rear end of the grommet or tape.

Some of the steps in the process were out of order, and my numbering was off by one level. All this was easy to see once I printed it out. I usually mark up the printout by hand and make adjustments electronically. Typically, this correction and improvement cycle occurs two or three times. I consider the process completed once it meets my criteria that the instructions are correct, the wording is in Simplified English, and it is properly formatted (each document has its unique format).

The process illustrated here was written for a maintenance manual, so Section 3, Connector Assembly, follows Section 2, Connector Disassembly. The jack in this case can be also referred to as a "connector." Many of the processes use connectors, so this nomenclature is maintained for these standard sections.

I originally didn't have the correct layout of the table in Section 3. The contact information should be included in its own table at the beginning of the process for total parts identification and referenced later as necessary. This was all modified as shown in the figure "Rough Draft: Page 2."

This example of the finished product and rough drafts should provide a good representation of what is involved in writing technical aircraft electrical process specifications. Technical writing is as varied as the thousands of industries that depend on documentation. Each industry has standard practices that reflect the needs that are specific to an engineering field.

B. J.

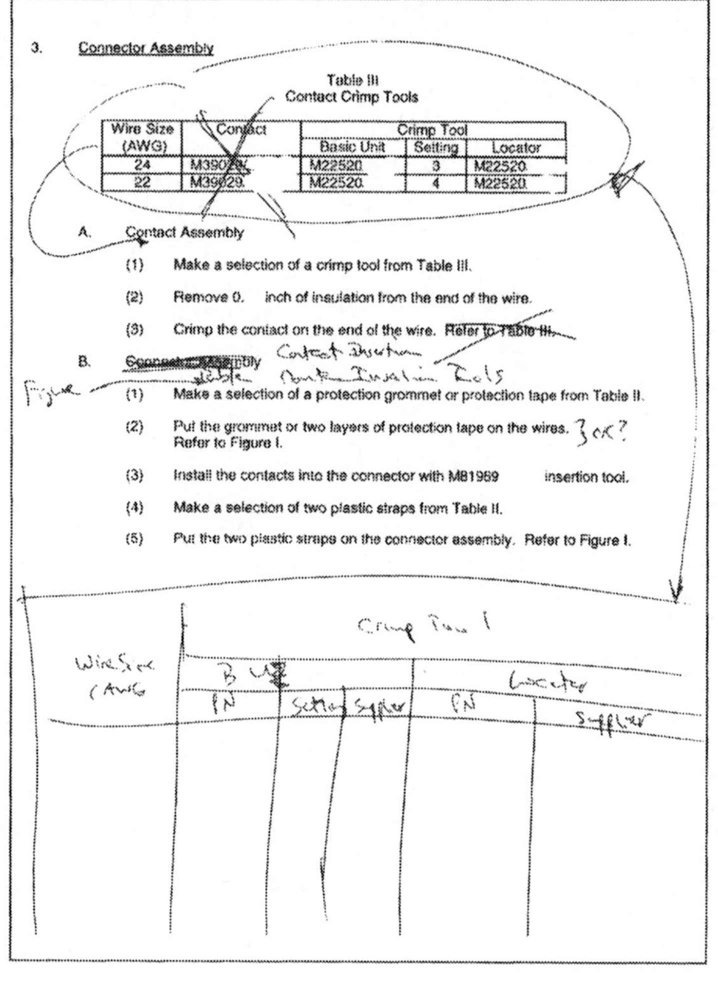

3. Connector Assembly

Table III
Contact Crimp Tools

Wire Size (AWG)	Contact	Crimp Tool		
		Basic Unit	Setting	Locator
24	M390[illegible]	M22520.	3	M22520.
22	M39029.	M22520.	4	M22520.

A. Contact Assembly

(1) Make a selection of a crimp tool from Table III.

(2) Remove 0.   inch of insulation from the end of the wire.

(3) Crimp the contact on the end of the wire. ~~Refer to Table III.~~

B. [illegible]bly

(1) Make a selection of a protection grommet or protection tape from Table II.

(2) Put the grommet or two layers of protection tape on the wires. Refer to Figure I.

(3) Install the contacts into the connector with M81969   insertion tool.

(4) Make a selection of two plastic straps from Table II.

(5) Put the two plastic straps on the connector assembly. Refer to Figure I.

Rough Draft: Page 2

The Samples

Comments

Introductions

The full-page samples of introductions reflect several popular types. Sample 4.A, concerning the spectrograph, opens with a definition and then shifts to an overview of what the paper will discuss. The author presents a line-itemed outline in the second paragraph. Conventional sentences are more commonly used to state a plan of activities (4.B) but the outline is equally effective in a running text. Sample 4.C is the opening page of a manual of assembly instructions for bench-tech assemblers. It develops a statement of importance for an introduction.

Notice that the samples are in the objective style and contain few pronoun references such as, "I will explain . . . ." The first sample refers to the project but not the author: "The discussion will deal with . . . ."

Conclusions

Conclusions are drawn if appropriate. One sample offers final thoughts concerning a laboratory project and concludes with recommendations. The other samples contain checklists and the summaries.

Outlines

The outline samples are vivid because they are topographically presented so that subsets are indented. (see Samples 4.G and 4.F). This feature is as important as the alphanumeric method that is generally used. Key points organize the sequence, generally by priority or occurrence. Evidence and important discussion points are identified in the subsections.

Summary

Introductions

- Use the outline introduction as a minimum starting point.
- Consider adding a statement of importance.
- Use alternative introductions to emphasize statistics, questions, incidents, history, or opposing views, if necessary.
- Place the outline introduction *after* any other type of introduction that is used.

Conclusion

- Use a summary conclusion, if needed, to highlight key points of a document.
- If you write a technical project, you may simply stop rather than draw conclusions, but recommendations and checklists are common conclusions also.

Pronouns

- Limit the use of *I* or *we* to introductory comments.
- If you write a laboratory narrative, you may use these pronouns throughout.
- Use the word *you,* as needed, for instructions.
- Avoid excessive use of passives.
- Use commands as a practical way to write instructions.
- Use the third person *he* or *she* as needed or nouns that identify the person: "assembler," "designer," "builder."

Outlines

- Develop outlines for supervised projects or team efforts.
- Use outlines for your own purposes as practical organizational tools.

Activities

Read the introductions in the models found in Chapters 5, 6, 7. In a memo to your instructor, briefly explain the different types of the introductions you identified.

Browse through some technical magazines and see if you can identify the types of introductions the authors used. Copy three or four different types. Attach them to a cover page in which you briefly explain them in a memo to your instructor.

Write three or four different types of introductions for projects. The introductions may be based on old papers or old lab reports or on projects you think you might develop. They may all be based on one project. Develop one of the introductions in the outline style.

Write conclusions for several of your recent lab reports or other technical reports that were developed for your program that were not structured in the introduction-body-conclusion format. Develop one conclusion as a summary. Develop the other conclusion as recommendations.

Share a Project

Work with two other members of the class to practice and discuss the effectiveness of introductions and outlines. Each member will need two recent technical projects from other courses to use as project models. An upcoming writing project will also work quite well. Using these projects, complete the following:

- *Develop a one-page* topic outline *of one project. Construct it in words and phrases. Save the outline but develop another one that is designed in complete sentences. Give the topic outline to the team members and ask them to explain your intentions. Then give them the* sentence outline *to look at. Briefly discuss the results.*
- *Develop* outline introductions *for both projects.*
- *Develop appropriate* alternative introductions *for the same projects and place each alternative introduction above the matching outline introduction. Show the introductions to the team members and discuss.*
- *It will be helpful if you have three extra copies of your outlines and introductions so that each member (and the instructor) has a copy of the project.*

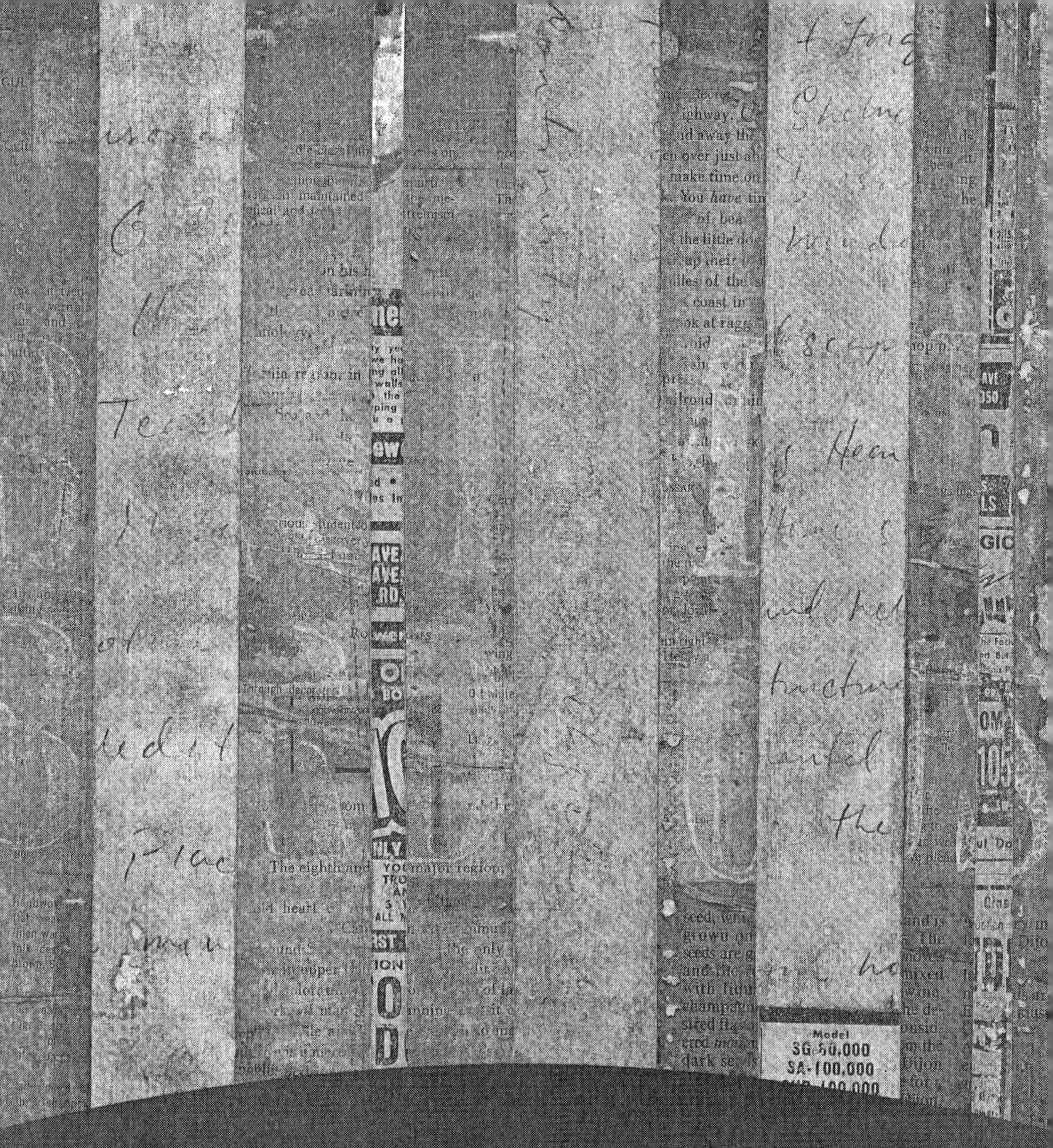

PART II

Document Prototypes

Work in Progress

5. Research

For many corporations, the standard operating procedure is that the engineers do the research. This is usually done by reviewing various specifications. The most common research at Genus involves comparing a customer drawing and a Genus drawing and creating procedures that will match both of the drawings and any other technical requirements. Various Military Standards (MIL-STD) are reviewed because Genus products are used on military vehicles and must meet governmental requirements. Nonengineering departments do research also, especially if something has a direct impact on their areas.

The distribution process I was studying is part of a production process for end items that are subsequently built into various parts that are sold to both customers and the government. That means there are non-Genus requirements that must be met, which partly explains the delay in getting the red-lined documents from the departments. The existing documents had references to customers, so additional agreements had to be arranged with manufacturing and government sectors before the changes could be agreed to.

After I updated the distribution process change as much as I could, I ordered the drawings, specifications, and a MIL-STD to understand why these were involved and to see if the proposed change would meet these requirements. During the review of these documents I paid special attention to items regarding meter stock and found that the MIL-STD 130 had specific instructions concerning the marking and identification of parts. This proved to be important because it explained how items are to be labeled.

The existing documents did not specify how parts should be labeled. I then went to several departments and noted that they had different types of labels for the parts. Some were using an "Accept Tag," which is generated by Inspection and identifies acceptance of a part; however, the tags did not identify the part as "meter stock," that is, there was no description of what the part was. Other departments were using other labels. Needless to say, no uniform labeling method was used, probably because the existing process document for the procedure did not specify how labels should be used, and it did not specify that all departments should use the same method. I was almost certain I had found the real culprit.

When I received the red-lined transaction information from the stockroom issuer, I placed this information in the applicable section of the document and modified it to conform to MIL-STD 130. I also received the red-lined production control information from the production planner and included this information in the applicable section of the document. The department changes standardized the distribution, and I added label standards to standardize the label practices.

S. B.

Texts Modeled by the Subject Matter

So far you have examined the generic layout and three primary formats of technical documents. Although the actual documents that a large corporation may generate can assume a huge number of shapes, you should be able to assign a substantial number of these products to basic categories. If you simply look at *utility,* you will find that you can take large chunks of the output and say "All this material can be developed according to such and such a procedure." The most conspicuous volume of writing is correspondence, and all this material more or less looks the same. It is a matter of convention, a matter of convenience, and a matter of efficiency to standardize writing practices. Often the subject matter, as noted in the last chapter, will shape the document, also.

Sample 5

CONSTRUCTING A BASIC WEB PAGE

Audience: First-time users of Microsoft FrontPage who are familiar with internet-browsing software and the Windows operating environment and who wish to learn how to create a basic web page.

Objective: To provide step-by-step instructions on how to create a basic, web page containing text, simple graphics, and links. The information is provided in a format that implies the existence of chapters that come before and after, but are contained within the same body of work titled *How to Design and Build Web Pages Using Microsoft FrontPage.*

Basic Placement and Testing of UTP CAT 5 Cable under EIA/TIA Guidelines

Audience: Assistant Network Systems Administrators and entry level Network Technicians interested in aspects of network infrastructure.

Objective: This paper is designed to aid in the understanding of the procedures that are involved in the installation and testing of UTP CAT 5 following the TIA/EIA standards.

Laser Welding Troubleshooting Matrix

Audience

The following document is designed for Manufacturing Engineers and Manufacturing Engineering Technicians who at some point will troubleshoot and subsequently repair a Solex Solstar Laser.

Purpose

This document provides a comprehensive explanation for the troubleshooting of alarms/error messages and the corrective action to be taken to repair a Solex Solestar Laser. This ensures that the laser has been optimized to function within its proper operating parameters and is ready for use with the laser welding system.

Categories of extremely commonplace documentation that are of concern here are those that engineers and technicians will encounter, either because they must frequently read and understand them or because they must occasionally write them. At least five basic types are critical:

1. Instructions	**(how to do it)**
Procedures	**(how it is done)**
2. Product Explanations	**(how it works)**
3. Definitions	**(what it is)**
4. Catalogs	**(the organization of products)**
Product Comparisons	**(the evaluation of products)**
5. Causal Analysis	**(if this occurs, the result will be . . .)**
Troubleshooting	**(if this is detected, do the following . . .)**

Among the five prototypes a total of eight common missions will generate a great deal of the copy in the engineering and technical departments of a corporation. The next two chapters will examine each type of document to explain how it is usually structured and why. Once you understand the structural necessities or other dictates that emerge from the subject matter, you should be in a position to develop any one of these primary documents. Models of each type of document accompany the text.

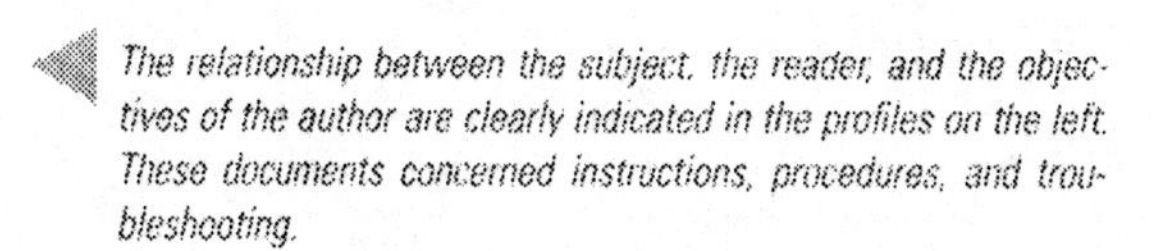

The relationship between the subject, the reader, and the objectives of the author are clearly indicated in the profiles on the left. These documents concerned instructions, procedures, and troubleshooting.

Model 5.A (1)

LABORATORY PROCEDURES MANUAL:

PROCEDURES FOR JUVENILE SALMONID STOMACH ANALYSIS

This section discusses the standard procedures for analyzing juvenile salmonid stomach contents

Audience:	Aquatic Biologists, Taxonomists, and Laboratory Technicians
Objective:	Section 2 of a 3-part manuscript that documents standard operating procedures for the PENTEC laboratory.

Name:	Kathryn C.
Date:	March 15, 200X
Course:	Engineering 231
Instructor:	Dave R.
Institution:	North Seattle Community College

Pacific Environmental Technologies, Inc.

Edmonds, Washington

Pentec

Instructions and Procedures

Instructions play a prominent role in most corporations. Broadly speaking, *instructions* include any type of itemized list intended to help a user or client or employee achieve some *mission*. It often is a *task*, of course, so instructions are a significant part of corporate paperwork. The client or employee may also read a document simply to see how others manage the task in a situation where perhaps twenty people actually are involved in the activity. In this case, the intent is rather different. These general *procedures* are not intended to be taken as instructions for the reader. They are usually somewhat more generalized and are oriented toward explaining how the group achieved the mission (see Model 3.C on p. 100). Included in both of these categories are a number of other somewhat similar materials—regulations, specifications, standards, labor contract agreements, and so on.

Readers probably do not perceive procedure documents as the sort they would pick up and "use" as instructions. They are more often points of reference. Nonetheless, they serve to help someone achieve some mission in accordance with prevailing practices or rules. In this sense, the building code books—*The National Electrical Code, The Uniform Building Code (UBC)* and others—are "instructive" even though they are not instructions. Readers consult these texts as regulatory references for proper outcomes more than for methods of completing the task.

On a practical level, let's leave the legalities and regulations and specifications for the appropriate staff. *Your* practical needs will be, more likely, either instructional in a specific sense relating to your field of interest or procedural in a general sense. The popular term for laypersons' instructions is "how-to" articles. The expression is helpful and clearly indicates the plan of action: how to do it. General procedures do not have a similar handle, but it is easy enough to see procedures in the how-to perspective: procedures explain "how it is done."

You can assume from the discussion of formatting (in Chapter 3) that there is one specific way to write instructions: in outline format. Certainly, the outline prevails throughout business and industry for developing instructions. It is worth noting that in popular literature, on the order of the *Reader's Digest* or *Time-Life* repair guides and the popular west

For the next three chapter we will be looking at models consisting of a number of pages of text. The pages are selected from longer texts in order to illustrate the discussion. Many pages are omitted.

Model 5.A (2)

INTRODUCTION

Fish stomach samples are analyzed to determine not only the diet of a fish but any change in fish diet due to environmental conditions. Laboratory analyses are performed primarily in PENTEC's biological laboratory in Edmonds, Washington. In addition, taxonomic support comes from Columbia Science in Royston, B.C., Canada, and the University of Washington, Seattle. All completed forms are permanently retained at PENTEC, with copies distributed to program files at the appropriate contract or subcontract offices.

Fish stomach samples returning from the field must be checked in according to the PENTEC Standard Operating Procedures for checking in samples, as outlined in the section on Sample Log-in Procedures—Receipt of Samples.

RESCREENING

Samples will be stored in labeled polypropylene or glass jars in the laboratory storage cabinets before and after rescreening occurs. Unless the field preservative used is alcohol, the samples must be rescreened before analysis can begin. If the specimens are preserved in formalin, samples should be rinsed or rescreened for at least a half day or preferably overnight to remove any odor of formalin. After the samples are rinsed, they should be stored in 70% EtOH until ready for analysis. If you have rescreened samples, be sure to sign off under the "Rescreened" column on the SampleTracking Log. All data forms are to be recorded in permanent-ink marker pen unless otherwise indicated.

coast *Sunset* homeowner's remodeling and repair guides, outlines of instructions are not always the preferred style for popular consumption. Publishers usually avoid the outline style unless the readers they address are fairly advanced and the document is intended as workaday instructions. The outline is favored for obvious reasons at all advanced levels of usage. There are reasons for this preference.

First, the outline can tick like a clock. Because instructions are obviously chronological and sequential, the outline structure and the instructional detailing are perfectly matched. Second, the material is usually clipped. Paragraphs are short, if they are used at all, and this brevity works quite well with instructions. Third, the instructions can be referenced quickly even if the ring binder is six inches thick. Whether employees of a local auto garage or a large engineering firm, the users of large volumes of instructions (or regulations and the like) depend on quick reference as a measure of the effectiveness of the document.

Distinct tasks come into play in the process of writing a set of instructions. Oversights are quite common in documents that provide instruction. These problems are frequently the result of a failure to visualize the task and target the readers properly. The categories for a set of instructions must include a series of activities of which the reader must be aware:

1. **Predoing tasks:**
    - **tools and equipment**
    - **parts needed**
    - **environmental conditions**
2. **Undoing tasks:**
    - **proper disassembly**
3. **Doing tasks:**
    - **the actual process of the project, step by step**

or 4. **Redoing tasks:**
    - **repairing an object is not the same as building a new one.**

Even though instructions and procedures are often presented in an outline format, the first page will usually be written in paragraphs. The paragraphs are used to introduce the reader to the text of the outline.

Model 5.A (3)

SAMPLE PREPARATION

After the sample has been rescreened, it is ready for analysis. Samples are processed by analyzing all or a subsample of salmon smolts from one jar at a time, per sorter. Data forms needed for analysis are the Sample Tracking Log, the Predator Stomach Analysis Form, and the Taxonomy Form.

Sample preparation includes the following steps:

1. Find the Sample Tracking Log for the current samples to be processed.
2. Record your initials under the "Processed" column for the next sample to be analyzed.
3. Note the storage location of the sample on the tracking log; check to see whether the column "Rescreened" has been signed off, and locate the cabinet in which it is stored. If the sample, has not been rescreened, it will have to be rinsed of the field preservative before analysis can begin.
4. Remove the sample from the storage cabinet and check the sample identification number on the jar to be sure that this number matches the sample I.D. number you signed out on the tracking log.
5. You will analyze five fish from each selected sample. This may mean processing all or a subsample of the salmon smolts from each jar. If you are subsampling, select five fish and gently rinse them through a number 60 screen in the sink with the fume hood on. Keep the fish you have rinsed in a petri dish or jar of water until you are ready to dissect one.
6. Determine the species of salmon in the sample jar (see Figure 1) and cross reference with the species I.D. indicated on the sample receipt log and jar. It is important that the correct species I.D. is made in all places where it is recorded.

Oncorhynchus gorbuscha

CHUM SALMON

Young without parr marks. Blue to greenish in color along back, sides silvery.

Oncorhynchus keta

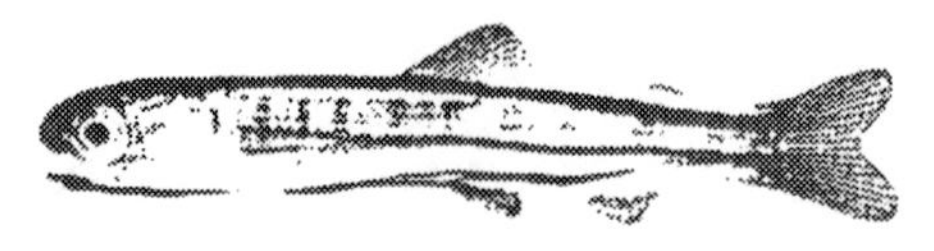

PINK SALMON

Young with parr marks as slender bars, scarcely extending below lateral line and with green iridescence on back.

The focus here is on "how to do it." You must provide the reader with all the correct details. If there are omissions, they will create problems.

Be sure to open a set of instructions with a conventional introduction developed in paragraph form. The introduction may be only a paragraph or it may take pages, but do not omit it unless your instructions are a section of a larger manual. Explain, even if briefly, the intent or purpose of the instructions. Notice the first page of the text that Kathryn developed for Pentec in Model 5.A (2).

Tools and Equipment

Once you have begun the outline, be sure to develop a section at the outset of the text to help your readers *prepare* for the task. Tools and equipment may be generic in a few industries, but technicians are highly specialized employees, and they will need to know *exactly* what they will need for an on-site repair, for example. If a needed item is not in the truck, it is an expensive oversight in the manual. The same logic applies to parts. You cannot depend on inventories. You need to itemize the materials needed. The environment may need to be specified also, whether it involves a dry day with no wind or a dust-free or temperature-controlled setting. A section of a parts list follows:

Suggested hardware components for building a basic PC

- 12X or faster IDE CD-ROM
- 32 MB to 64 MB 72-pin EDO RAM or 168-pin SDRAM (preferred)
- Socket 7 TX-chipset–based motherboard (AT or ATX)
- System case with a 200 W or greater power supply
- 3 GB or larger Ultra-DMA hard drive
- PS/2 compatible mouse with scrolling feature and mouse pad

Instructions often combine graphics and lists. The graphics represent the subject of the instructions (in this case, specimens). The instructions identify the activity to be undertaken.

Model 5.A (4)

PREDATOR STOMACH ANALYSIS FORM

The Predator Stomach Analysis Form is used to record pertinent data from the fish to be sampled from each jar. See Figure 2.

To fill out a stomach analysis form:

1. Record the project number and name at the top of the page under "Project #."
2. Record an S under "Sample Type."
   The first fish stomach you process will be stomach 1. Each sample should contain repetitions 1 through 5 to represent the 5 stomachs processed. If there are fewer than 5 fish in a sample, then you may have fewer than 5 repetitions. Start numbering again with 1 for each new sample processed.
3. Record the sample I.D. number from the jar and tracking log under "Sample No."
4. "Stom/Rep" refers to the number of repetitions of the sample.
5. Look up the NODC code in the code book and record it under "NODC code." Species have a 12-digit NODC code; therefore, all the blocks in the code should contain a number.
6. Use the calipers or the length board to measure the fork length of the fish under "Length (mm)." Note that this measurement is in millimeters rather than centimeters.
7. "Type" refers to the type of length measurement taken. Record an F here.
8. After measuring, blot the fish dry on a paper towel until no more moisture soaks into the towel. Do not let the fish dry out.
9. Place a weigh boat or a piece of Parafilm in the middle of the analytical balance and tare the scale so that it reads 0.000 g. If you have never used the balance, review the procedures prior to weighing.
10. Position the whole fish in the center of the weigh boat and close the glass door on the scale.
11. Record the weight under "Weight (gms)."
12. All weights used in processing are wet weight and should be recorded as W under "Type."
13. All fish processed for this project are juveniles. The juvenile code from the code sheet is 2. Record the code under "Life History Stage."
14. Write in your initials and the date under "I.D. by."

Warnings

Of equal importance are any warnings that may be appropriate. This is a very commonplace feature of technical documents and may appear at the suggestion of legal counsel at times. The following example is typical.

System Requirements

There are minimum system requirements in order for this product to work properly. It is recommended that before you purchase or install the AIGotcha!, compare your system with the list of requirements, as seen in Table 1, to verify compatibility. This product is also available for the Macintosh platform, as well as Microsoft Windows 3.1x, however, this paper is designed for Windows 95 (or later version) users.

Table 1: Minimum System Requirements

- IBM—PC Pentium 133 or greater
- Microsoft Windows 98 (or later)
- Minimum 16 MB available RAM
- Minimum 100 MB free hard disk space
- Parallel Port
- SVGA video card, 16-bit color
- CD ROM Drive

There might also be unique considerations. For example, compatibility is a special consideration in computer-related fields.

Disassembly

Second, consider disassembly as equal in importance to any other dimension of instructions. Obviously, this feature concerns technicians who "repair" or "correct" and not, for example, contractors who create *new* buildings, or engineers who design *new* bridges. The operating assumption of many manuals is that the reading audience is professional. This view seems to affect disassembly instructions more so than the rest of a manual. You will often see vast directives on how to get the job done and nothing on how to remove the damage first. Be sure that disassembly is properly tackled. It should be quite thorough if, in your judgment, the disassembly instructions are critical.

The instructions do not need to follow an elaborate outline unless such a technique is called for. Frequently, a simple numbering of steps is adequate and clear.

Model 5.B (1)

TRUSS ANALYSIS

Audience:

The primary reader of this paper will be second-quarter students of Mechanics in the Construction Engineering Drafting Technology program.

Objective:

The objective of this paper is to explain procedures involved in truss analysis by the joint method using the laws of physics, trigonometry, and algebra.

Anthony V.

April 25, 200X

Technical Writing

Dave R., NSCC

Instructions

Third, proceed with the instructions proper. Focus on three elements in particular:

CLARITY

STAGES

THOROUGHNESS

Every feature must be *clear* to the reader: the processes you describe, your vocabulary, and your graphics. Staging the instructions in correct *order,* and vividly, is the second concern. Remember that in the case of instructions, your reader is also a "user" who will employ what you write. Any omissions can be problematic. There can be none, so the document must be *thorough.*

Finally, note that *assembly* and *repair* are two distinct skills. Doing is one task. Redoing is another. Repair, for example, will call for troubleshooting in the service-related fields of electronics. In the construction industry, the builders of homes are a distinct group that seldom connects with remodelers, who reconfigure homes. The disciplines are quite distinct: construction and reconstruction. Maintain a sharp distinction between instructions that originate and instructions that restructure (or repair) because engineers are likely to be the readers of the former and technicians of the latter. They are different audiences.

The Scenario

If at all possible, develop instructions by actually doing what has to be done. It will greatly contribute to the success of the project if you do *not* imagine the job on the paper but, rather, walk the job through by penciling in each stage as you *do* it. Appendix B of this text, for example, was checked repeatedly for errors. Andreas, an expert in such matters, would suggest one way of phrasing the instructions. I would suggest another. We repeated the instructions on the keyboards at least six times over a period of a year, and rephrased the computer tips again and again to make them clear.

If you have difficulty finding an audience target for your projects, you may be overlooking the handiest one available: the peer group. Engineering Tech students and coworkers are convenient readers to address.

Model 5.B (2)

TRUSSES

1 BASIC PRINCIPLES

A truss is a structure generally formed of straight members to form a number of triangles with the connections arranged so that the stresses in the members are either tension or compression. Trusses are very efficient structures and are used as an economical method to span long distances. Typical depth-to-span ratios range from 1:10 to 1:20, with flat trusses requiring less overall depth than pitched trusses. Spans generally range from 40 feet to 200 feet. However, some wood-trussed rafters are used to span shorter distances.

Some typical truss types are shown in Figure 1. Generally, roof loads on a truss are transferred from the decking to purlins, which are attached to the truss at the panel points to avoid putting any bending stresses in the top chord of the truss. If concentrated loads are placed between panel points, or uniform loads are applied directly to the top chords, the member must be designed for the axial loading as well as for bending.

Trusses act much like beams in that there is usually compression in the top chords and tension in the bottom chords, with the web members being either in compression or tension depending on the loading and type of truss used. As with a beam, the forces in a parallel chord truss increase toward the center. In a bowstring truss, on the other hand, the chord forces remain fairly constant because the truss depth varies from a minimum at the supports to a maximum in the center.

In a trussed-roof design, trusses are placed from 10 to 40 feet on centers, depending on the loads and the spanning capabilities of the purlins. In residential and light commercial construction, truss rafters made of 2″ × 4″ or 2″ × 6″ members are often placed 2 feet on center. Open-web steel joists are usually placed 2 to 3 feet on center for floor construction and 4 to 6 feet on center for roof construction, depending on the spanning capabilities of the roof deck. Since trusses are thin and deep and subject to buckling, they must be laterally supported with bridging along the bottom chord. In some cases, diagonal bracing is required along the top chords of pitched roofs if the roof deck is not adequate to act as a diaphragm.

Audience Checks

Ideally, you should have a representative of your audience read a sample of the project. The best person is the user-reader, someone who will put the document to use. Have the reader or readers indicate any point of confusion they see, and revise accordingly. Perhaps you have seen the *Complete Idiot's Guides* to car repairs from John Muir Publications. Having volunteers work on their cars in the publisher's parking lot greatly refined these books. The volunteers used the books, and the books were then edited for clarity and thoroughness whenever there was a misunderstanding. In this case it was very important to use people who really did not know much about automobiles. As one of the editors explained, "When an 'idiot' tester gets too savvy, we have to get a new one." On a far greater scale, the 800-numbers that you use to call in questions to computer and software corporations serve as the audience checks for computer corporations. Subsequent additions, of software manuals in particular, can be revised—if the next generation isn't online already!

The Walk Through

Instructions are intended to be performed, of course, and the acid test *is* the performance. If the instructions are "used," you will be able to see areas that need revision. Procedures pose less of a problem. Procedures may not be as detailed as instructions, and there may be no easy test for thoroughness, since a walk-through is not an appropriate use of the procedures. In the case of instructions, remember that you have more than a reader; you have a user-reader. Accordingly, a number of typical technical writing documents carry a responsibility not seen in other writing areas.

As a final note, observe the tendency people have to trust instructions. They tend to take an orderly list and assume that the order is correct. The author is liable for the document and could end up in court. For example, poorly made exercise equipment has led to bodily injury and some rather major court suits in the fitness industry. I would imagine that the assembly instructions could sometimes make or break the case for either the defense or the prosecution.

Notice that this model is handled in a running text of paragraphs. Because the project is a discussion of procedures, the paragraph form is acceptable.

Model 5.B (3)

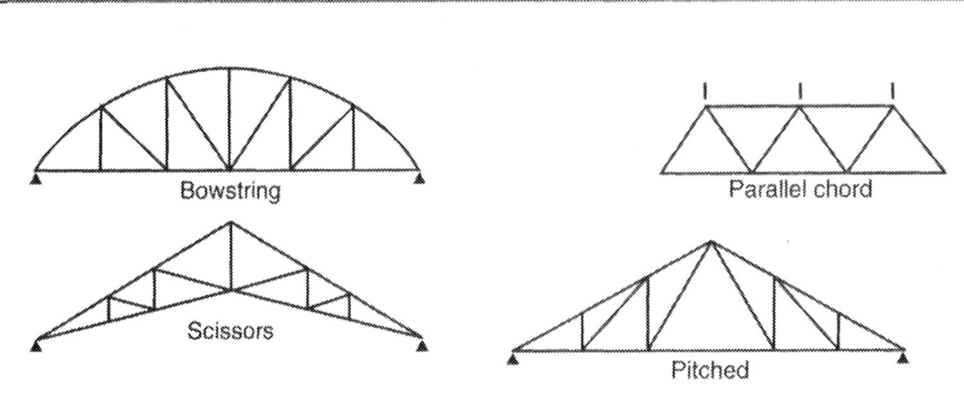

Figure 1. Types of Trusses

Individual truss members are designed as columns if they are in compression. If in tension, they must have adequate net area (after deducting for the area of fasteners) to resist the unit tensile stress allowed by the material being used. If concentrated loads or uniform loads are placed on any chord member between panel points, the member must also be designed to resist bending stresses.

When designing steel trusses with double angles, as is usually the case, the engineer can find allowable concentric loads and other properties for various double angle combinations in the American Institute of Steel Construction (AISC) Manual of Steel Construction. By knowing the compressive load, the length of the member, and strength of steel, the engineer can determine the size and thickness of a double-angle combination.

For members in tension, the net area must be determined. This is the actual area of the member less the area of bolt holes, which is taken to be 1/8 inch larger than the diameter of the bolt.

Regardless of material, truss members should be designed so they are concentric, so the member is symmetric on both sides of the centroid axis in the plane of the truss. Thus, steel truss members are often built with two angles back-to-back separated by 3/8-inch or 1/2-inch gusset plates, with tee sections or with wide flange sections. See Figure 2.

Regulations and Specifications

If you are developing regulations or specifications, you will have to determine the organization of the material based on the specific needs of the subject matter. Whereas the instructions and procedures are chronological, there is no clock ticking in a book of regulations or specifications. Remember that outlines are, nonetheless, sequential and that they are subsectioned. The sequence can, for example, be used to *rank*. What should come first? What is most important? What is less important? The subsections similarly ascend and descend in importance. A document can greatly clarify confusion when details are subordinated to some other consideration. Rely on the greater than/lesser than nature of the outline logic. Be sure to indent subsections uniformly so that the ladder logic is visual.

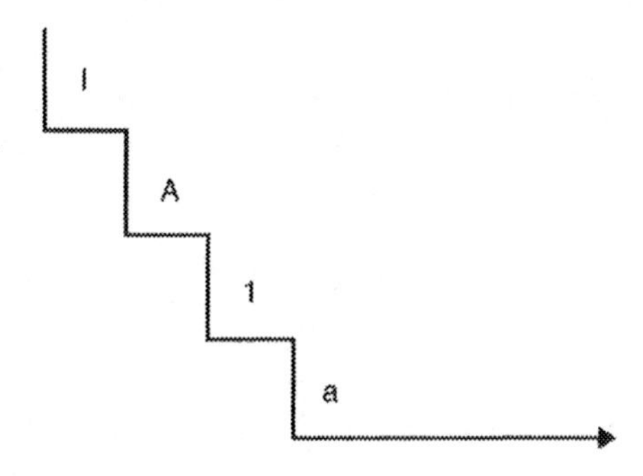

It is partly the topography of the outline that allows a volume of regulations to be referenced with speed and efficiency. The method of enumeration is the other tool that makes large texts of this sort less daunting, whether they are alphabetical lists, numerical lists, or alphanumeric lists. For example, some variation on the outline style of numbering is used to codify state laws. Whether a single volume of specifications for a BMW motorcycle or fifty volumes of federal importation laws, the material is easily consulted because of very basic devices: outline topography and uniform numbering systems.

The text explains how trusses are used from a general perspective. There is no effort to provide a set of instructions for a specific application.

Models 5.A and 5.B

Comments

Document model 5.A concerning salmon analysis was designed for a hands-on situation. Kathryn developed the project for her company as a set of instructions. Model 5.B is distinctly different in being a discussion of procedures. Anthony developed this project to explain how truss systems are selected for a variety of different structures. His explanation of the process is not intended to be viewed as a set of instructions.

The difference between the two processes is quite apparent in that procedures tend to explain how a task is accomplished but not with the idea that you are supposed to do it. To put it simply, instructions explain how to do it (by you, for example) and procedures explain how it is done (by them).

Instructions and Procedures

Project game plan

- Determine whether the text should be developed as instructions or as a description of a procedure.
- Open the presentation by stating the value (importance) of the instructions or procedures.
- Briefly outline the plan of the document.
- List or otherwise explain all necessary materials, tools, and environmental considerations.
- Select the format that will be appropriate (Chapter 3): outline, block paragraphs in a running text, headings with paragraphs, or a mixture of these.
  - There is a general tendency to use outlines for instructions. The instruction document is easy to follow if it is step-by-step and has an outline topography.
  - The paragraph style is most appropriate when discussion is intended to be a substantial part of the presentation.
  - Headings can be used to identify key divisions in either an outlined text or a text of paragraphs.
- Develop the text using a chronological start-to-finish format.

Model 5.C (1)

A Basic Audio Amplifier and Its Operation

Audience: First- or second-year electronics students with a strong understanding of DC solid-state theory.

Objective: To establish an understanding of the AC operation of a basic audio amplifier by conducting a stage-by-stage analysis.

Damon J. H.

EGR231

Project #2

North Seattle Community College

Product Descriptions

The promotional and technical documents that explain a product line or describe mechanisms are certainly as important as the instructions. A corporation does not sell its products by simply watching them move off the shelves of stores or warehouses. The sales representatives present the devices to various wholesale and retail sectors. They may have the genuine items in their pockets, but not if they are refrigerators, forklifts, or jet engines. Writers devote a substantial body of documentation to presenting the products in a more practical way—on paper.

In this sense, writing creates a substitute or a model of the products. A sales rep may impress his clients with his computer chip cuff links, but usually the paperwork will do a better job. When the products are too small or too large to present to the marketplace, paper representations will do the job more effectively than any other method of promoting a product, with the exception of demonstration videotapes.

To write a product into existence on paper, you must bring to the drawing table a physical reality that you do not have to deal with in the writing of instructions. A reader must "see" the product. If the product is nonmechanical, the challenge is not too great. For example, if your company sells floor tile, you can photograph the tiles in color and not worry about what the tile "does." But if the device is mechanical, you have a more difficult product to represent. A mechanism that involves moving parts is difficult to "create" on paper, photograph or no photograph. If the salesman with the chips on his cuff links also wears a digital watch that he distributes, how will he explain the mechanism? All watches keep time, so he must explain what is unique about his particular oscillating crystal. Yes, it is mechanical in some sense, because the crystal is "performing" a task at a nuclear level. A photograph will not show this, but how do you render the mechanical nature of the movement in writing?

In truth, product descriptions can easily be made more vivid than the products themselves. You cannot magnify, amplify, qualify or otherwise change the products in your briefcase or at a trade show. What you *can* do is present a written description of the product

There is often an important proviso in the audience description in technical documentation. This author explains that the text is for second-year students, but specifically, the readers must have adequate knowledge of DC solid-state theory in order to understand the device under discussion.

Model 5.C (2)

Analysis

This text deals specifically with the AC analysis of the amplifier. Due to the elementary nature of the DC quantities and their calculations, it is assumed that the reader possesses the knowledge needed to calculate these values independently.

Like any other multistage amplifier, the audio amp is composed of multiple independent stages of amplification. In order to understand the overall operation of the audio amplifier, we need to analyze the amp stage by stage. In the case of the amplifier shown in Figure 1, there are three stages. Each stage consists of a transistor, or transistor pair, and the necessary support components. Figure 2 graphically isolates each stage of the amplifier. Notice that each stage is fully self-sufficient and can operate independently of the others.

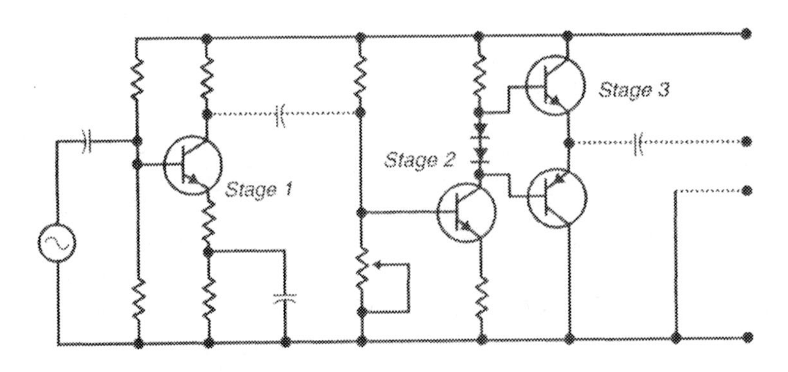

Figure 2. The three stages of the studio amplifier shown as independent transistor stages brought together by coupling capacitors.

In Figure 2, we can see that stage one is a small-signal class A amplifier, stage two is a large-signal class A amplifier, and stage three is a class B push-pull emitter follower.

or mechanism that can sit beside the sample. This representation of the merchandise can be distributed by the millions in a cost-effective manner. The point of writing a product description is the convenience of *replacing* the product, but it is also an opportunity to represent the product in a way that is appropriately developed to meet the needs of a given readership. For example, no one is going to benefit from watching an I beam resting on two corner-post beams. This "demonstration" would be as exciting as watching paint dry. But with a drawing and a short text, the way in which the beams are a mechanism can be explained. If readers watch the beams *do* the job, they will not see much getting done. But with a diagram, a few arrows to represent force, and a short description of statics, you can bring the beams to life and explain what they are "doing."

In this respect, if you are in a program related to construction engineering fields, be particularly alert to the fact that the focus on "mechanisms" does not exclude architectural form. The distinction that is usually made between "static" and "dynamic," or between "potential" and "kinetic" energy, is an issue that concerns movement in some sense. The absence of movement does not change the mechanism to a nonfunctional object. A building is, in my eyes, a mechanism; all its parts are mechanical because they are doing a job. I say this to suggest that static effort, even more so than mechanisms that clank and chug, calls for a little explaining so that readers understand the product. "Passive" energy products—insulation, for example—call for detailed product explanations to increase consumer understanding.

Procedure

To describe a product, particularly a mechanism of any sort, divide the device into three major categories:

Function

What does it do?

Parts

How is it made?

Principle

How does it work?

Observe the role of graphics in this model. Each page calls for an illustration of some sort. Since this project focused on principles, the in-depth nature of the topic called for schematics.

Model 5.C (3)

3

When dealing with audio amplifiers, we need to consider two fundamental characteristics, namely, the amplifier's fidelity and its AC voltage gain. The fidelity of an amplifier is essentially the quality of its sound reproduction. Fidelity is a very complex subject, dependent on many diverse factors far beyond our scope of analysis. The AC voltage gain is the ratio of the output voltage to the input voltage and is represented by the symbol A. It can be thought of as the amplifier's amplification factor; the higher the gain, the greater the output voltage for a given input voltage. Equation 1 states the expression for gain.

$$A = \frac{V_{out}}{V_{in}}$$ *Equation 1*

From Ohm's law, we know that voltage equals current times resistance. If we substitute current times resistance for voltage in Equation 1, we get Equation 2.

$$A = \frac{r_c i_c}{r_e i_e}$$ *Equation 2*

However, because the current in the emitter is approximately equal to the current in the collector, they cancel each other out, leaving only the resistances. We can see this in Equation 3. Note that the emitter resistance is the total emitter resistance. This includes the swamping resistor, r_e if one is present.

$$A = \frac{r_c}{r_e + r'_e}$$ *Equation 3*

r'_e represents the AC resistance of the base-emitter diode of the transistor and is calculated by the equation given in Equation 4.

If you discuss all three categories thoroughly, you will define the item or device with precision. You will *always* need all three levels of discussion to adequately represent the product. The text can be left in this general format at times, but writing gives you flexibility to do more with the project.

I noted that your document can magnify, amplify, or otherwise modify the desired focus you seek for the item under discussion. You might take that to mean the use of a host of graphics, such as cutaways or exploded views or photographs of objects, and so on. Illustration has a flexibility that the real object does not have, but so does the *text* of the product description. You can move the three major parts around, and you can also tailor the discussion of procedures, parts, and principles.

Consider who will be reading the product literature. There are three major categories of readers that more or less correspond to the three categories you should develop in documents that concern products. Suppose you have imported 5000 smart typewriters. They are digitally designed to compete with word processors, but there is no CRT screen. What literature will you need? First, you will need an operator's manual that must go out with the typewriter. An attractive glossy booklet with many close-up photos of the exterior parts will do the job. But since the booklet is for secretarial staff, will it make much sense to devote equal time to functions, parts, and principles? As you realize, the clerical staff needs a document that will focus on *functions*.

What will be the demands of the repair technicians? They will need interior illustrations and not exterior illustrations. Assembly drawings will replace the photographs, and the literature will be much more *parts* oriented. For engineers, the primary concern will be theoretical *principles*. If you think about mechanisms in terms of principles, you realize that theoretical explanations will be the most technical literature you design to support the instrument at this level.

Readers will dictate their priorities, which, in turn, will help you see how to prioritize your document. (See Chapter 8 for a more detailed analysis of readerships.) In general, assume that most of the document can focus on *one* of the three categories—functions, parts, or principles—but do not omit the other two. In other words, the secretaries who

The use of mathematics is a practical dimension of engineering documentation. A project dedicated to operating principles is quite likely to involve mathematics. The boxes you see here are unusual but perhaps add clarity to the page layout.

Model 5.C (4)

10

In stage two, we also have the 5 kΩ variable resistor. The purpose of this resistor is to fine-tune the base bias of the transistor to avoid distortion due to saturation or cutoff. The value of the resistor has been predetermined at 1.92 kΩ for optimal performance.

Figure 6 displays the important AC quantities for stage two.

Quantity	*Description*	*Equation*	*Value*
r'_e	AC emitter resistance	$\frac{25\text{ mV}}{I_E}$	27.3 W
$Z_{in(base)}$	Input impedance to the base	$\beta(r'_e + r_e)$	102.7 kΩ
r_c	AC collector resistance	$Z_{in(Stage\ 3)}$ $R_C + r'_e(\text{of diodes})$	3.46 kΩ
A	AC voltage gain	$\frac{r_c}{r'_e + r_e}$	3.37

Figures 6. Important AC quantities for stage two

Stage Three

The third and final stage of the audio amplifier is a class B push-pull emitter follower. Figure 7 shows the single-stage equivalent schematic for stage three. Remember that a class B amplifier requires two transistors to amplify the full sinusoidal waveform.

will use the typewriter need to know a little about the parts. They often make office adjustments or repairs of a minor sort. And a *little* theory may help so that they have patience with digital processes and do not expect miracles, or so that they do not put refrigerator magnets on the chassis to post reminders. The case of the technician is more obvious. Today's engineering technicians may be concerned about repairs, but their need for theory is the reason why technicians all need college degrees now. Thus, the use of theory, explicit or implied, will be essential for the repair manuals.

To write the text, consider beginning with the two elements you will *not* emphasize. If the documents are for technicians, you might open with a brief discussion of function and theory. Use these two areas of interest for background, and then devote the text to *parts* and drawings and so on. If the document is for operators, you might open with a discussion of parts and perhaps a brief note on principles; then follow with a focus on *functions*—what the typewriter will do.

Graphics play a key role in product descriptions for obvious reasons. The product under discussion in the document is a physical object. Instructions describe activities. As a result, the instructions are easy to sequence though perhaps a little difficult to visualize. In the case of product descriptions, you will find the opposite to be true. Product descriptions are easy to visualize—particularly with graphics—but a little more difficult to explain in writing. Use functions, parts, and principles as your guidelines. Keep in mind the fact that mechanisms "do" something. The document will have to explain both spatial needs (illustrations) and how the product performs tasks in time (chronology).

The logic structure of the document will magnify one of the three categories of concern for the readership. Because there is very likely to be a strong emphasis on the physical makeup of an object, the text may take on a pattern of development that is quite unlike instructions. Instructions are sequential and organize activities in a time sequence. Descriptions of mechanisms are likely to shift attention to component parts and component organization, and this creates a shift to spatial sequences in which the logical continuity is controlled by the graphics. The paragraphs of the text are then ordered according to the necessities defined by the graphics.

Tables often play a role in a technical project. Note that the amplifier is being divided into stages for discussion. The table provides data in support of the second stage of the discussion, according to the one-sentence introduction above it.

Model 5.C (5)

To perform the calculations for this stage, we will need to assume a value for the output resistance. Typically, output resistances can vary greatly, depending on what type of device is connected across the output terminals. For convenience we will assume a resistance of 00 Ω across the output terminals.

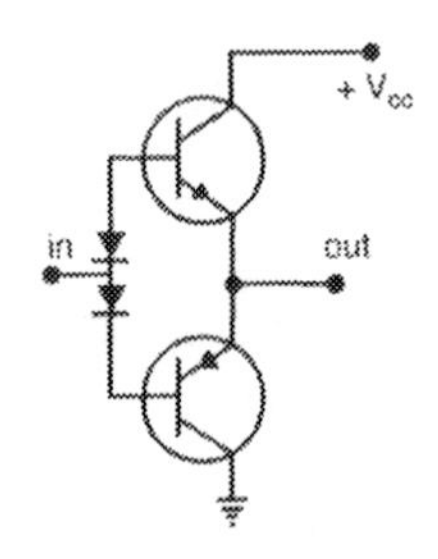

Figure 7. Stage three, a class B push-pull emitter follower.

Figure 8 displays the important AC quantities for stage three.

Quantity	*Description*	*Equation*	*Value*
r'_e	AC emitter resistance	$\frac{25\text{ mV}}{I_E}$	27.3 Ω
$z_{in(base)}$	Input impedance to the base	$\beta(r'_e + r_e)$	12.7 kΩ
r_c	AC collector resistance	$r_c = R_L$	100 Ω
A	AC voltage gain	$\frac{R_L}{r'_e + R_L}$	0.785

Figure 8. Important AC quantities for stage three

If the mechanism performs a task with movements, you have a unique consideration. In this instance, the product description will have to be both spatial and temporal. Parts are used to describe the object; simulated movements explain what it does. Graphics are indispensable for these descriptions and explanations, but the text is also important because it will have to include both spatial and temporal elements and explain or amplify the graphics. (For discussion of the controls that keep the graphics and the text aligned, see Chapter 11.)

Damon developed the audio amplifier project (Model 5.C) with some very good equipment to help him. Computers have greatly improved our ability to render our own graphics. Since the project explains audio amplifers, the computer proved to be a handy way to generate graphics that visualize the mechanism. The original graphics are in color. Damon used the following computer products:

The equipment and software used to create this document include:

- Gateway Pentium 750 computer
- Epson Stylus color printer
- Hewlett-Packard laser printer
- Hewlett-Packard desktop scanner
- Windows 98
- Microsoft Office (main document and several graphics)
- Aldus Photoshop (to create and modify graphics and title)

This page illustrates the important point that technical writing is a multimedia and multilingual undertaking. These systems are called into play to explain many dimensions of engineering, including in-depth product descriptions.

Model 5.C

Comments

The audio amplifier project is a discussion developed primarily through an examination of operating principles. The first page, omitted here, began with a statement of importance based on the usefulness (functions) of the amplifier. Parts are examined at many points of the project, but primarily in the context of the discussion of the operating principles.

Early on in the project there are schematics indicating the structure and its components. These graphics are tools that allow us to enlarge important parts of the discussion—visually and in text length. Like a dissection or an examination under a microscope, any discussion or visualization of assembly provides a unique view of an object. From this point on, the author focuses on the mathematical principles involved in the design and operation of the amplifier.

Product Descriptions

Project game plan

- To examine a device or product, identify its functions, its parts, and its principles.
- Determine the needs of your audience in relationship to the product.
- Introduce the subject that is to be defined and explain the importance of the subject.
- Briefly describe the two components of the discussion that are of less interest.
- Thoroughly explore the third component that is being addressed to the specific readership.
  - ✔ For users, focus on function.
  - ✔ For repair personnel, focus on parts.
  - ✔ For designers, focus on principles.
- Make sure the analysis satisfies the reader's needs.

Model 5.D (1)

UNDERSTANDING MEDICAL DIAGNOSTIC ULTRASOUND

Audience: This paper is intended for people who have had at least high-school-level physics, including wave theory.

Objective: To explain, in basic terms, the process of ultrasound scanning.

by

Dawn E. S.

May 15, 200X

Instructor: David R.

North Seattle Community College

Definitions

The process of explaining or describing a product is, in effect, the process of defining it. Because the object is physical and has a utility, we noted in the previous section that the practical parameters of definition are functions, parts, and principles. There are also other types of definitions and other situations to define. Virtually anything that calls for a substantial explanation is defined by that explanation. Dictionary definitions will not do the job. A simple sentence is hardly a definition. It may take four hundred pages to define a pulsar.

We understand every aspect of reality by *defining* it. An apple is a fruit, a cabbage is a vegetable. A tomato (is it a fruit or a vegetable?) is a point of conflict when the parameters of the definition do not quite fit the object. Definitions are a point of reference and a matter of convenience, but they can never embrace all our world. The early European explorers who dared to journey across the oceans often brought back specimens or tales of a host of new animals, from the platypus to the lungfish, and these creatures did not fit the known realities *defined* by the animal orders of Europe.

The pursuit of *specimens* among botanists and zoologists reflects the most desirable definition: the object itself. Lacking the specimen, the early explorers and scientists, as I noted earlier, kept vivid descriptions and occasionally were accompanied on their voyages by an illustrator. The learning experience of a child's classroom includes a very old tool: show and tell. The method is scientific. The challenge for the child is the description. An author faces a similar challenge. To develop a definition a writer must focus on methods that describe and therefore define. Illustration is the most obvious of the tools of definition.

Ideally the best approach to defining a thing is presenting the thing itself. In the absence of the object, the most practical substitute may well be a visualization or graphic rendering. Once the image is presented, it can then be amplified, cut away, rendered as an exploded view, or otherwise explained. Essentially, the object of the definition must somehow be divided for analysis. As long as there is a *physical* nature to the object, the analysis may simply follow a breakdown of the parts—a figurative or literal dissection.

Extended definitions may seem far less common than instructions or product descriptions or descriptions of mechanisms. In truth, they are valuable and popular but lack the immediately identifiable characteristics of the other types. Anything can be the subject of a definition.

Model 5.D (2)

Ultrasonics

Medical diagnostic ultrasound is an important tool in the hospitals and clinics of today. It is one of the few **noninvasive**[1] diagnostic tools available.

X-rays pass right through some types of tissue, like skin, but are largely reflected by others, like bone. Medical diagnostic ultrasound is based on the same idea. The ultrasound waves pass through some tissues easily, and other tissues reflect most of the signal. Unlike X-rays however, which are harmful in high doses or long-term exposure, ultrasound has not been proven to be harmful to humans.

Ultrasound provides information on tumors and cysts that cannot be obtained by conventional X-ray studies. Echocardiography, the study of heart motions by ultrasonic means, is another medical application. Ultrasonics plays a significant role in observing pregnancy problems as well and is often the first screening examination for many medical conditions that affect children. An ultrasound machine and scanheads are shown in Figure 1. Learning what ultrasound is and understanding some of its common uses is the best introduction to understanding how the medical diagnostic ultrasound scanning process works.

---

[1] *All words appearing in bold are defined in the glossary on page 14.*

If the object of the definition is energy or theory or anything on the order of what might be called phenomena, demonstrable or hypothetical, the "object" or subject focus must be scrutinized in its appropriate terms. Gravity is popularly compared with electromagnetism, for example, even though gravity is like the tomato and does not satisfactorily fit the known categories. If a particle physicist wants to demonstrate the existence of bosons, she cannot. She can provide only a demonstration of "events" in which the boson was involved. Similarly, a black hole is the star that was. The traditional star charts, from the red giants to the white dwarfs, had no provisions for stars that are not. Theoretical conditions such as these are constructed as models in astrophysics. There will be no pictures of a black star. There is only a sort of aura bordering its horizon or edge.

Apart from illustrations and explanations based on the conditions of the object to be defined, there are many definition techniques to help define the parameters of what something is. There are no rules, in a way, and a definition can be established in any terms that serve a writer. For technical work, the usual approach, of course, is to focus the definition on the needs of the reader. Notice that this suggests the relative nature of definitions. People often assume that definitions are fixed. In truth, mutual agreement on a definition is often in question—particularly in the sciences, exactly where the layperson might assume that definitions are absolute. The social sciences may be an even better example, since, for example, economists frequently disagree about economics, both in theory and practice. The science of psychology, as well, is prone to remarkable disparity—with well over one hundred "schools" of practice or theories of what's what.

In physics, for example, there are still many matters for conjecture—from neutrinos to bosons to gluons. Engineers argue over the proper ways to produce power, recycle toxic chemicals, process sewage, and so on. Often these conflicts concern definitions of terms. What is a toxin? When is it toxic? What is toxicity? What is the standard measure? The chemical engineer may be delighted to point out the toxicity of "organic" compounds, such as pyrethrums, that organic gardening buffs construe as safe only because the compounds are "natural." At federal hearings, a European engineer will proclaim the risks of EMR (electromagnetic radiation), and an American engineer may deny the risks. As you can see, definition is often the point of conflict in many confrontations, scientific or otherwise. The gluon is theorized to exist. Maybe it does not. Then again, the planet Pluto was theorized to exist, and it does.

The first page of this discussion of ultrasound includes a brief explanation of the process, a brief explanation of its uniqueness (it is not an X-ray), and highlights a few of its uses.

Model 5.D (3)

3

There are four basic divisions in the ultrasound scanning process: wave production, wave transmission, wave interaction, and wave reception. In the first part of the process, the ultrasonic waves are produced. This is accomplished inside the **scanhead,** or **transducer,** as shown in Figure 2. The scanhead is the hand held part of the ultrasound machine that actually touches the patient. Inside the scanhead there is a **piezoelectric crystal.** When a voltage is applied to this crystal, as shown in Figure 3a, it produces vibrations at a 90° angle to the applied voltage. The precise frequency at which the crystal vibrates is determined by the size of the crystal. In the same way, if the crystal is squeezed, it produces a voltage at a 90° angle to the applied pressure, as seen in Figure 3b.

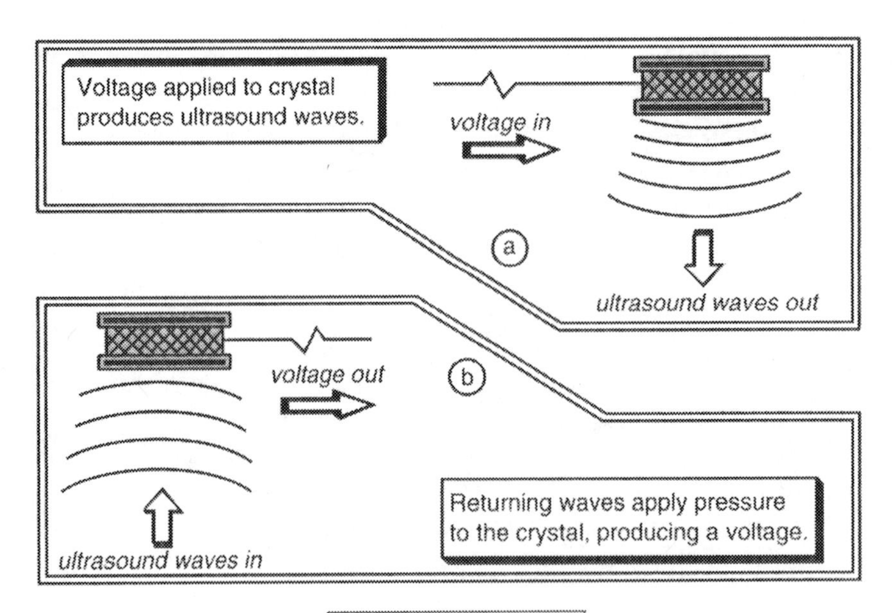

Voltage applied to crystal produces ultrasound waves.
voltage in
a
ultrasound waves out
voltage out
b
Returning waves apply pressure to the crystal, producing a voltage.
ultrasound waves in

Figure 3b. Piezoelectric Crystal

Analysis

There are other tactics for defining, apart from samples, illustrations, and simulations. I would head the list with an analysis of parts in the general sense, by observation, disassembly, dissection, or whatever method suits the object. A general explanation is, in effect, often the analysis of parts. This is frequently a first order of preference for a definition. At times it is often the only preference.

Interpretation

The conditional viewpoint is always a possible approach, but it is likely to be a potentially subjective or personal perception. If a doctor defines your headache as psychosomatic, that is a definition. Perhaps it was stress related, or sinus related, or vision related. Interpretation is likely to reflect the interests of the beholder. An artist will not see the Eiffel Tower with the eye of an engineer. If either of them explain this steel oddity, he or she might declare it a masterpiece, but their differing reasons for respecting Eiffel's genius would be apparent when they explained their perceptions.

History

Anything can be defined by its *evolution* or by its *past* or by its *growth*. It is rare for me to receive a paper from a student about Marconi or Tesla or some other interesting historical figure in the development of electronics. I will receive several hundred papers in total, but few will be historical. Across campus in the anthropology department, it will be different. Paper after paper will concern evolution. In the geology department, endless projects will have a strong historical concern for eons of formations in, let's say, the Columbia River basin. In the biochemistry lab, the issue of organism growth will be a focus. In all these areas of scientific inquiry, history is a major component of perception and definition. Notice that all the strategies for definition work well, but the subject matter of a given discipline is often dominated by certain techniques of definition. History is simply not a common or practical approach in electronics.

As you might guess from the first sentence of this page, the project will discuss the process of ultrasound by dividing the technology into four subject areas of discussion.

Model 5.E (1)

LAN NETWORK ADMINISTRATION

Audience: This paper was written for I.S. department employees of Swedish Health Plans.

Objective: LAN NETWORK ADMINISTRATION is for those who wish to gain an understanding of what network administration entails.

John T.

01/29/00
EGR 231
Project 1
Instructor: David R.

Comparison

A traditional technique for establishing what something is is to state what it is similar to. This is a method of taking "known," or defined, conditions of the world that are understood and establishing comparisons for practical purposes. What you know can then contribute to your understanding of what you do not know. The popular observation, "The mastodon was very much like an elephant" is a very handy definition. In the case of an extinct branch of the animal kingdom, such as in this instance, the comparison helps define the subject.

Contrast

Equally common and equally effective is the approach that examines dissimilar situations. If you have an animal that shares nearly all the conditions of being a mammal, but not quite, you have another tomato. The dissimilarities are critical to definition. If the critter lays eggs, for example, you need a distinct new category for the type. Such creatures are gathered into their own family and given a new name, for example, "platypus." The unique way in which platypuses are contrasted to other mammal groupings gives people a better perception of them. In a similar way gravity behaves much like electromagnetism, but there is a critical dissimilarity in gravity's lack of polarity. This dissimilar element is perhaps *the* defining feature of gravity.

Elimination

Another way to attempt to define something is to describe what it is *not* by using a process of elimination. You can go a long way toward explaining a phenomenon by systematically removing possibilities of what it could be. At the practical level, this was the focus of the famous Project Bluebook, which debunked the UFO theories of the 1950s. Through a process of elimination, sightings were explained as weather, balloons, imagination and, so on. This meant that, through elimination, whatever the sightings were, *they were not* UFOs. Because investigators will not know what a UFO is until they have one, it was the only logical strategy to take. By eliminating *known* factors, the Air Force hoped to define the limits of the unknown—the parameters of which they evidently wanted at about zero.

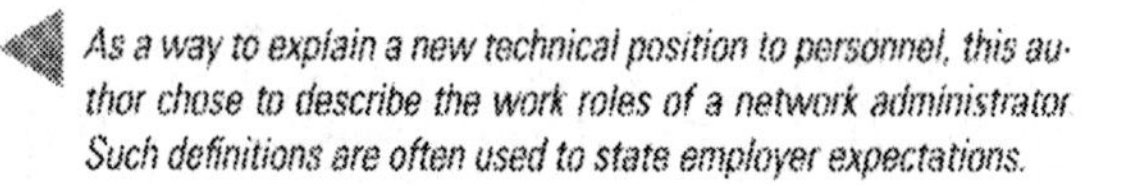

As a way to explain a new technical position to personnel, this author chose to describe the work roles of a network administrator. Such definitions are often used to state employer expectations.

Model 5.E (2)

The network administrator is responsible for network design, implementation, and daily network management. Additional duties of the LAN administrator include record files, security, building the directory structure, installing the applications and files, and establishing drive mappings so that they are both easy to use and practical.

Record Keeping

Without proper recordkeeping, the rest of the LAN administrator's job is made that much more difficult, if not impossible. Careful recordkeeping facilities, efficient support, and effective problem determination make it possible to provide an overall high quality of network administration. Proper recordkeeping does not mean an elaborate filing system or on-line database. It simply means that any information that may be needed at some future date should be written down and remain retrievable. A simple paper system should suffice until the need for something more elaborate arises.

Data Integrity and Security

Since the data on a network is typically shared, it is important to secure it in directories where only authorized users have access. Netware security consists of access restrictions, privileges, and file conditions. This way you can store data in certain directories on the server disk without having to worry about someone accessing the data. The Netware security model consists of four layers:

1. **Login/password security**
2. **Login restrictions**
3. **Access rights**
4. **Attributes**

To some extent, the only way mortals can understand quasars, pulsars, black holes, worm holes, and similar phenomena is by elimination. They are not what is *known*. New phenomena call for new theories. New theories account for new phenomena. Like the explorers of a former time who challenged flat worlds, it is as though we are about to set out on a new voyage—always at the expense, or the elimination, of the old order.

Do not confuse the process of elimination with the odd and popular use of the following logical fallacy. If someone says,

> "Ghosts exist, and you cannot prove to me that they do not,"

do not be fooled. This is not an elimination at all. My failure to prove that ghosts do not exist has nothing to do with the assertion that they do. It cannot be argued that they exist only because no one has proven that they do not.

A popular variation of this is a scientific convenience that well-trained investigators will allow themselves. Reputable sources will say,

> "The Sasquatch does not exist. There is no evidence that it does."

Investigators will make this comment as a matter of practicality. The problem is that they have eliminated the evidence and not the Sasquatch! Thus, this second comment is *also* a fallacy. The absence of evidence is not necessarily the evidence of absence. If you have no evidence, well, you have no evidence—nothing more. You have not proved that this legendary fellow is not out there bounding through the woodlands.

Classification

I consider classification last only because it is a transition to the next type of technical document. Classification is, in fact, one of the primary methods of definition. A variation on definition by analysis is to develop a perception of how—let's say a coot—fits into the scheme of things. A coot is an aquatic bird with slightly webbed feet, but it is *not* a duck. Certain contrasts and similarities can be used to establish *categories* or *groups* into which the object

This model is designed to explain the role of the network administrator in detail. It defines the administrator's tasks. The text is for experienced employees, and so it quickly moves into specific details concerning structure and files.

Model 5.E (3)

Tape Backup and Archiving

The backup and archiving of network files on the file server is a critical part of maintaining the integrity of any network. Without proper backups, the smallest problem can turn into a disaster. In order to assure that backups are available after a catastrophe, such as a flood or fire, a copy of every system backup should be maintained off-site. It is advantageous to run backup sessions during periods of low network activity when most or all of the files will be closed and available for backup. Some backup products provide a means to schedule timed, operator-free backups.

In addition, the administrator is required to perform preventive maintenance and timely network troubleshooting. Keeping the system operational requires continual management and maintenance. Listed below is a summary of some of the more crucial maintenance responsibilities.

Tape backup and archiving
Record keeping
Data integrity and security
Hardware maintenance
Software maintenance
Monitoring file server performance

of a definition will fit. Fundamentally, this is taxonomy, the principle of scientific classification by type. An entire description of a PC, for example, could be handled strictly by an analysis within the groupings of PCs by relevant categories concerning memory or cost or speed or similar conditions. The method is a simple matter of creating groups.

The taxonomic classification system is the source of Latin names for plants and animals. For example, I have a rose called *Rosa damascena bifera.* To understand the practical applications of taxonomic classification, look at the concept in the following drawing:

The idea is to observe or find the categories that will define an object, such as my rose. Usually the approach involves finding the group to which it belongs, any larger family it fits into, and the distinguishing features of the object within the family. In other words, my rose is a "damask" type within the larger rose family, and it is unique because it is twice flowering (the meaning of *bifera*). So it is a *rose,* a rose that is a *damask,* and a rose that is a damask and twice flowering. It is a *Rosa damascena bifera.*

The second page of text shifts attention to the larger challenge of the position. It divides the administrator's tasks into divisions, each of which is then analyzed.

The technique of defining by developing categories is quite ancient, but the Latin system of names was created and developed by Linnaeus (his original name was Karl von Linné, 1707–1778), a Swedish scientist. The science of optics doubtless gave exceptional credibility to the concept of categories. Magnification allowed scientists to vastly enlarge their perceptions of greater or lesser categories. The analytical methods of many sciences appear to have been profoundly altered by the telescope and the microscope. *Macro*analysis and *micro*analysis are the two types of observations that allow me to describe my damask in terms of categories greater and lesser than itself. Botany and zoology have always been greatly indebted to the sheer power of observation. Optics simply unveiled the incredible potential for investigating new categories far greater and far smaller than any that had previously been known.*

In summary, remember to define by your readers' needs. Our discussion suggests that the object to be defined somewhat dictates the nature of the definition. Use the different methods of definition as needed. Circumstances will suggest the approach to definition that you should explore. In research, for example, writing will take one approach; whereas corporate documents may be directed more toward definitions that fit such blunt realities as consumer interest. Promotional material often "defines" by explaining a product or service (see Model 4.A, p. 140), and the limits of those definitions seem to be very practical bottom-line concerns. Computer memory (MB) for the dollar, the best buy, and so on are often perceptions that define.

In the end, definitions are endless in shape and form. The subject, popular perceptions, personal interests, and even commercial needs dictate the definitions. They are relative, since we construct worlds of understanding to serve our own ends. The fact that definitions appear to be true is the marvel of the methods we use to define. Perhaps there really is no *one* definition of anything.

** The microscope was widely used before the twentieth century but simply did not stand at the center of the kind of epic event that came to be associated with the telescope. Galileo is sometimes credited with designing and constructing both the microscope and the telescope. Other authorities deny him either invention. What is important is that in or around 1610, the design and use of both devices were recognized. The Dutch biologist Antony van Leeuwenhoek made hundreds of microscopes in the seventeenth century. In the United States, John Prince, a precision instrument maker, began to manufacture the first American microscopes in 1792. The implications of what was seen with the instruments are significant to us here: there appeared to be no end to greater and lesser categories in our world.*

Models 5.D and 5.E

Comments

The development of an analysis of parts is one of the most popular strategies for defining. Samples 5.D and 5.E both involve the divide-and-analyze tactic. In the case of the project concerning ultrasound technology, the author focuses on the apparatus. The text systematically divides the instrument into its components, which are then amplified and explored.

The LAN administration presentation was written for an area hospital, and you can see the analysis strategy at work. The author sets out to define an administrative position. He begins by dividing the job description into categories. He has both outlined the project for the reader as well as created the plan of the analysis that follows. Subsequent pages detail the work involved in the position.

The Extended Definition

Project game plan

- Identify the subject to be defined and explain its importance.
- Prioritize the definition techniques. You can begin with the least important—perhaps a history—if you want to build an introductory discussion.
- For the body of the text, shift to more important defining strategies, particularly those appropriate to your discipline or those appropriate to the subject to be defined.
- Thoroughly amplify the definition to enlarge the reader's perspective.
- Detail the definition according to the reader's needs and your intent.
- Remember that illustration is a type of definition. A graphic may be a valuable sample of the object of your definition. Certain types of graphics, such as the exploded view and cutaways, will also allow you to create a visual analysis of parts.

Activities Industrial Applications: I

You are now ready to compose a technical writing project.

---

1. Instructions and procedures are easy to develop because the sequence logic of a set of instructions or procedures is the same as the sequence logic of your general pattern of conversations. Decide on a subject, then decide whether you want to develop instructions or procedures. A technical topic from your major study area is the ideal choice. Equally effective is a topic that will have value where you work. Consult with your instructor at the college or your supervisor at work.

Using the Instructions and Procedures Project Game Plan as your guide (see p. 175), develop a project of 1500 words or more.

---

2. Your second option for a project involves the description of a device of some sort. Many of the community and junior college technical programs are very product oriented in a general sense: devices are all around you. Computers, testing equipment, and other pieces of equipment are critical to the tasks of engineering techs. Select one such device or a product or mechanism from your workplace for the project.

Using the Product Descriptions Project Game Plan as your guide (see p. 187), develop a project of 1500 words or more.

---

3. Consider the possibility of a project that defines. If you describe a device or mechanism, you define it, but you can also define a system, a theory, a process, or many less tangible features of your work world. Remember that the technical, engineering, and scientific disciplines seem to favor distinctly different defining methods. Determine which methods are most appropriate for your area of interest, and proceed.

Using the Extended Definition Project Game Plan (see p. 202), develop a project of 1500 words or more.

Part III of this volume can be very helpful to you. Skip to Part III and read on while you are writing your first project. If you are really rushed, at least read Chapters 10 and 11.

Share a Project

Work with three other members of the class to discuss and edit the first set of technical documents. Your group will be most efficient at technical editing if all the members are from the same engineering technology program.

- *Before holding the meeting, you should read the peer review discussion in Appendix A (pp. 495–497) and study the first three editing checklists concerning instructions and procedures, product or mechanism descriptions, and definitions (pp. 498–501).*
- *On the day the projects are due, have each member of the editing group explain his or her project in terms of objectives, intended audience, and project development.*
- *Hand around the papers for a critical reading and editing by each member as time allows.*
- *Have each member edit the texts for both writing errors and technical errors.*
- *Have each member write a one-paragraph critique at the end of each paper.*

6. Coworker Responses

I went to visit the employees who provided the red-lined documents because I still had questions on the data they had given me, and I had to explain the label problem. I mentioned earlier that those who work as part of a process know a certain stage in detail and perform tasks with which other employees are unfamiliar.

My report said what should be done but did not describe how and what to enter at the system prompts. John, the stockroom issuer I visited, explained the items to me, but I was still uncertain, so I asked him if he would be doing the transaction that day. He said he was just getting ready to do one. To actually see a transaction (or any function) being done is about the best way to get detailed instructions. I made notes as I watched, and after he was done I had him print the screen, which showed what he entered at each prompt, and the steps of the transaction.

I felt I now had enough information to update that portion of the process. You may have noticed that I had to ask if he was going to be doing that transaction soon; he did not volunteer the information. He was not being rude. The stockroom is an information-sensitive area due to the items held there and he was probably trained accordingly. I have to respect these areas and understand that some information may be confidential.

Since I had not received the red-lined document from Inspection, I paid them a visit also. I found out that the delay was due to an increase in orders of meter stock that was going through their area, and the inspector, Cheryl, had not gotten the non-Genus documents (from customers) to review the requirements that must be met. The inspector stated that they were now having to do additional switch-contact resistance testing on new batches of meters. She was going to send me procedures for the additional testing that had been assigned. She was able to give me the figures (graphics) for this procedure so that I could include them in the production process description as a meter stock requirement of a certain customer. One of the figures I received is shown here.

S. B.

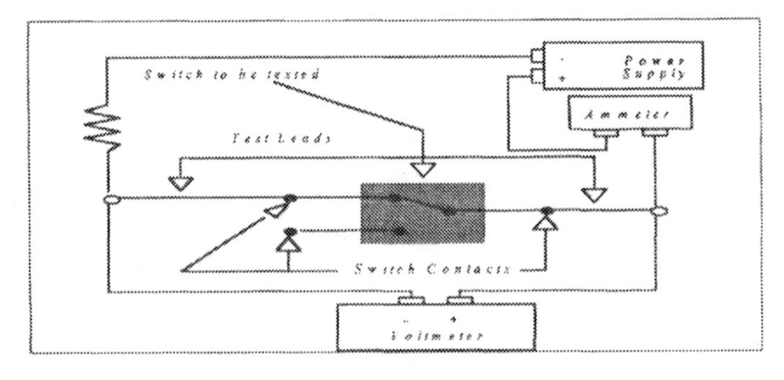

Switch Contact Resistance Test Setup Figure

Product Evaluations and Cataloging

Cataloging is not a task that you will likely undertake in most circumstances. Large catalogs, in any case, are seldom the work of individuals. However, there are other uses for catalog writing skills that may interest you. A catalog is a device that designs and builds classification. Usually catalogs are *descriptive;* that is, they pass no judgments. Consumer catalogs usually brag a good deal, but that is part of the sales routine. Otherwise, the catalogs are, by intent, purely descriptive. There *is,* however, a popular variation on the catalog concept that is immensely popular in trade magazines and finds its way into corporate documentation as well. It is this second approach—the product evaluation—that deserves your attention.

Model 6.A (1)

ASTRONOMICAL TELESCOPES

PART 1: DESIGN AND FUNCTION

PART 2: CHARACTERISTICS AND PERFORMANCE

Audience: the beginning and intermediate telescope user

wishing to purchase an astronomical telescope

Mark a. S.

June 5, 200X

Technical Writing

David R. Instructor

A classification analysis can be *evaluative*. The word *catalog* is probably not appropriate, since the evaluations usually involve small numbers of products. These products are *compared,* and the comparison includes an evaluation that usually suggests which product is superior or which product best serves certain circumstances. If you see an analysis of six CD players, it will probably be evaluative. If a construction engineering report compares four new roofing materials, the analysis is likely to focus on which applications will best utilize each product. A discussion of new chip technology is likely to discuss the characteristics and applications of comparable chips and speak directly to their strengths and weaknesses.

The most popular document that is always consulted for evaluative analysis is *Consumer Reports*. The general trend toward consumer awareness has greatly enlarged the evaluative analysis spectrum, and such analyses as those concerning generic medicines or cameras or insurance policies have become increasingly popular. You will see "single-product evaluations" in many periodicals, but they often avoid comparison. Without analysis by comparison, there are fewer significant benchmarks for a single-product evaluation. Even if you do see a multiple-product analysis in popular magazines, you must realize that the advertising copy in the magazine probably is just as reliable. In other words, the evaluations will be a wash. An auto magazine cannot be completely truthful if it depends on millions of dollars of Detroit advertising. Only *Consumer Reports* has no ad copy. The evaluations are *the* business of the magazine.

You will notice that every issue of *Consumer Reports* makes extensive use of tables. The table *is* the text to some extent. The text repeats much the same information you find in the table. Evaluative articles often contain a table, and the redundancy is noticeable. Of course, without the text there is no article, but there must be a more logical cause for writing it. The text explains the primary concerns of the evaluation. Also, the text must explain what criteria were used for the analysis and why. Finally the evaluation of "best" or "most appropriate" has to be explained in the text. The tables are usually not fully understandable until the text is read. Readers need the text in order to learn to use the table, but seldom thereafter do they return to anything but the table.

The emphasis on a consumer magazine should not be misleading. It is simply an easy example you can study to see how comparison articles that include tables are constructed. In truth, this variety of technical documentation can serve many purposes in both commercial and academic applications. Any group of three or more items of concern—from genes to ailerons to tires to alloys—can be approached and studied as a classification analysis. In fact, although our emphasis here concerns manufactured products, the subjects of an evaluation can include industrial, engineering, or scientific considerations of a wide variety. To make any evaluative analysis, the criteria that will suit the needs of the

This model is part of a large project that was developed as two major papers. If you have an interest that is too large for one paper, consider the possibility of dividing the project into parts.

Model 6.A (2)

INTRODUCTION

Today's astronomical telescopes are the result of innovations in design and materials. Refractor objectives made from calcium fluoride and fluorophosphate crystals have virtually eliminated the chromatic aberration that used to plague every refractor. The introduction of nonspherical curves to primary and secondary mirrors has improved image quality. Coated optics have increased the brightness of telescope images. Convex secondary mirrors have compressed long focal length reflectors into short, manageable tubes. Cat scopes have further refined the compound telescope by adding a large correcting lens and enclosing the optics in an instrument capable of high power and resolution.

No single telescope will satisfy everyone. Technological innovations have expanded the variety of commercial telescopes and raised the expectations of consumers. With this rise in expectation have come increases in features, size, and price. Buying a telescope is no minor purchase. For most it requires planning and informed decision making.

The following information is provided to help the consumer evaluate and choose which type of telescope suits his or her needs. Part I provides basic information on the function and components of all astronomical telescopes. Five characteristic attributes of astronomical telescopes are defined. The three major types and how they work are described in detail. Part II address the strengths, weaknesses, and performance of these telescopes.

project must be determined. The criteria may concern an unlimited set of conditions. The *best* in relation to different products may mean

the fastest	**the tastiest**	**the quietest**
the cheapest	**the biggest**	**the most powerful**
the most durable	**the easiest to use**	**the most trouble free**
the most efficient	**the most deadly**	**the least damaging**
the lightest	**the smallest**	**the most functions**
the longest	**the safest**	**the least fattening**

Commonly, the *best* is less of an issue than *the most appropriate* selection for a set of conditions. Tables allow you to graphically render as many criteria as you need for an analysis. As you can see in the following example, the text is the important link that *explains* the table. The table may or may not be self-explanatory, and the priority of your criteria usually will call for discussion.

Table 21. The 7400 series family.

Generation	*Power Dissipation*	*Power Supply*	*Input Load H/L (U.L.)*	*Output Load H/L (U.L.)*	*Propagation Delay*
7400	10 mW	5V	1/1	10/10	10 ns
74L	1 mW	5 V	0.25/0.1125	2.25/5	33 ns
74S	20 mW	5 V	1.25/1.25	12.5/25	3 ns
74LS	2 mW	5 V	0.5/0.225	5/10	10 ns
74ALS	1 mW	5 V	0.5/0.125	5/10	4 ns

Note: Unit Load (U.L.) Standard is referenced with respect to standard TTL device loading.

To see the larger implications of the evaluation model, let's look at several samples that do, in fact, analyze items of interest outside the arena of products or devices. The table on p. 216, for example, was designed to examine energy. That table and the following one are designed to examine characteristics unique to industry or engineering. You can see that the criteria in the energy table are identified in the leftmost column. Notice that the table of metals reverses this procedure and places the products, metals in this case, on the left.

Notice that the second paragraph introduces the basic criteria: features, size, and price. The next paragraph introduces the additional considerations: five characteristics and three major types.

Model 6.A (3)

7

THE CATADIOPTRIC TELESCOPE

The **catadioptric** or cat scope (9:2) is a hybrid of the refractor and reflector designs that uses both lens (refraction) and mirror (reflection). A correcting lens in the front end of the tube refracts incoming light, which is further corrected by a figured primary mirror. The lens physically supports a separate secondary mirror or replaces the secondary with a silvered circle on the interior of the lens. Like the refractor, cat scopes are closed optical systems.

The Schmidt-Cassegrain-Telescope

This design is based on the photographic telescope developed by Bernard Schmidt in 1928 (5:13) and the Cassegrain telescope (see Figure 8). On the interior surface of the correcting lens is a convex secondary, mirror. Light is refracted through the correcting lens reflected from the primary mirror and reflected again by the convex secondary mirror, exiting the tube through a hole in the primary. A tapered, tubular baffle projecting forward from the primary mirror prevents light from the correcting lens from entering the eyepiece directly.

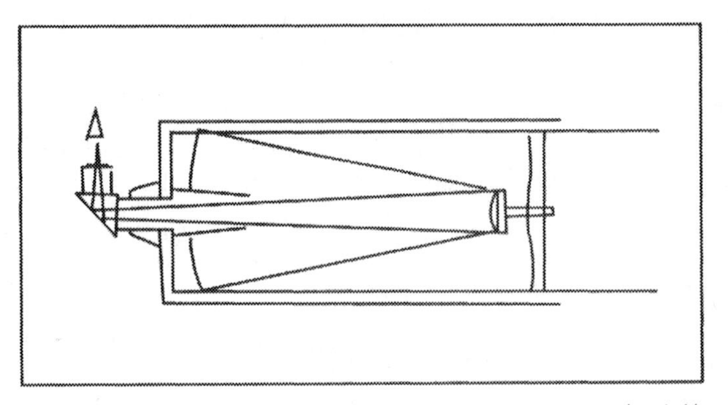

Figure 8. ***The Schmidt-Cassegrain telescope*** *A full-aperture correcting lens holds the secondary suspended in place while closing the optical system.*

The authors used different graphic methods because the papers were written by using different organizational methods. The energy paper was written to discuss desirable energy properties, and the text moves from one property to the next. The paper concerning metals was handled in the opposite manner. It developed a discussion of each metal in turn. The characteristics of each metal were handled in seven separate subsections.

Table 1. Properties of metals.

Metal	Boiling Point (°C)	Melting Point (°C)	Color	Density (g/cc)	Conductivity	Malleability	Ductility	Resistance to Corrosion	Symbol
Iron	3000.0	1535.0	Silvery/ white	7.674	Poor	Yes	Yes	No	Fe
Gold	2807.0	1064.4	Bright/ yellow	19.32	Poor	Yes	Yes	Yes	Au
Copper	2563.0	1064.9	Reddish/ orange	5.96	Yes	Yes	Yes	Yes	Cu
Lead	1750.0	327.5	Bluish/ gray	11.34	Very	Yes	Yes	Yes	Pb
Aluminum	2520.0	660.5	Silver	2.535	Yes	Yes	Poor	Yes	Al
Tin	2270.0	231.9	White	7.200	Poor	Yes	Very	Yes	Sn
Mercury	356.58	36.87	Silver	13.54	Yes	Yes	No	Yes	Hg

Once a table is designed, the text can be developed either in terms of the objects of the comparison or the criteria (the variables). Often the conditions, characteristics, or other features are used as the controlling factors for developing the project.

This is a page from the middle of the project. It is part of a discussion of one of the three major types of telescopes.

Model 6.A (4)

For the sake of brevity, some comparison factors have been deleted or abbreviated in the following tables. Temperature effects, for instance, are more complicated than indicated. The vulnerability of telescope to thermal air currents varies a great deal. The simple act of walking in front of a telescope and temporarily warming the air can cause transient image distortion until the air cools down again. This type of secondary information, although useful and interesting was sacrificed to explore other, more basic data.

	REFRACTOR	*REFLECTOR*	*CATADIOPTRIC*
OPTICAL ABERRATIONS (3:64, unless otherwise noted)			
Chromatic (color fringe around observed objects, also known as a secondary spectrum)(4:29)	■ Achromatic refractors exhibit this aberration, especially at fast f-ratios ■ In apochomatic refractors this is neglible.	■ None	■ None
Spherical (fleeing of star image when the eye is shifted off the eyepiece, pronounced at f/7 and below)(5:153)	■ This is found in less expensive telescopes with poorly made optics.	■ This is found in less expensive telescopes with poorly made optics. Especially noticeable in short focal ratio newtonians (4:35).	■ This is found in less expensive telescopes with poorly made optics.
Coma (stars at edge of field have comet-like tails that point away from image's center)	■ Low	■ Coma is prominent in short focal ratio neutonians (4:35) and in Cassegrainians (4:36). Several companies produce a coma-correction lens that can be inserted into the focusing tube. It will increase the telescope's magnification by 15% or less (4:121).	■ Low
Curvature of field (images at center of field focus in one eyepiece location while those at the edge of field focus in another) (5:287)	■ Low	■ Low	■ Yes
Astigmatism (seen as irregularly shaped star discs in poorly made optics)	■ Low	■ Astigmatism can be present in short focal ratio Newtonians and Cassegrainians. A coma-correcting lens (see coma above) can correct some forms of astigmatism.	■ Low

As you know, I always encourage engineers and technicians to outline first. However, a catalog or an evaluative classification is the exception; make a table first! The table virtually *becomes* the paper by listing the criteria and the products to be analyzed. The text follows suit by discussing the need for the evaluation, the criteria, the products, and the outcomes. Generally speaking, the analysis will contain a modest number of items to be compared, so there are likely to be two or three paragraphs of discussion devoted to each item in the group. If the text is in pursuit of a specific judgment, it can be stated clearly if the criteria lead to an apparent outcome. This may not always be the case, and you may instead define a group of adequate or appropriate outcomes and leave the final selection to other parties. To develop a table, locate the table templates on your computer (see Appendix B).

Catalogs are usually far more complex than a product evaluation. Few authors are likely to write one but the concept of the catalog deserves our attention. Catalogs are a linchpin for any corporation—and they are "written." Readers do not think of catalogs in terms of writing—and certainly not in terms of creativity—but catalogs consume enormous energy in the marketplace. The catalog is very important as the major representation of the goods and services of a company. A catalog is often glossy and costly, and every effort is devoted to making the products move out of the warehouse as a result of catalogs. There are usually aftermarket catalogs, also. They are filled with replacement parts and the like and are usually devoid of pictures, and they are often generated in fine print and ten-digit product numbers. The glitter is gone, although there are still profits to be made.

Catalogs (including Internet offerings, e-mail versions, junk mail disks, and so on) can be viewed from two perspectives, both of which contribute insight into structures for technical writing. First, let's take the scientific approach, and then let's look at popular consumer applications. You have perhaps looked through an *Audubon Field Guide* or a *Peterson Field Guide*. These books are a practical point of reference because they "catalog." They are not consumer catalogs, of course, and authors seldom use the word *catalog* to describe the more scientific efforts to organize and record, for example, flora and fauna. In the guides, you will notice that graphics are critical. Bird guides, for example, absolutely depend on details concerning bird shape, feather colors, shape of bill, and so on. You will also notice that there is a predictable graphic tool—the grid—that helps assign order to these distinctions. The table is a critical complement in catalogs of all types.

I noted earlier that Linnaeus's system of nomenclature (the names) of plants and animals is based on categories. In a sense this nomenclature positions any object in a series of related categories. Often, similar categories can be visualized, as you see in field guides or product catalogs. If a group of birds or a list of resistors is tabulated by some criteria, this approach creates a unique way of "seeing" the object in a context—in the environment of other similar things.

Toward the end of the project, the author brings the types of telescopes together in one of a number of tables that are presented and discussed.

Sample 6.A

Table of Alternate Energy Sources and Their Properties

	Solar	***Biomass***	***Wind***	***Ocean***	***Hydro***	***Geothermal***
Cost	\$3–\$8 per gigajoule	5¢–10¢ per gigajoule	3¢–4¢ per kilowatt hour	NA	NA	1.5¢–6¢ per kilowatt hour
Best use	Heating	Heating	Electricity	Electric/Heat	Electricity	Heat/Electric
Best conditions	Hot, sunny climate	Anywhere	Consistent 20+mph winds	High tidal flow/Depth	Large water flow, valley	Geothermal area
Power output	0.4 exajoules	8–10 exajoules	1.7 exajoules	0.7 exajoules	100 gigajoules	3 exajoules
Present state of development	Commercially ready	Commercially ready	Demonstration stage	Technically ready	Commercially ready	Commercially ready
Resource/raw materials availability	Excellent	Excellent	Excellent	Good	Very good	Fair
Fossil fuel replacement potential	Good	Excellent	Good	Small	Good	Fair
Cost-competitiveness with fossil fuel	Excellent	Excellent	Very good	Fair	Excellent	Good
Degree of social acceptability	Excellent	Excellent	Very good	Good	Good	Fair
Environmental acceptability	Excellent	Very good	Excellent	Excellent	Very good	Good

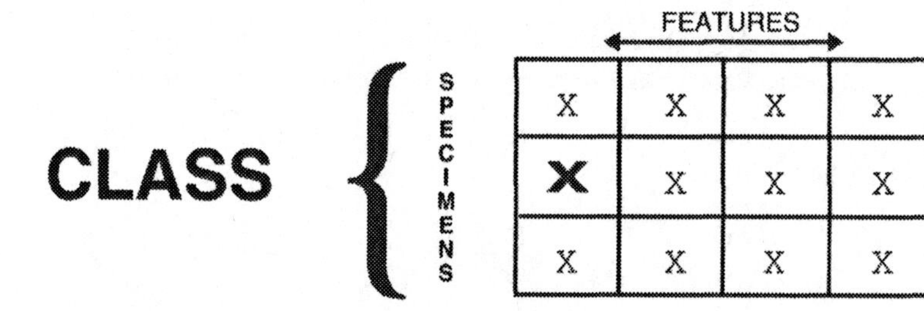

The definition provided by a Latin name focuses on the *particular* (or one) and the combined Latin words are taken to be one of something (the species). Tables focus on the general—the *total* (the group). The table card catalogs combine *all* the relevant definitions. The nomenclature more or less gives us a way to remove *one* item from the table and give it a name based on the features, the class of specimens, and the family.

In a table the vertical column usually represents a criterion used for assigning categories found in the table. The vertical columns are usually explained above the graph. The horizontal column represents one of the groups under scrutiny. The table is usually filled in with comments or graphics. The matrices are often boldly drawn, also. In other words, the focus is on the *categories* and not a single moment or quantity. The table is used to find what you are looking for, but the focus of the table is not, let's say, a specific parrot, but a family of parrots—the group. In sum, tables plot an enclosed world of whatever groups they set out to analyze. Precisely this technique of using matrices to plot groupings is what allows us to use them to "look up" everything from wildfowl to computer chips in guides and catalogs.

Graphs differ from tables. Graphs focus on points of interest. Tables focus on fields of interest. In addition the ordinates of a graph are read from the bottom up and the categories of a table are often read from the top down.

Here we see another sample table from another project. In this sample notice that the criteria are placed to the left and are examined horizontally.

Sample 6.B

Configurations	Dell Inspiron 7500	Gateway Solo 9300-se	Toshiba 4100XDVD
Processor	Pentium III 600MHZ with SpeedStep[1]	Pentium III 500MHZ	Mobile Pentium II 400MHZ
Memory	64MB SDRAM[2]	64MB SCRAM	64MB SDRAM
Memory Min/Max	64MB—512MB SDRAM	64MB—288MB SDRAM	64MB—192MB SDRAM
Operating System	Windows 2000 Professional	Windows 2000 Professional	Windows NT 4.0
Video System	2x AGP ATI Mobility P w/8MB SGRAM[3]	2x AGP ATI Mobility-P w/4MB SGRAM	Trident Cyber9525 with 2.5MB video memory
Display Size	15" SXGA+[4]	14.1 XGA Active Matrix TFT Color Display[5]	14.1 Active Matrix TFT Color Display
Audio System	3D Sound with Wavetable	Integrated 16-bit sound	16-bit stereo Sound Blaster Pro compatible
DVD or CD-Rom Drive Hard Drive Size	6X DVD-ROM 6.0 GB Ultra ATA	6X DVD-ROM 6.0GB Ultra ATA	2.4 DVD-ROM 6.4GB EIDE
Battery	79WHr Lithium Ion Battery	12 cell Lithium Ion Battery	Lithium Ion Battery
Modem	Internal PCI 56K V90 Fax Modem	3COM MHz 56K PC Card	Integrated V9.0/K56flex
Dimensions (l × w × h)	12.9" × 10.5" × 2.5"	12.7" × 9.8" × 1.59"	12.2" × 10.2" × 1.6"
Weight	9.1 lbs	7.35 lbs	7.0 lbs
Price	XXXX.XX	XXXX.XX	XXXX.XX

[1]Intel Pentium with SpeedStep Technology provides automatic or user switching between performance modes based on the power supply (AC vs Battery). Maximum performance is 600MHz while connected to AC and 500 MHz while on batteries.

[2]SDRAM Synchronous DRAM, a new type of DRAM that can run at much higher clock speeds than conventional memory.

[3]SGRAM Synchronous Graphic Random Access Memory, a type of DRAM used increasingly on video adapters and graphics accelerators.

[4]SXGA+ Super Extended Graphic Array, capable of displaying 65,536 colors at 640 × 480 resolutions or 256 colors at 1,024 × 768 resolution.

[5]Active Matrix TFT, a type of flat-panel display in which the screen is refreshed more frequently than in conventional passive-matrix displays. The most common type of active-matrix display is based on a technology known as *TFT (thin film transistor)*. The two terms, *active matrix and TFT*, are often used interchangeably.

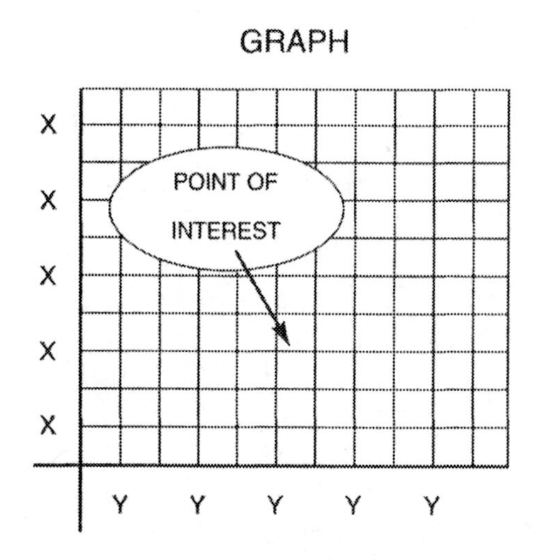

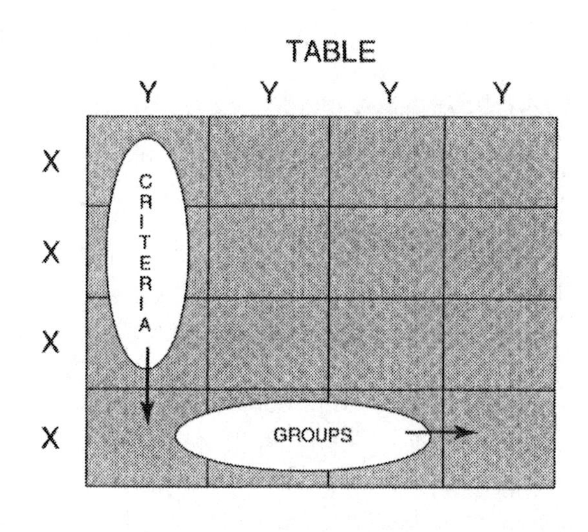

Consumer catalogs are designed in much the same way as technical catalogs: a clothing catalog, an electronics catalog, an auto parts catalog. There are often tables in both technical and consumer catalogs. I keep a copy of the Newark Electronics product catalog in the office, and it contains well over 5000 tables. Consumer catalogs usually avoid the tradition of using tables extensively unless they want to "look" scientific. Tables of down sleeping bags and running shoes are common in sporting goods catalogs. There is, by the way, a fairly identifiable historical moment when consumer catalogs, both technical and commercial, began. They were rare until the nineteenth century. Thomas Jefferson ordered seeds from lists he received from England. A hundred years later, Claude Monet was ordering seeds from Thompson & Morgan Seed catalogs for his famous gardens at Giverny north of Paris. More or less between the seed list and the garden catalog, the U.S. patent office opened in 1836. Product protection was an important step in product manufacturing, and before long, the mass production of patented items led to the creation of the catalogs that listed the products. In addition, mass production was linked to mass consumption, and catalogs were the link.

Manufacturing and its bigger brother, mass production, did more than turn the blacksmith into just a horseshoe artist. Such features of the American landscape as the general store and the hardware store began to emerge. More and more gadgets meant more and more inventory and a demand to organize and control the items. The catalog was and remains the solution both to marketing merchandise and to maintaining inventories of products. Only catalogs allow us to reference the array of items in our various disciplines. The Sweet's catalog, which is consulted by architects, is measured in many feet of thickness. The most famous of them all, the Sears catalog, was begun in 1891. By 1907 the catalog had reached a circulation of 3.5 million. It was already 500 pages long and heavily illustrated by the turn of the century.*

**Richard Sears created his first catalog for a mail-order business in pocket watches. Circulation ballooned to millions as the catalog grew in its variety of products. He was in the right place at the right time. Rural Free Delivery was established in 1896, and the catalogs were suddenly brought straight to every farmer's mailbox. Parcel Post was introduced in 1913, and then the packages started to show up at the mailbox, also! Amazon.com may be admired as a cyber phenomenon, but the idea behind it is an old one: the catalog.*

The table on the left represents a popular use of the device of product comparisons. Notice the details in the table are explained below.

Model 6.A

Comments

The comparison of telescopes was created by a commercial artist specializing in science fiction illustration. An avid fan of astronomy, Marc developed this project in two sections because the overall project was lengthy. It was easier to divide the project into two areas of concern: design and performance. Both projects contain several pages of tables to make the analysis easier for the reader to understand.

You will notice that the text is based on an analysis of three "products." Each type of telescope is presented for discussion, and each is visually rendered so that the optics can be clearly understood by the reader.

The tables in his project, one of which is included in the model, highlight relevant criteria that are explored in the text. There are judgments in the second project that emerge from the tables, but final decisions depend on the particular needs or preferences a reader may have.

Product Evaluation

Project game plan

- Develop a table that will structure the evaluation of several products that are "equal" in some basic respects such as price or power.
- Select the criteria to identify distinguishing characteristics of the products.
- Develop a presentation based upon the table.
- In the presentation, introduce the reason for the evaluation to the readers.
- Introduce the products.
- Discuss the criteria of the evaluation.
- Follow the discussion points of the table for each evaluation.
  - As one strategy, develop and detail each product in the remaining text.
  - As an alternative strategy develop the discussion by criteria, using each criterion as a point of departure for evaluating the group of products (or topics). Follow the columns of discussion points of the table for each evaluation.

Model 6.B (1)

HOW HIGHWAY NOISE BARRIERS WORK

Audience:

Highway Design Technicians

It is the intention of this report to introduce important developments in reducing environmental noise. An introduction to the basic concepts of noise barriers and how they function will assist in responsible future highway design.

Janine B-R

5-25-200X

Instructor: Mr. R.

Project

EGR 231.01

North Seattle Community College

Causal Analysis and Troubleshooting

The study of "caused" relationships is very much at the heart of modern technology. Whether we are washing dishes with a phosphate or building an aircraft part with a titanium alloy, we are using products based on their apparent predictability. We expect outcomes or effects from everything we use. The focus is, of course, utility, but the practicality is an outcome of the long road to invention. Any reaction, atomic or otherwise, is a *reactive* response. There is an initiator. Phosphates clean.

The study of these causal relationships is the study of causes; the process is called *causation*. There is no mystery to doing dishes, and in your technical discipline there is no mystery to your use of causation. In electronics, in particular, troubleshooting is an integral part of all the technical disciplines within the field and at the heart of the activities of technicians in many fields. Troubleshooting is the bread and butter of service and repair departments, whether the object is a digitized gyroscope from a satellite or a component of an electrocardiograph or a fuel injection circuit on a car. Troubleshooting is also pure causal analysis.

The first important consideration is to realize that there are two relationships and not one. If you have a cause, you can pursue its effects. It was understood early on in the nuclear century that radioactive compounds were quite dangerous. Two of Pierre and Marie Curie's chemists died from radiation-related complications that were directly connected to the radium compounds in the lab. A decade later, the commercial application of radium resulted in the deaths of many workers in Europe and North America. There were many symptoms. The question has always been, what are the *specific* effects of radiation poisoning? The question of radiation effects is now studied in such terms as long-range cellular or genetic damage. In other words, there is often a *known cause,* and investigators pursue the outcomes.

In the case of troubleshooting, there is an opposite situation. When a technician analyzes a mainframe computer system that is down for unknown reasons, she will look for *unknown causes*. The system is malfunctioning. She has an *effect,* and she can localize the source of the effect in a drive unit. That is the particular effect. She then pursues the cause. A general physician spends much of his workday "troubleshooting," only that word is seldom used in medicine. In essence, the doctor is observing effects and looking for

Notice the title and the objective. The concern is how noise barriers function, and the approach is causal. Noise barriers cause the effect of sound control.

Model 6.B (2)

INTRODUCTION

This report will introduce the concept of noise barriers and how they function. The first section introduces the fundamentals of highway noise, beginning with how noise is generated, defining special terminology associated with environment noise control, expanding further on how sound moves, and discussing how it is measured. The second section describes the physical characteristics of each type of barrier and introduces basic barrier concepts. The next section deals primarily with the physical mechanics of barriers. Special problems in barrier design are explained, including how sound can diffract over or around a barrier, how the mass of a barrier affects the sound transmission path of noise, why certain locations of barriers are more effective than others, and how weather can affect noise and the performance of a barrier. The conclusion discusses the importance of accurate future noise predictions and the role of acoustical recommendations.

FUNDAMENTALS OF HIGHWAY NOISE

Noise is defined as a sound of any kind, particularly when loud, but specifically when unwanted or undesired. *Sound* is defined as a pressure variation in the air that is detectable by the human ear. For example, sound is produced when a vibrating piston first moves outward, compressing the layer of air particles surrounding it and pushing against others in an outward progression forming a compressional wave front. Small molecular displacements transfer energy in the form of a traveling wave from air particle to air particle. As the piston moves inward the closest surrounding air is sucked inward leaving a slight vacuum (rarefaction) that also travels outward behind the compression, resulting in alternating rarefaction and compression waves that move in sympathy with the vibration of the piston.

In the following subsections, specific examples of how highway noise is generated will be discussed, and point and line noise sources will be explained. How noise is measured will be explained and special terminology associated with highway noise will be defined.

possible causes. If you have a fever, you have the general effect identified on the thermometer. Your systems are down. If it is a low-grade fever and your white cell count is up, the doctor might see a sharper picture. Perhaps next he will try to determine if the bug is viral or bacterial. He pursues the cause. In other words, both the technician in a service division at IBM and the doctor at a nearby clinic develop an analysis based on *known effects,* and they then pursue the causes.

The patterns observed in causation, therefore, go in either of two directions. The result is a logical analysis from causes to effects or from effects to causes.

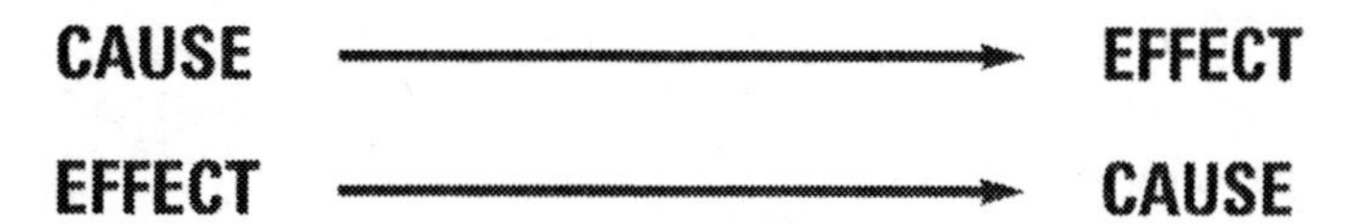

These two methods of analysis somewhat direct the way engineers and technicians ply their trades. Engineers rely on the former for design. Technicians rely on the latter for repair. Model 6.B is Janine's analysis of the way in which noise barriers can be developed to control highway noise pollution. This is the engineering approach to causation in that the barriers *cause* noise reduction.

Consider a book of engineering tables, perhaps one concerning I or T beams. All the tables are the results—the effects—of carefully determined calculations derived from engineering research. In the lab, the engineers may add pound weights to a certain beam until the point at which microfractures appear and show stress. From these calculations, verified repeatedly, predictability can be determined, and standards are established. Observe that the engineer *created* the *cause* for the effect.

It is interesting to note that a "corrective" response to the cause of a problem involves both the cause-to-effect pattern and the effect-to-cause pattern. Whether a technician troubleshoots an alarm system for a bank or a doctor prescribes tetracycline for an infection, both situations are derived from the two methods of causal reasoning. For example, look at the following diagram.

First, some "bug," electronic or biological, enters the system and creates a dysfunction, mechanical or warm-bodied. Then you set about in dealing with the bug.

This introduction is structured in the outline style. The paper promptly turns to a discussion of terms and principles, which is often a necessary first step in technical projects.

Model 6.B (3)

This allows the barrier not only to reflect the noise but also to absorb it. Absorptive material is typically porous and is adhered to the side of the barrier facing the noise source.

HOW BARRIERS FUNCTION

With the concept of sound-wave motion in mind, imagine a drop of water falling into a motionless pond. If you placed a straightedge in the pond, you would see how the straightedge interferes with the wave motion caused by the drop of water. Wave motion is diffracted around and behind the straightedge, creating a shadow zone behind it. Sound moves in a similar way. The wave motion bouncing away from the straightedge is similar to the reflection of sound.

In the next section, discussion of the wave motion of sound in barrier design will begin with diffraction over and around barriers and additional reduction acheived through vegetation. Further discussion will include barrier mass and sound transmission paths, as well as effective barrier location, and will conclude with the effects of weather on sound propagation.

Diffraction by Barriers and Diffusion by Ground Cover

Noise can bend around and over barriers that are smaller than the sound wavelength in the same way that light can bend around objects (see Figure 5 on the following page).

Noise diffraction is a very challenging factor in barrier design. It can be reduced partially by the location and by forming a slight angle at the top of the barrier, which can reduce noise by up to 3 dB. Another method of coping with noise diffraction is to place trees or bushes adjacent to the barrier in order to "scatter" or diffuse the sound in its path to the receiver.

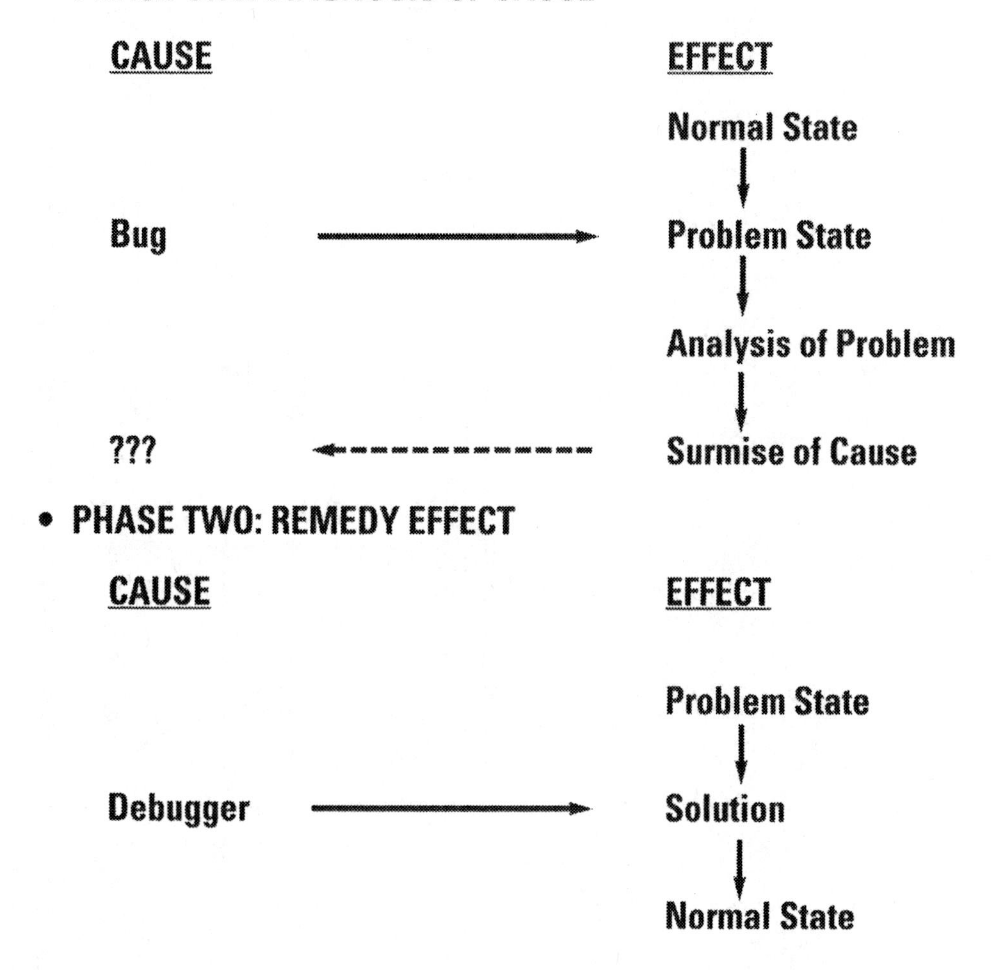

The first go-round is the investigation of effects to determine cause. The second go-round is the initiation of a cause (cure) that will correct the problem. Infection is a cause with an effect. An antidote or a solution is also a cause with an effect. An antibiotic, or a repair, will effectively remove the original cause of the problem.

A very substantial discussion is developed before we get to the specific cause-and-effect issue of barriers. The limitations are part of the discussion and define the extent of the effectiveness of the structures.

Model 6.B (4)

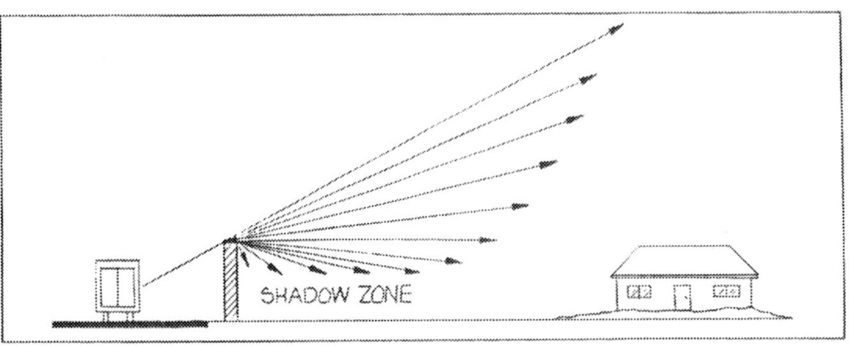

Figure 5
Diffraction by Barriers

How Barrier Mass Affects Sound Transmission

Sound travels along a sound-transmission path and is recognized as a measure of oscillating force. When this force is obstructed by a barrier, the barrier that receives the sound energy will vibrate, which in turn will radiate sound to the receiver. The weight (mass) of a barrier is very significant in the reduction of noise. It is understood that the amount of sound transmitted through a barrier is inversely proportional to its mass. For example,

If Force = Mass × Acceleration,
and Force = 4,
then 4 = 2 × 2.

If the mass is increased to 4, then
4 = 4 × 1.

Notice that when the weight (mass) of a barrier is doubled, the acceleration or amount of vibration transmitted is halved. The minimum mass configuration used for a highway barrier is material weighing at least 4 lb/ft^2. This is another important consideration in noise reduction.

It should be noted that the identification of cause can be very problematic. Even in electrical systems such as those found in a car, it can be difficult to determine causes because events can occur in series or in parallel, so to speak. If A causes B and B causes C, then fixing B is going to create an angry client once A triggers B again a week later. Complex systems, such as the human body, are frustratingly more complicated. Researchers simply do not quite know what causes Alzheimer's disease or Parkinson's disease or many cancers. The causal relationship is either too complex or inadequately understood. Even a troubleshooting checklist on your clipboard can be a problem. It looks simple enough, but suppose you find that there are unclear "problems." You conduct six tests and only four are decisive. The result might look like this:

✓	**A. OK**	**(NO CAUSE)**
✓	**B. DEFECTIVE**	**(TRUE CAUSE)**
✓	**C. OK**	**(NO CAUSE)**
?	**D. PARTIAL DEFECT**	**(PROBABLE CAUSE)**
?	**E. CAN'T TELL**	**(POSSIBLE CAUSE)**
✓	**F. OK**	**(NO CAUSE)**

If you use the word "cause," you can examine the technical problem clearly to explain the shades of gray. Notice that there are probable and possible causes that are the result of such considerations as serial problems or, perhaps, intermittency. In addition, in some environments problems have *multiple* causes. If a social worker sends a delinquent to his aunt and uncle's for a "structured home life," this strategy may overlook the other possible causes of his crisis: peer pressure, education, television, substance abuse, and so on. Causation seldom is as absolute as the method that appears in a troubleshooting procedure. It is a system of analysis that is limited by our own limits, the limits of our instrumentation, or the complexity of the subject under investigation.

Flowcharts

One contemporary technique for representing troubleshooting on paper is the flowchart. This is a very modern device, so you are not likely to see flowcharts in books dating to much earlier than the beginnings of the postwar transistor and chip technologies. Because a flowchart is very graphic, it is perhaps the most vivid way to develop troubleshooting documents. Each shape has a meaning, and it routes you to the next task. Like Monopoly,

Barrier technology involves more than a blocking effect. Mass must absorb sound and not deflect it. Here the principle is explained.

Sample 6.C

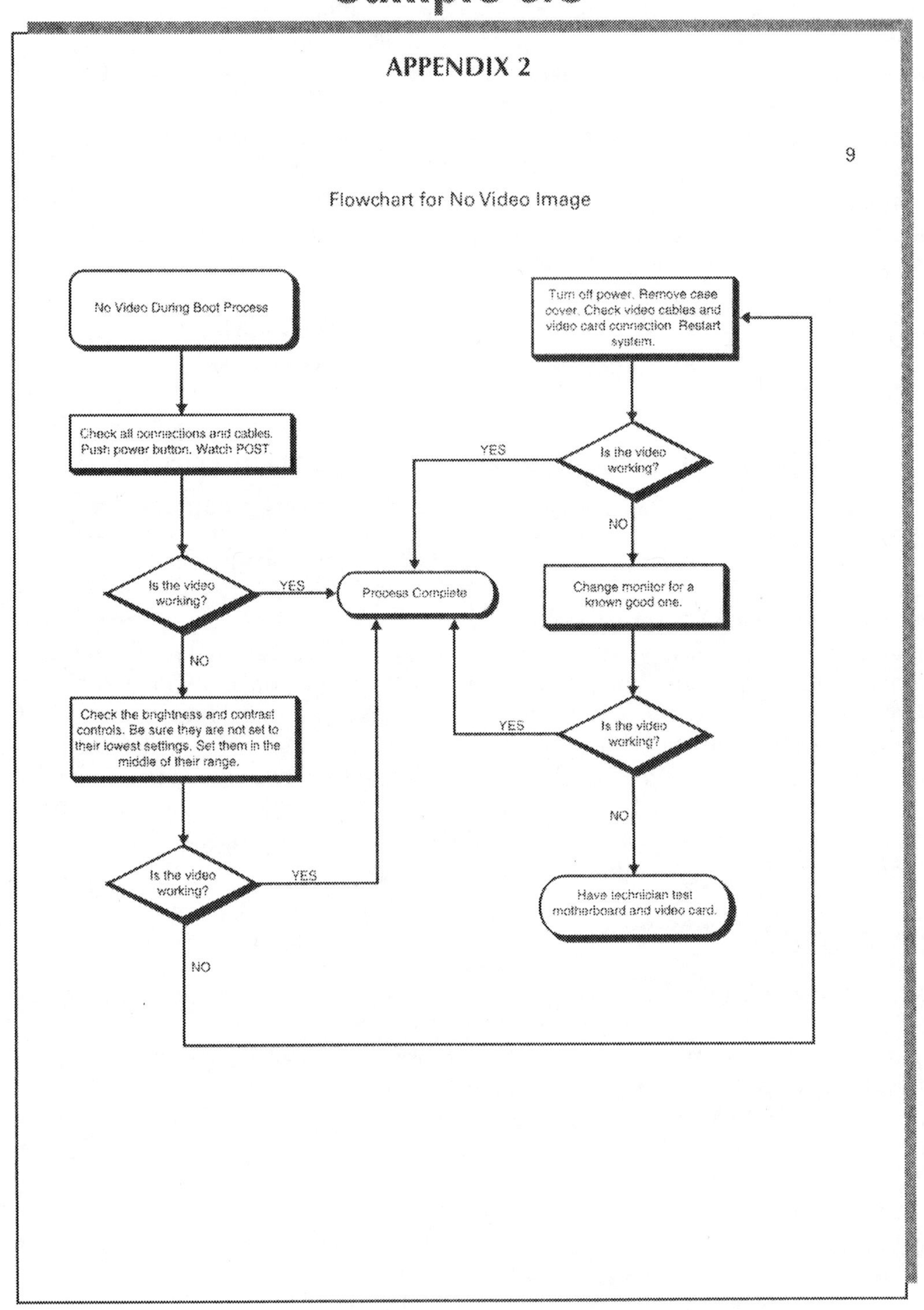
APPENDIX 2
9
Flowchart for No Video Image
No Video During Boot Process
Check all connections and cables. Push power button. Watch POST.
Is the video working?
YES
Process Complete
NO
Check the brightness and contrast controls. Be sure they are not set to their lowest settings. Set them in the middle of their range.
Is the video working?
YES
NO
Turn off power. Remove case cover. Check video cables and video card connection. Restart system.
Is the video working?
YES
NO
Change monitor for a known good one.
Is the video working?
YES
NO
Have technician test motherboard and video card.

flowcharts have all the makings of a game—except that a technician is trying to get "Chance" off the game board. The basic symbols are shown here:

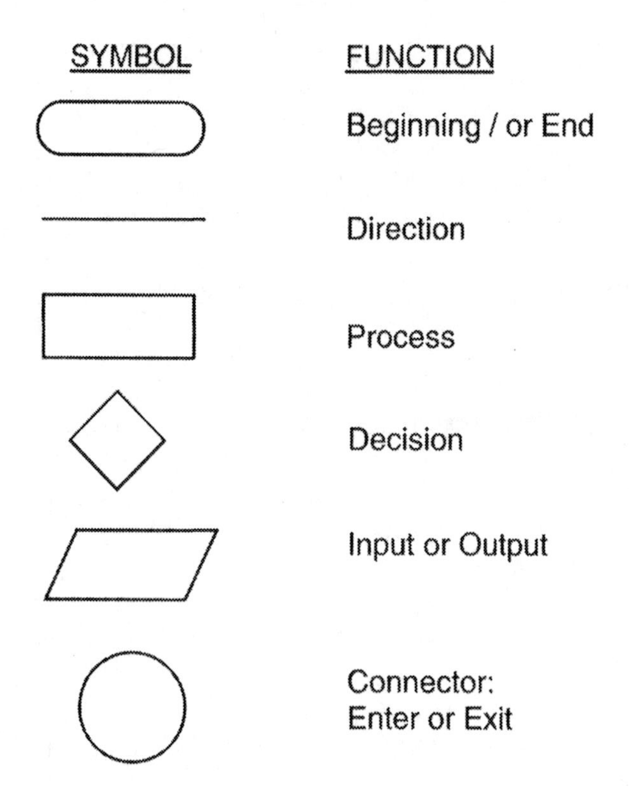

A flowchart is an unusual example of spatial design, symbology, chronology, and language all rolled into the logic structure of a chart. It is visually symbolic, unlike the outline, which uses only topography to indicate priorities. Also, the use of the decision box allows a reader to see how a "path" can lead in one of many directions. On a flowchart, this is a literal image: lines emerge from the decision box and head off to options. Pointing out directions is probably one reason for the popularity of the flowchart. Flowcharts also indicate *alternatives*. Outlines give a reader the feeling, whether it is true or not, that there is only a certain way to do a task. What seems to be a lack of options may not be intended in an outline. By contrast, paths are *very* visually apparent on a flowchart, as in Sample 6.C.

Notice that flowcharts provide the process but not the instructions. If you need instructions, a very practical way to strike a balance is to place the flowchart on a page of text that also contains appropriate guidelines for walking through the chart and perhaps visuals of the actual equipment the flowchart was designed to investigate. Usually graphics are placed above or below the text, but a flowchart works particularly well if it is placed *beside* the text.

Flowcharts are vivid but cryptic. The trade-off is of no concern to experienced technicians but may be a problem for inexperienced readers.

Sample 6.D

- The analyst must be aware of the safety precautions given at the end of the document.
- All volumetric glassware must be class A.
- All glassware must be acid washed and rinsed before use.
- The water used for preparing samples must be Type II, with a resistance of greater than 10 megohms.
- The reagents used for sample preparation must be Trace Metals Grade or better.

WATER SAMPLE PREPARATION

Acid Digestion Procedure For GFAAS analysis

- Shake sample well and transfer 100 mL to a 250 mL beaker.
- Add 0.5 mL of concentrated HNO_3.
- Add 2 mL of 30% H_2O_2.
- Cover the beaker with a watch glass and heat on a hot plate for 2 hours at 95°C.
- Cool sample to room temperature.
- Adjust the sample volume to 100 mL with deionized distilled water.
- Allow sample sediment to settle overnight before analysis.
- A flowchart for this analysis is shown in Figure 2.

Acid Digestion Procedure For ICP/OES Analysis

- Shake sample well and transfer 100 mL to a 250 mL beaker.
- Add 0.5 mL of concentrated HNO_3.
- Add 5 mL of concentrated HCL.
- Cover the beaker with a watch glass and heat on a hot plate for 2 hours at 95°C.
- Cool sample to room temperature.
- Adjust the sample volume to 100 mL with deionized distilled water.
- Allow sample sediment to settle overnight before analysis.

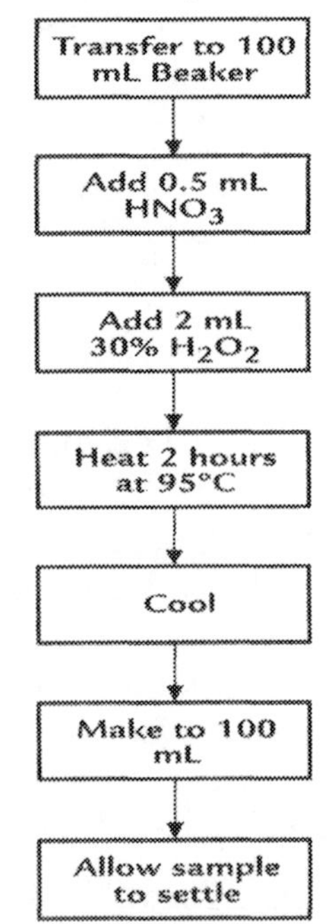

Figure 2 A Flowchart for the Analysis of a Water Sample by GFAAS

Because a flowchart flows down the page, much like the wording of the text itself, the two can go hand in hand. If, in addition, the text is outlined, the items are quite easily seen if the flowchart is adjacent to the text. A component that is parallel to the text and runs the length of the page in a box is usually called a *sidebar.* Sidebars are fairly common, and very vivid, but a little difficult to incorporate. Here is a simple example from a student project.

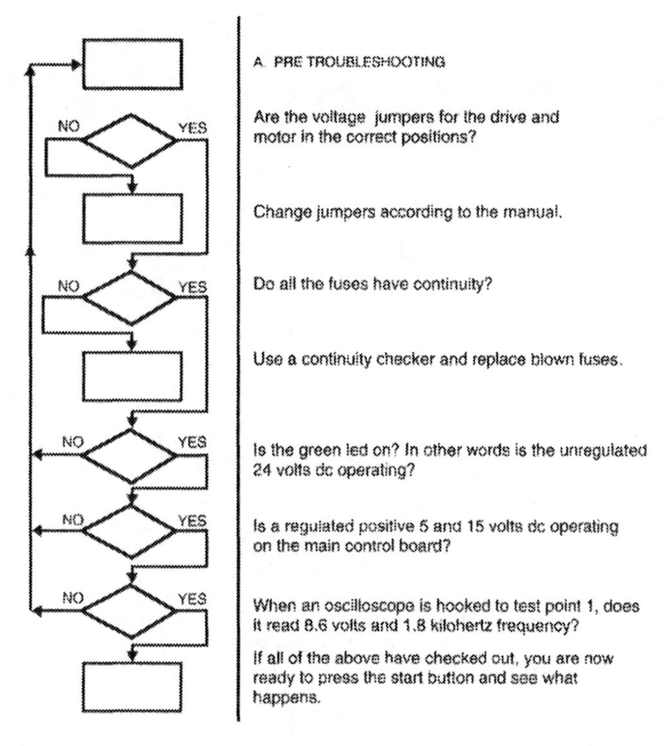

A. PRE TROUBLESHOOTING
NO
YES
Are the voltage jumpers for the drive and motor in the correct positions?
Change jumpers according to the manual.
NO
YES
Do all the fuses have continuity?
Use a continuity checker and replace blown fuses.
NO
YES
Is the green led on? In other words is the unregulated 24 volts dc operating?
NO
YES
Is a regulated positive 5 and 15 volts dc operating on the main control board?
NO
YES
When an oscilloscope is hooked to test point 1, does it read 8.6 volts and 1.8 kilohertz frequency?
If all of the above have checked out, you are now ready to press the start button and see what happens.

Notice that the "flow" of what needs to be observed is made graphically clear. Not only do the symbols cue the reader, but the arrows indicate the range of if-then possibilities. As a result, the flowchart can virtually stand on its own, without a text, if it is certain that a technician knows what to do (see Sample 6.C).

Usually the flowchart symbols contain enough information to convey instructions without further comment—*if* the reader knows the subject. The preceding sample is unique in that each decision box is placed beside a question that precisely identifies a step in the troubleshooting procedure. Another example of a flowchart is on p. 230. Notice that the cryptic format of the commentary inside the symbols on that example calls for a great deal of knowledge on the part of the reader.

Do not be discouraged from using the flowchart concept because you are unfamiliar with the various shapes of the symbols. The device is quite commonly used without the many symbolic shapes, particularly if the chart is not a troubleshooting tool. I included the

The flowchart frequently appears without any designated function assigned to the shapes that are used to construct it.

Model 6.C (1)

TROUBLESHOOTING

REFRIGERATION SYSTEMS

AUDIENCE:

FIRST-YEAR HVAC STUDENTS IN THEIR SECOND QUARTER. STUDENTS WILL BE FAMILIAR WITH THE REFRIGERATION PROCESS.

OBJECTIVE:

PROVIDE GENERAL DISCUSSION OF SOME OF THE MOST COMMON PROBLEMS ENCOUNTERED IN REFRIGERATION SYSTEMS.

NAME: MICHAEL K.

EGR 231.01
INSTRUCTOR: DAVID R.
03/09/200X
PROJECT NO. 3

NORTH SEATTLE COMMUNITY COLLEGE

sample on p. 232 to illustrate the idea of the parallel text. Arthur's chart (Sample 6.D) avoids symbols and simply indicates a testing procedure in chemistry.

Because flowcharts are symbolic and may be very cryptic, be sure to develop a few paragraphs to lead your readers into a group of any such charts. The same is true of troubleshooting tables. Briefly explain them before you present them. There is no substitute for explaining the use of a document to a reader. If you develop a flowchart or a troubleshooting table for a large manual, you may not need explanations, but if the flowchart document stands by itself, add an introductory paragraph or page and explain the chart to the user-reader.

Troubleshooting Tables

The more traditional method of rendering a troubleshooting logic process is to create a table. The formats are conventional, and they are derived from the logical order of causal reasoning. At a minimum, the troubleshooting table consists of a list of *symptoms* (effects) and a list of *remedies*. A little more thorough model will include symptoms, *causes*, and remedies. An even more thorough model will have symptoms, *location*, causes, and remedies. Other columns of data may be added as needed.

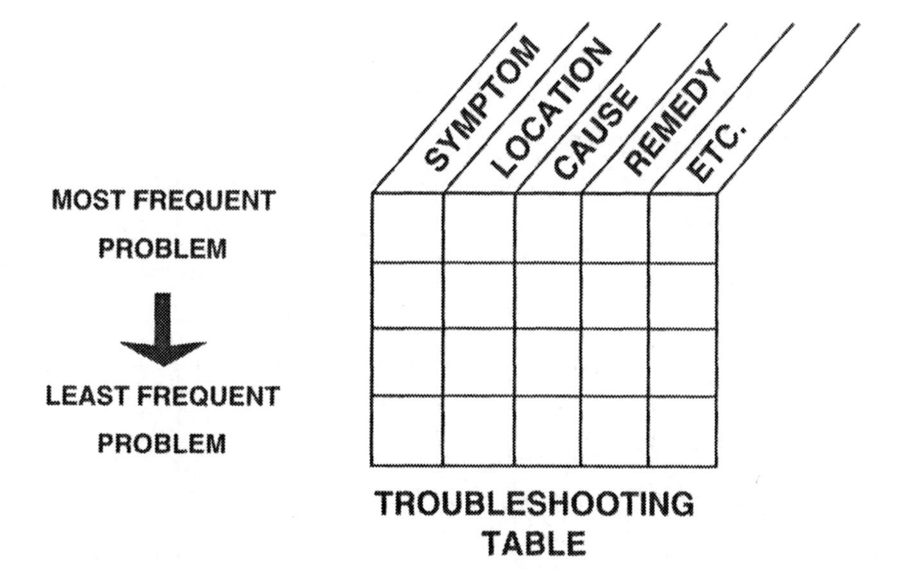

Many two-year engineering technologies programs involve troubleshooting, and this activity is often a major skill that must be developed by engineering techs.

Model 6.C (2)

Troubleshooting Refrigeration Systems

There are many different types of refrigeration systems: some are designed for high-temperature applications such as room air-conditioning, some are designed for medium-temperature applications, such as nonfrozen food preservation, and still others are designed for low-temperature applications such as ice making.

Regardless of the type of system, some common operating problems will be encountered, and the service technician, like a doctor, must be able to recognize the symptoms, diagnose the cause, and take corrective action. A diagnosis is made through the process of elimination of several possible causes, each of which may be the source of the problem in the refrigeration system.

Problem Categories

A refrigeration system is a heat-energy-transfer device. It only is capable of absorbing heat from a heat source and rejecting the heat into a heat sink. The rate at which it does this is dependent on the amount of heat available and the rate of heat transfer. The rate of heat transfer will depend on maintaining the proper temperature difference between the

There is one major distinction between the comparison tables for product evaluation and the one in the following prototype. The comparison table is read in either direction, whereas flowcharts and troubleshooting tables are always read top down.

The importance of problems or frequency of problems is used to prioritize the tables, and they are read from the top. Usually the tables are constructed somewhat like this:

Table 3. Troubleshooting Approach to Upgrading.

The Complaint	***What's Happening***	***How Do I Fix It?***
Takes a long time to load programs	Your computer is having a hard time reading all the data, processing it, and storing it in RAM.	• Defragmenting the disk may increase performance somewhat. • A faster CPU will speed up the data reading. • A wider, faster bus can move a wider flow of data more quickly.
Takes a long time to redraw graphics or complex screens	May not have the RAM needed (4 MB minimum for basic operation with Windows). CPU may not be fast enough to compute new screens quickly. Bus may be too slow to move data to video quickly.	• More RAM can help. • A new CPU or motherboard with wider, faster bus can move data more quickly.
Hard drive is constantly running in Windows	There is not enough free RAM, so Windows is continually writing to virtual memory.	• Streamline system configuration. • Install more RAM.
Takes a long time for word sorts, database queries, or mathematical computations	The CPU is slow at processing data.	• Install new CPU or motherboard.

Early in the text, the author introduces the idea of "categories" so that the subject can be divided into manageable areas of discussion.

Model 6.C (3)

-2-

refrigerant and the material from which heat is to be extracted or into which it is to be rejected.

The problems of a refrigeration system can be divided into four categories:

1. Heat source
2. Heat sink
3. Refrigeration
4. Electrical.

Heat Source

All refrigeration systems must have available to it a sufficient heat source to satisfy the capacity of the system. If a system is capable of transferring 12,000 Btu of heat in an hour's time with 20°F ΔT, for example, this amount of heat must be available. If air is the means of carrying the heat from the product to be cooled to the evaporator for extraction, the correct amount of air must pass through the cooling coil. If an insufficient amount of air is being supplied because of dirty air filters, a slow-pumping fan or blower, dirty coil fins, or an improperly placed product that impedes the flow of air, the amount of heat absorbed is reduced. A reduction in air flow reduces the quantity of heat absorbed, lowers the coil boiling point, and lowers the suction pressure and, therefore, system capacity.

The concept is quite simple and remarkably consistent in thousands of manuals. In practical terms, the result will appear as you see in the sample on p. 241.

I have already discussed the effectiveness of tables in technical writing. Tables can be consulted quickly and easily, and they can hold a large body of information in an orderly and graphic way. When I suggested that you construct a table to use as a basis for writing a product evaluation, I was recommending a visual layout in preference to an outline. The outline is somewhat graphic—or topographic at least—but the table and the flowchart are far more visual. If you want to develop a troubleshooting document, design the table (or flowchart) first. Follow the table to develop the paper. Begin by explaining the causal problems to be analyzed and resolved. Then, using a problem-by-problem approach, discuss the available remedies. Prioritize the problems, either from most frequent to least frequent or, possibly, least costly to most costly, or some similar ranking.

As in the case of the flowchart, you usually *will* need a text. The table becomes more clear if explanatory text is wrapped around it. The tables are self-explanatory only to the experienced reader, as you can observe in the table concerning upgrades. The following table concerning network troubleshooting is taken from a project that included elaborate tables, but the presentation depended on a well-structured text as well.

A multipage troubleshooting table, such as the following one, becomes a visual tool and a quick reference guide. The design is the secret to the clarity. The logic is orderly, and the layout is geometric. You could not ask for more from a document. Notice that the table is reasonably thorough in comparison with a flowchart because there are no space restraints. The boxes can be as large as the author desires. On the other hand, this table is for the experienced reader, and text support would be important for anyone who might be unfamiliar with the networks and their care and maintenance.

The table is an excellent example of the way in which technical documentation is not exactly technical writing. The table is largely a design concept and it functions visually. Seldom do two sentences even link together in this context. The result may well also include illustrations.

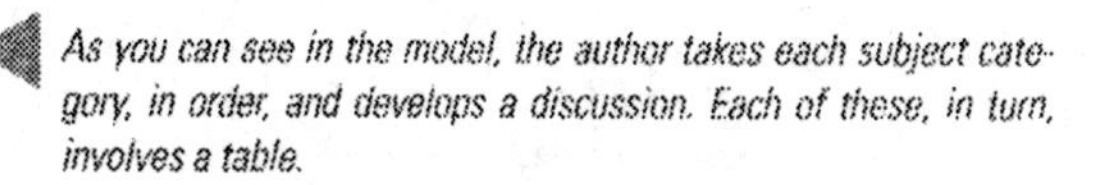
As you can see in the model, the author takes each subject category, in order, and develops a discussion. Each of these, in turn, involves a table.

Model 6.C (4)

-3-

Heat Sink

Problems in the heat sink area are usually easy to diagnose because the change in the system becomes more radical with a change in the operating conditions of the heat sink. When the amount of air through the condenser is reduced, head pressure and compressor amperage draw go up, and system capacity drops. The effect on capacity is not as great, however, as a change in load on the evaporator, so this problem usually exists and grows until a radical departure from the norm occurs.

The most frequent causes of problems in air-cooled condensers are shown in following the table.

Table 1. Most Frequent Causes of Problems in Air-cooled Condensers

Symptom	*Problems*	*Causes of problem*
Reduced system capacity	Insufficient air flow through the coil/blockage	Dirt, leaves, etc., on coil surface Restricted air inlet or outlet Prevailing winds Unit located in restricted airflow area
	Insufficient air flow through the coil/mechanical	Blower bearing too tight Incorrect fan or blower rotation (especially on three-phase motors) Incorrect fan or blower speeds Dirty blower blades Motor bearing too tight

Tables can be readily constructed by your computer. The table below was conveniently constructed by the author using Word and you do not have to use Excel in order to design tables of the sort illustrated in this chapter. Troubleshooting tables will seldom involve extensive use of narrow columns, and the tables usually develop thorough material in three, four, or five columns. If the author of the table below had decided on a fifth column, the usual choice is the addition of a column that will identify the "location" of the problem. The horizontal list of problems or symptoms is a very different matter, and at times a table can run continuously for many pages. If you develop a table that continues from page to page, place the word "continued" in the lower right hand corner immediately below the table. Similarly, the word is often found in the upper left corner of subsequent pages of the table.

NetworkingTroubleshooting Chart for Lab1650a

Symptom	***Cause***	***Remedy***	***Steps***
You cannot see the other computers on the Local Area Network.	The wrong protocol is being used to view the network.	Check the configuration tab in the properties of Network Neighborhood and make sure that you have NetBEUI installed.	Right click on "Network Neighborhood" and click on "Properties." Highlight "Client for Microsoft Networks" and click the "Add" button. Highlight "Protocol" and click the "add button. Highlight "Microsoft" in the Manufacturer's table and then highlight "NetBEUI" in the Network ProtocolsTable and click on the "OK" button.
You can see the other computers on the Local Area Network but your computer is in the wrong hierarchy in the Explorer window.	Your computer is in the wrong workgroup.	Check the other computers in the room to see what the name is of the workgroup they are using and change your Workgroup Name to match them.	Right click on "Network Neighborhood" and click on "Properties." Click on the "Identification" tab. Change the workgroup name to "1650a-classroom."
You cannot see any of the network while getting a message on the desktop that states "Your system cannot access the Network Interface Card."	1.) One of a variety of software and firmware errors and omissions is causing a failure in the NIC Card. 2.) The NIC card is not fully inserted into the bus.	1.) Run "Hardware ConflictTroubleshooter." 2.) Properly seat the card.	1.) Start the Hardware ConflictsTroubleshooter and follow the steps to resolve the software or firmware error or omission. 2.) Have the instructor call the technical support staff at the help desk and do not open the computer cases in the lab.

(continued)

The text explains the table. The table visually organizes the subject matter of the text in such a way that it functions as a ready reference upon first reading, and thereafter.

Model 6.C (5)

Refrigeration System Problems

If the refrigeration system is correctly charged to the point at which the evaporator is fully active, and the condenser contains a sufficient quantity of refrigerant to produce subcooling, and the compressor is pumping the required amount of vapor, the system has to perform as designed. Any deviation from these requirements will show up in the test pressure and temperatures that are taken on the system.

In refrigeration systems, the most common problem is refrigerant shortage, which results in failure of the refrigerating effect. Problems of refrigeration system malfunction are varied and can arise from the refrigeration system itself or from an electrical cause or a combination thereof. Following is a list of common problems that can be encountered in the refrigeration system. Listed under each problem are possible causes.

Entire System

Evaporator temperature too high

Evaporator load improper
Refrigerant shortage
Ambient temperature too low
Head pressure too low
Metering device misadjusted or blown out
Oil clogged coil
Refrigerant lines undersized
Undersized unit

Evaporator temperature too low

Evaporator load improper
Improper wiring
Control contacts overloaded
Metering device misadjusted or blown out
Ambient temperature too low
Oil clogged coil
Improper unit location

Liquid line extremely hot

Head pressure too high
Refrigerant shortage

Engineering Tables

Unless you are involved in research and development, you are not likely to contribute to the development of tables that calculate "standards" of any sort. Engineering tables are derived from *much* researched demonstrations. These predictable outcomes are then evaluated and adopted by a wide variety of agencies and professional groups that are responsible for the integrity of national standards. The tables themselves are somewhat the reverse of troubleshooting. The tables identify correct practices; these specifications and regulations then "cause" quality, safety, strength, and other such effects.

Authors tend to place the symptoms or effects on the left side of a troubleshooting table because they start by looking for *trouble*. In contrast, readers start by looking for the *solutions* (the effects) when they consult engineering tables. They want to *avoid* trouble; they consult the tables for standards that cause (or are the cause of) a trouble-free application of the engineering calculations provided for them. If you look for problems, you begin with *troublesome* effects. Engineering tables are keyed to *desirable* effects and provide recommendations (causes) for your needs. If you engineer the project correctly, you use proper regulations and standards that *cause* a safe and predictable quality in your product at the outset.

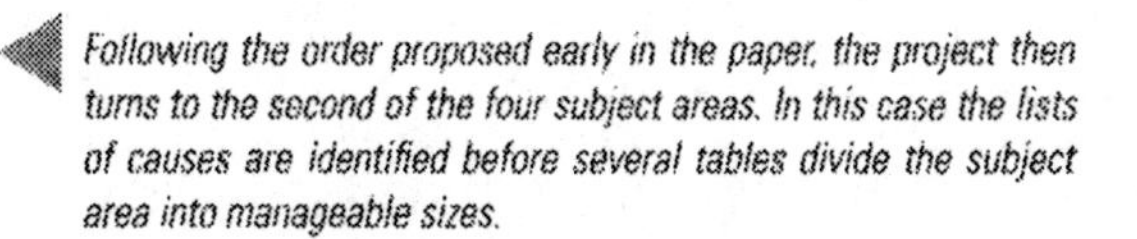

Following the order proposed early in the paper, the project then turns to the second of the four subject areas. In this case the lists of causes are identified before several tables divide the subject area into manageable sizes.

You have looked at several brief narratives concerning projects in which authors have explained computer applications. There are countless combinations of equipment and software that are available to help construct these projects. It is probably helpful to see what goes on behind the scenes now that the old methods of the writing process have been digitally remastered, so to speak. The troubleshooting project in Model 6.C was developed by Mike for a heating technology program. Mike provided a narrative about how he constructed his paper with modest computer equipment. Some of the other authors have had access to excellent high-end computer equipment, but you do not need expensive gear. Mike explains how his project came about.

> Because I have such patchwork equipment I had to be creative in preparing attractive-looking finished products.
>
> The projects were prepared using the following hardware.
>
> 1. Toshiba laptop
> 2. School computer--Pentium based PC linked to file server via LAN
> 3. IBM Pentium at photocopying store
> 4. IBM laser printer at photocopying store (600 dpi?)
> 5. Photocopying machine at photocopying store
>
> The software involved was Word Perfect and PageMaker. All the text was generated using Word Perfect, and all the graphics were done using drawing tools, from PageMaker due to limited access to other more potent drawing tools, such as AutoCAD. The text was imported to PageMaker, layout was established, and drawings made. When I needed to change the size of the graphics, I simply used a photocopier and cut and pasted.
>
> Word Perfect has a potent table-generating tool. I used it to construct tables fairly quickly.
>
> When I borrowed illustrations, I used a different process. I generated all the text and layout with the word processor. All the figure boxes were empty when the initial text was generated, even though the captions were in place. The graphics were photocopied from books and pasted into a draft copy.
>
> Mike K.

Models 6.B and 6.C

Comments

We have examined several samples that explored the different approaches to analysis through causation. The first project concerns highway engineering and acoustic engineering. The presentation is a cause-and-effect analysis concerning sound barriers and their effectiveness. The text explores sound transmission and examines methods of control, such as diffraction, diffusion, and absorption.

The second project is a discussion of troubleshooting for mechanical heating systems, and it includes four traditional tables, one of which you see in the model.

The acoustic engineering project creates a cause-to-effect analysis that will allow us to "construct" (the barrier). The troubleshooting analysis creates an effect to cause analysis that will allow us to "repair" (the heating system).

To see the heavy emphasis on tables that is typical of troubleshooting documentation, please see the final model in this text on pp. 482–484.

Causal Analysis

Project game plan

- Develop a table or flowchart to represent the analysis being conducted.
- Such a table is generally read from the top down.
- Rank problems in a troubleshooting table in the left column. Begin with the most significant problem at the top.
- In the presentation, introduce the causal problems so that readers see the point of your project.
- Introduce the table by explaining the purpose of each vertical column to be used for analysis.
- Develop the analysis by following the itemized list of symptoms in the left column. Develop each horizontal line of data in a brief discussion of a problem and its solutions.

Activities Industrial Applications: II

You are now ready to compose another technical writing project.

An evaluation of a group of similar objects is a very useful document. Everything imaginable, from medical insurance to car radios can be compared. Use a technical focus of interest to you, and develop an evaluation. Compare three to six similar items and develop a table after deciding on the features or concerns that should be investigated. Try not to repeat data from an existing document that has already accomplished a similar mission. Redundancy is a waste of company time. If you determine your interests and focus on a specific and unique audience, you will avoid repetition.

Consider alternatives to comparing devices or mechanisms. You could look at the merits of three installation procedures, four alloys, five toxic disposal processes, six ways to control population growth, and so on. Finally, do not compare motor vehicles; the subject is too complex for a 1500-word project, which is your goal. Build a table for the project and use the Product Evaluation Summary as your guide (see p. 500).

If you are familiar with a particular troubleshooting process, develop a presentation to explain it. Use a novice, or perhaps someone at work, as the target audience. Include at least one table or one flowchart. Locate the table templates on your computer. The text should run 1500 words or more. Use the Causal Analysis Summary (p. 501) as a guide.

If you prefer, develop a text from cause to effect. Examine conditions that cause an outcome and explain the process. The topic can be based on your technical specialty, but there are endless causal topics of interest, from computer viruses to vitamin deficiencies.

Share a project

Work with three other members of the class to discuss and edit the second set of technical documents. If desired, use the same groups that were assembled for the first projects, or develop new groups.

- *Before holding the meeting, you should review the peer review discussion in Appendix A (pp. 495–497) and study the third and fourth editing checklists concerning comparison, causal analysis, and troubleshooting (pp. 500–502).*
- *On the day the projects are due, have each member of the editing group explain his or her project in terms of objectives, intended audience, and project development.*
- *Have each member edit the texts for both writing errors and technical errors.*
- *Have each member write a one-paragraph critique at the end of each paper.*

Work in Progress

7. Document Outline

I soon found that still other departments were experiencing problems with distribution. I explained to the department supervisors, Jan and Howard, that I was working on a process change to solve the problem and asked for their ideas. The supervisors and I went over the existing distribution process and labeling procedures.

After interviewing these additional department supervisors and receiving their inspection redlined documents a few days later, I was ready to prepare an outline of the project, even though I actually had pieces of it written. An outline is my favorite format because you can organize the items into sections, rank them, and put them into a sequence that will aid you in the flow of the document. After all, a document is not created for the writer; it is for the reader.

When assembling a document you may have to do so by taking portions from a variety of documents and nondocuments. A few of the nondocuments I often refer to are situations such as verbal instructions, watching a test be performed, watching a video, physically operating or taking apart something, or physically looking through archive boxes for historical data. This can be thought of as structuring rough-draft materials. Needless to say, this process is challenging.

Every work-related document can be a new challenge in structure. I mentioned that at Genus, documents should be uniform in appearance. I find pleasure in building order into an unstructured issue, especially if the result aids the user in performing a function.

I prepared my outline and needed to contact Tom Whitaker, the primary engineer of production, to set up a meeting to discuss the distribution changes I outlined. I arranged a meeting in which we reviewed my outline. As I expected, he had some useful suggestions and approved the outline with his suggestions. This meeting was beneficial in that he acted as a technical editor. When I am in the process of writing a document I get to a point where I have memorized it and might not be able to understand it from a "reader's" point of view. That is when an outside reader or a second set of eyes is helpful. Besides, I needed to have an engineer's perception of the issue.

Most documents used in schools and colleges are in a "paragraph essay style" format. I tend toward an outline style because I feel it is easier to read and refer to when trying to find references within the document. Of course, most procedure and instruction documents at Genus are in an outline format as end products, so the outline is a practical tool for me from start to finish.

S. B.

The Double Viewpoint

Research-oriented writing is usually distinct from the types of documents examined in Chapter 5 and Chapter 6. Obviously, the practical, day-to-day documents that emerge from a corporation are related to production. The writing supports the services or products of the company; the writing is product-driven (and consumer- or user-driven). For example, a set of instructions usually calls for an outline style and topography of necessity. Similarly, the documents of the R&D division of any corporation serve the ends of research and lack the applications orientation of instructions, at least initially. R&D documents are research oriented. Likewise, universities are usually research oriented. The documents developed in these research environments, in the form of "papers," are quite similar on both the corporate campus and the university campus, though perhaps they are far more confidential in corporate environments.

What goes on in this variety of technical writing? Basically, the material is likely to be much more analytical and much more driven by causal analysis. The inductive process I mentioned at the outset is probably most predominant in the R&D reports, in which the primary research must precede not only product development and applications but is likely to precede even the engineering modeling phase. Theory must first take the shape of demonstration. As I have noted frequently, technical writing is highly data oriented, and nowhere is this more obvious than in the research papers of corporations and universities.

You may not be planning a future in research and development, but you need to look at the analytical models for academic purposes if you are a transfer student. All the models discussed to date will be of value to you sooner or later—sooner if you are an engineering tech student, later if you are an engineering transfer student who must still finish upper-division courses at a university. In either situation, you will usually find your way to the marketplace, where applications-oriented writing is daily fare. However, if you are likely to develop papers for upper-division courses, you should look at a few of the models that will be useful for upper-division work. The industrial applications examined in Chapters 5 and 6 may not work successfully in the academic environment. To put it simply, if you *outline* upper-division papers with an eye to *brevity,* you may raise more than a few eyebrows. The writing procedures of universities may not coincide at all with the workday fare of Chapters 5 and 6.

This chapter's selection of models in the analytical area is limited to three prototypes. Along with the single-viewpoint narrative (discussed in *Basic Composition Skills*), the following three models are very likely to be structures of interest for upper-division college papers:

Comparisons	**(of theories, for example)**
Problems and solutions	**(a causal research format)**
Risk analysis	**(analysis of benefits and risks)**

Certainly there are many others. I have selected only those that involve a divided structure in the document so that you can focus on the technique for constructing more than one conceptual movement in the project. I assume that you have mastered the single-point-of-view perspective you found in most of the projects you examined in earlier chapters. Those eight styles or types are fundamentally designed as straight shooting without any "ands" or "buts." Each of the "industrial" types reflects a single mission that can be written in a straightforward fashion by the author. Only the product evaluation model suggests that there are other organizational problems, which are discussed in the sections of this chapter.

You need to examine academic or research projects in structural terms to decide what is usually unique to them or at least distinctly different from the practical, work-related writing you have seen so far. What is probably unique at the research end are academic writing conventions, depth of discussion, and double or multiple viewpoints.

Regarding the *conventions,* most college writing targets a professor. The document is very narrow in focus and style, since it reflects course work and because an instructor usually reads it only once. You might inquire after any formats your instructors prefer. You will find a wide disparity in what you might call "commitment to style," and yet professors will, nonetheless, expect high standards. This book will help serve your purpose, since it conforms to most professional conventions, both academic and corporate. I will point out any substantial variance, such as the issue of graphics and where to put them. You might ask to see a paper as an example. I keep dozens at hand because I believe strongly in the modeling concept. It will help if you are thoroughly familiar with the writing standards discussed in the *Writer's Handbook,* Volume 2 of the *Wordworks* series.

Depth of discussion is the second issue. Running text is likely to be the choice for analytical work. The standard writing style for academic work seems to be the running-paragraph style, with necessary headings. This practice is partly a matter of convention, but it also has to do with the preference for paragraph logic. Research is likely to have the most thorough or lengthy, discussions of any sort of technical writing. The reason is readily apparent in the mountain of data that must be scrupulously gathered to support a contention—even if it takes years or decades to gather it. Then the mission of writing the document is just as scrupulously attended to. The paragraphs—whether dozens of them or hundreds—are usually the focus of the discussion because the paragraph develops the logical analysis: the proof.

Quite simply, the paragraph is, in and of itself, a unit of logic. Most English composition books look at the paragraph as a subject-related cluster. I am less concerned about what a paragraph is *about* than what it *does.* It is a tool. In analytical work it is the point of analysis. That analysis is usually inductive and usually follows a pattern that is recognizably mathematical.

Topic Sentence	**Theory**	**Hypothesis**
Evidence	**Proof**	**Analysis of Data**
Discussion	**Discussion**	**Discussion**

I can state the preceding *process* in a dozen ways, but all of them suggest long paragraphs. The paragraph must *state* its purpose, *verify* its purpose, and *defend* its purpose. In other words, you open a paragraph with the intention and then devote the rest of it to two tasks, which often run together. You produce the evidence, and you discuss how the evidence proves the point. In sum, for upper-division courses you may find the documentation highly analytical, and the writing will usually reflect the needs of the analysis.

When you examined the dominant varieties of industrial applications for technical writing tasks, you looked at eight types of projects. It may appear that analytical papers are a single "type," but there are probably many varieties, including the three practical variations in the following discussions. A sample outline and two models are presented in support of each discussion. First, the sample outline shows you how to organize the logic of the structure. Then, one model is used as the usual running illustration, as in Chapter 5 and Chapter 6. However, papers are not, as I noted earlier, as structurally apparent as many of the trade and industrial documents, and so a second sample is presented, accompanied with commentary boxes to clarify the details of development.

Sample 7.A

Technical Writing
May 13, 200X

C vs. C++ Programming

Audience: Individuals with little or no programming experience who intend to learn a computer language.

Purpose: To acquaint beginning programmers with the differences in programming styles of two popular languages. The text will look at the advantages and disadvantages of each language.

I. History of C A. Developed in 1973 at Bell Labs B. The older language	**I. History of C++** A. Developed in 1983 by Strousop to extend C B. The subsequent language
II. Programming Orientation A. Procedural B. Emphasis on processing of data (Functions)	**II. Programming Orientation** A. Same B. Emphasis on data itself (Object)
III. Advantages A. Flexible, portable (Can move between platforms) B. Forgiving C. Small Few reserved words	**III. Advantages** A. Same B. Expanded capabilities (i.e., animation) C. Object-oriented programming Becoming the standard in industry
IV. Disadvantages A. Less capable B. Use of functions makes manipulation of data more difficult	**IV. Disadvantages** A. Less forgiving B. Harder to learn

V. Conclusion

A. When learning to program for general purposes, learn C

B. When learning to program for specific applications, learn C++

The Balanced Comparison Design

The third distinct feature of academic writing concerns the viewpoint or viewpoints under discussion. Apart from conventions and the length and depth of academic discussion, there is also the matter of *alternatives*. A great deal of research is relational in some way. That is, if researchers develop a theory, they must compare it with other theories. If a team develops an antidote for a disease, the team develops a solution to a problem. The problem and the solution are relational. If an engineering group proposes a particular solution to the need for a bridge over very deep water, the engineers must consider the advantages and disadvantages of the design, which are relational issues. And if the group has both a floating bridge and a suspension bridge on the drawing boards, there are advantages and disadvantages of *each* to consider. Still more relational material to analyze! These particular demands are not overly complex if it is clearly understood that the structure of the analysis is relational.

Essentially, you can take an architectural view of the complex projects just identified. You will observe, of course, that they basically are issues that involve two sides or two viewpoints. They are, as one engineering student said recently, binary. With that idea in mind, the documents are easy to structure. The first challenge with all three of the basic double-viewpoint approaches is to divide and conquer. The easy way to do this is to compartmentalize both the sectors involved and the issues you want to focus on in discussion. Essentially, you build a concept on the order of the following diagram.

The outline concerning programming is tightly structured and constructed in the fashion of a table. It could be developed either by program (the subjects) or by features (the criteria).

Model 7.A (1)

CONSTRUCTION PROGRESS SCHEDULES

A COMPARISON OF BAR CHARTS AND NETWORKS

AUDIENCE: Architects, Developers, Construction Managers

OBJECTIVE: To compare the two common types of construction progress schedules, particularly as they are applied as specific types of projects

Mike S.

Project 1

EGR 231.01

SCI

Cost Estimating • Value Engineering • Project Management

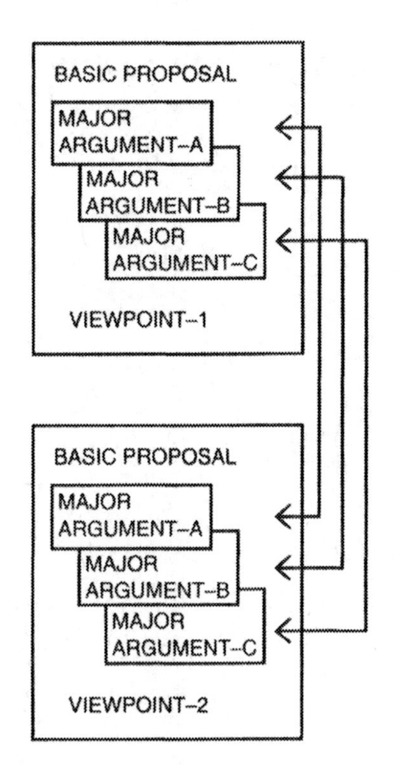

The next concern is to decide how to transfer the conceptual organization into a text. Depending on the topic under discussion, you might develop the paper by "sides" (proposals) or by "issues" (arguments). The document that does a clear job of covering one side first, and then the other, is easier to write. This may be more practical as a strategy for longer papers, as well. On the other hand, as one author explained to me, "Quality control is my bottom line." He wrote all his papers by issue because the issues, in his case, were all quality oriented. He was not surveying two of something because his company was concerned only about analyzing the *criteria* (or needs) that the company established for quality. Of course, by implication, his supervisors used the presentations he wrote to judge which products were the *best*. They indeed were *not* interested in a survey of products. They were primarily concerned about meeting company standards.

Because the comparison document has two of something to compare (two sides or two subjects), develop a two-sided outline to reflect the two interests involved. Try to keep both the vertical sections balanced (equal) and the horizontal sections balanced (equal). A symmetrical design will usually take shape (see Sample 7.A), although some issues will not cooperate. The world around us is not always in balance.

This model is an example of a comparison that involves systems or processes rather than products or other objects. Large construction projects are planned by using the two timetables under discussion.

Model 7.A (2)

INTRODUCTION

A construction progress schedule represents a properly planned and precisely developed system of operations for a building project. This paper will compare the two common types of schedules: bar charts and networks. The comparison will help architects, developers, and construction managers decide which type of schedule is best suited for each of their projects.

Bar charts and network schedules are graphic descriptions of the construction process (see Figure 1a and 1b). Each schedule is unique to a specific construction project. Both types show what work items must be done, how long each work item will take, and when each work item should be done in order to complete a project in a specific amount of time.

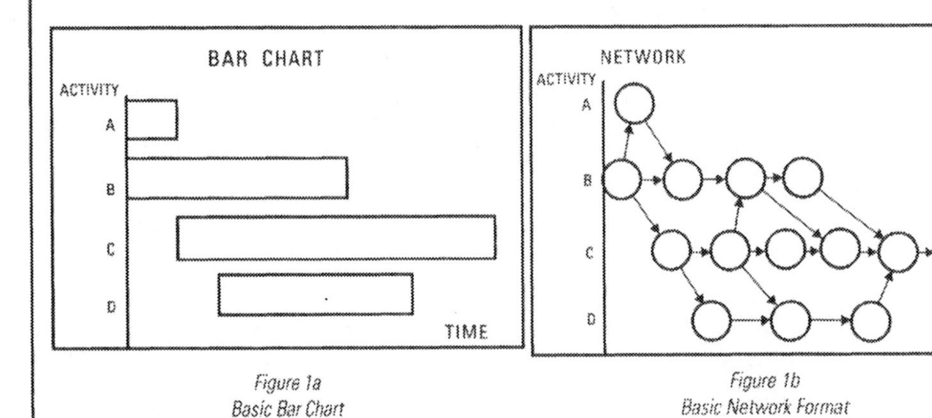

Figure 1a
Basic Bar Chart

Figure 1b
Basic Network Format

Bar charts and networks differ greatly in how this information is described. For many construction projects, one type of schedule is clearly preferable to the other.

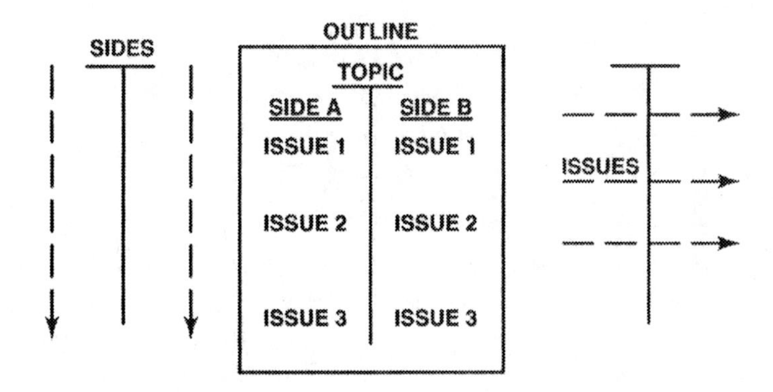

Then develop the text by viewpoint (sides) or by issue. You can can go either direction.

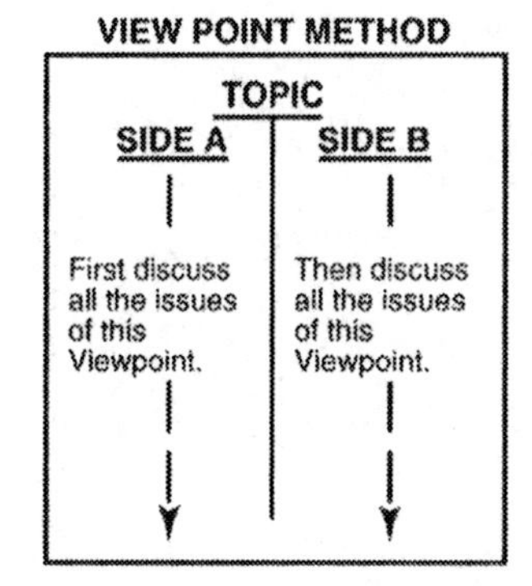

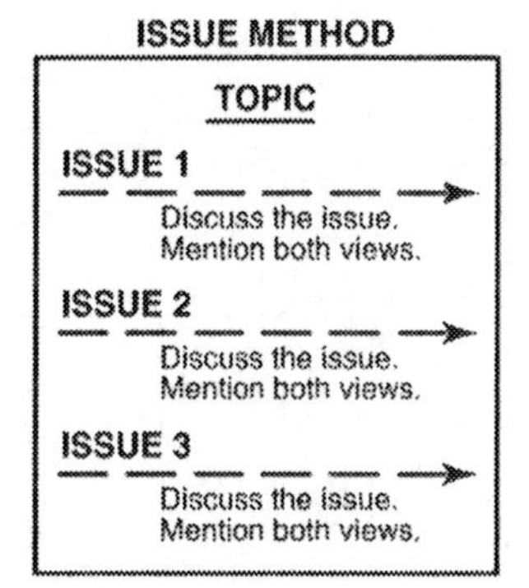

A comparison of two theories is structurally no different from a comparison of two brands of car wax. Certainly a discussion of plate tectonics and volcanism will take about 200 more pages than the car wax discussion, but the fundamental structure does not change. The task is one of balance.

The challenge is to research the "two" of whatever you choose because, along with that research, you must learn which criteria are significant to the analysis. The *criteria* are the points of comparison. They are indeed the measure, not only of the two sectors of discussion, but of your success at measuring. You must decide on criteria because the document is fundamentally conditional; it is only a product of the circumstances you assign it. If cost is no object, for example, you will have one document. If cost is the bottom line, you will most certainly have another. Determine the criteria for evaluation based on the subject, your perceptions, or the supervisor's requests. If this is a term paper, be sure you understand the exact criteria you are supposed to examine. Research may reveal the criteria, or the criteria may be driving the research. The circumstances vary. If the criteria function as "given" due to some condition, of a grant for example, the research will be approached from a very particular angle.

The text begins with an explanation of the two basic systems that are to be analyzed: bar charts and network formats.

Model 7.A (3)

2

DETAILED COMPARISON

APPEARANCE

A bar chart schedule has a logically ordered list of work activities along the left side of the chart. A time scale runs along the top and/or bottom of the chart. The time scale indicates the days and/or dates during which construction will be in progress (see Figure 3).

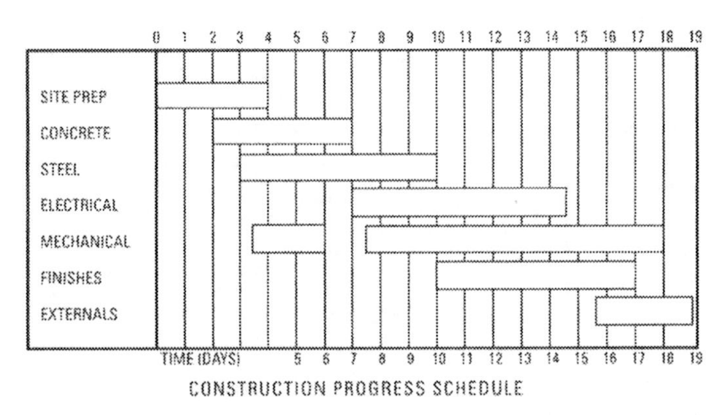

Figure 3. Conventional Construction Bar Chart Progress Schedule

Usually, the time scale is in terms of calendar days from start to completion of the project, although the time may instead be shown in terms of working days (no weekends or construction holidays). The time scale is divided into day, week, or month increments. Milestone dates may also be indicated for significant stages in the construction process.

Horizontal bars or lines are used to indicate the start and completion of each activity. The completed chart consists of staggered bars representing the anticipated steps from start to completion of the project.

In the following introduction the author lists the criteria for the discussion at the very outset. She could follow the list as an organizational device for his entire document, or she could apply the criteria to each network she discusses. Since the structure involves two subjects, one option is to develop them as two sections once she has established the criteria.

> **Network Considerations**
>
> Implementing either a client-server or a peer-to-peer network depends on many conditions. The distinction between server-based networks and peer-to-peer networks is important because each has different capabilities, and each serves many unique functions. The type of network to implement will depend on numerous factors, including the
>
> - Size of the organization—small, medium, or large
> - Level of security required—strict or tolerant
> - Types of business—heavy or light technology demands
> - Level of administration support available—people to support the network
> - Amount of network traffic—depends on many circumstances
> - Needs of network users—depend on the size of the network
> - Network budget—how much an organization is willing to invest in a network.
>
> Consideration of all of these factors can provide an idea of the type of network that is appropriate in a given situation.

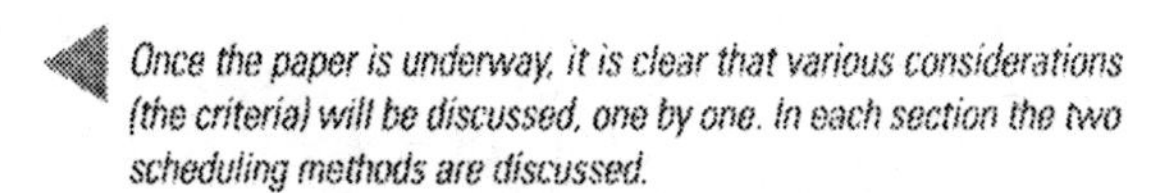

Once the paper is underway, it is clear that various considerations (the criteria) will be discussed, one by one. In each section the two scheduling methods are discussed.

Model 7.A (4)

DEVELOPMENT

Bar chart schedules are relatively easy to develop, which makes them convenient. The charts force project participants to do some preconstruction planning and, later, some monitoring of job progress in relation to the plan. They are widely used and accepted in the construction industry, largely because they are easily understood (see Figure 6, page 7).

Networks are, however, more costly and time-consuming to prepare and update. They are often intimidating in their complexity and just plain hard to read. Many construction superintendents and subcontractors simply cannot, or will not, work with networks. Networks are also less tolerant of changes at the worksite. Small changes may cause radical network revisions.

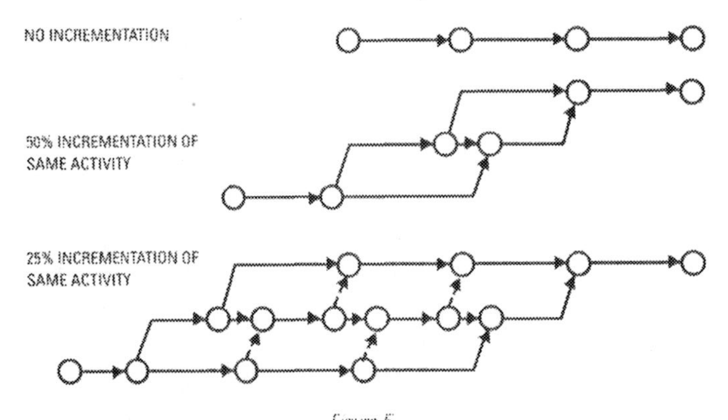

Figure 5.
Activity Incrementing within a Project

If you want to design a structure for a text that is developed by subject or viewpoint, the criteria become subsets of the major points of discussion. The structure of a comparison of continental drift and continental emergence might look something like the following one. A model that is presented by viewpoint begins on p. 265.

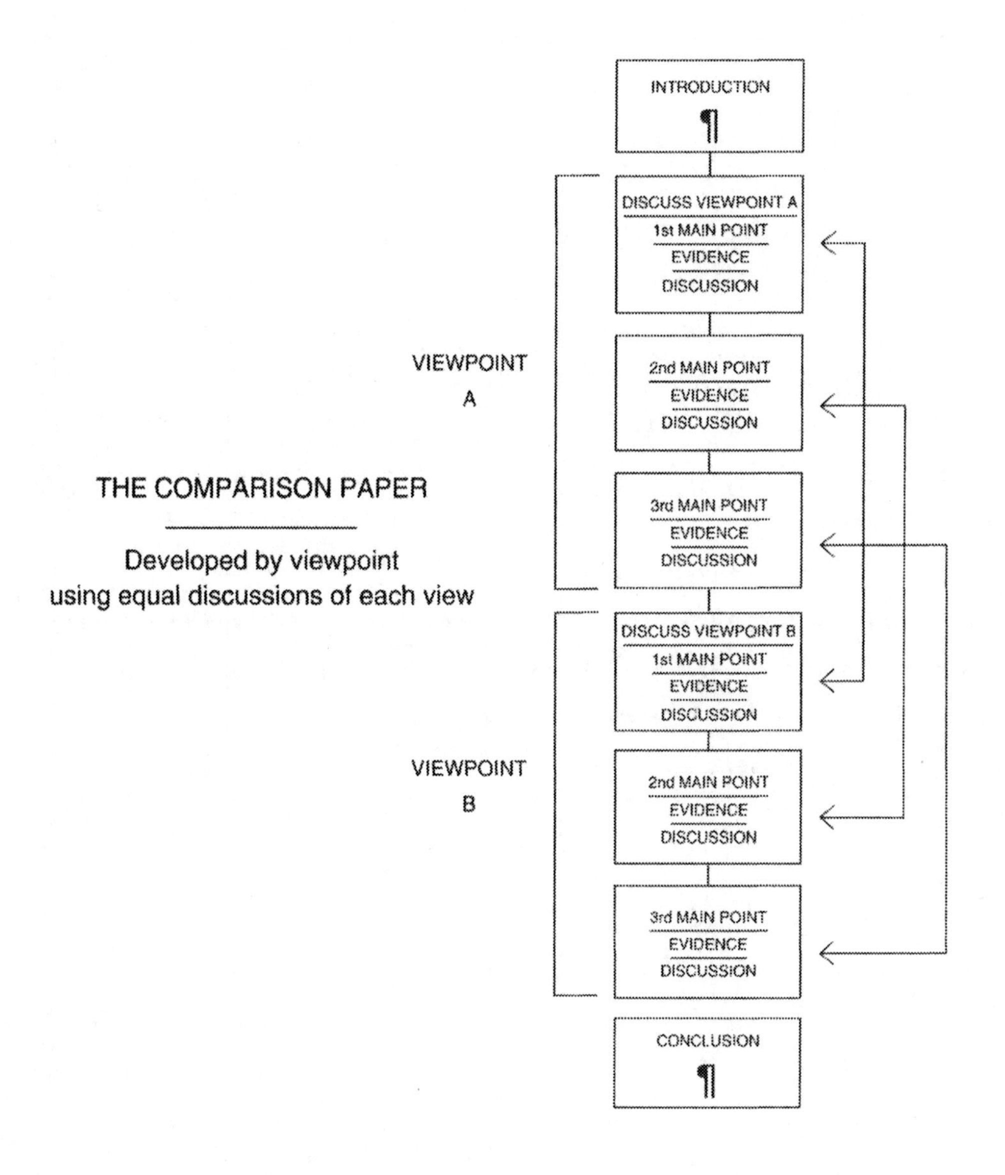

Notice the use of headings to identify the criteria on pages 2 and 3 of the authors' text. Each discussion begins with bar charts and moves to networks. Consistency is important.

Model 7.A (5)

4

APPLICATIONS

Bar chart schedules are typically used for smaller, less complex, and shorter-duration projects than are network schedules. Time, although important, may be somewhat less critical to the success of projects for which bar charts are utilized. They are also best suited for projects requiring minimal documentation, which often means private rather than public work. Bar charts are typically used for housing and small commercial projects. They are often used for government projects under a $1 million construction cost.

In contrast, network schedules are typically used for larger, more complex, and longer-duration projects than are bar chart schedules. Time is usually critically important to the success of projects for which networks are utilized. Timely completion of projects, such as research centers or dams, prevents costly or catastrophic situations.

Networks are best suited for projects requiring thorough documentation in minute detail, which often means public work. Networks are typically used for hospitals, utility plants, and municipal facilities. However, they are often found to be too cumbersome and costly for projects under a $1 million construction cost.

A commonly seen alternative is a document structured by issue or criteria. The project is symmetrical, as shown in the following figure. Note that this is an architectural "model" and not a reality. The document may take thirty pages to develop, but the basic design will seldom change. The issues organize the model. For example, the heading at the top of Model 7.A(5)—APPLICATIONS—is the issue discussed. The text then moves on to another issue. The issues are the criteria. Here is a structural diagram for a document developed by criteria:

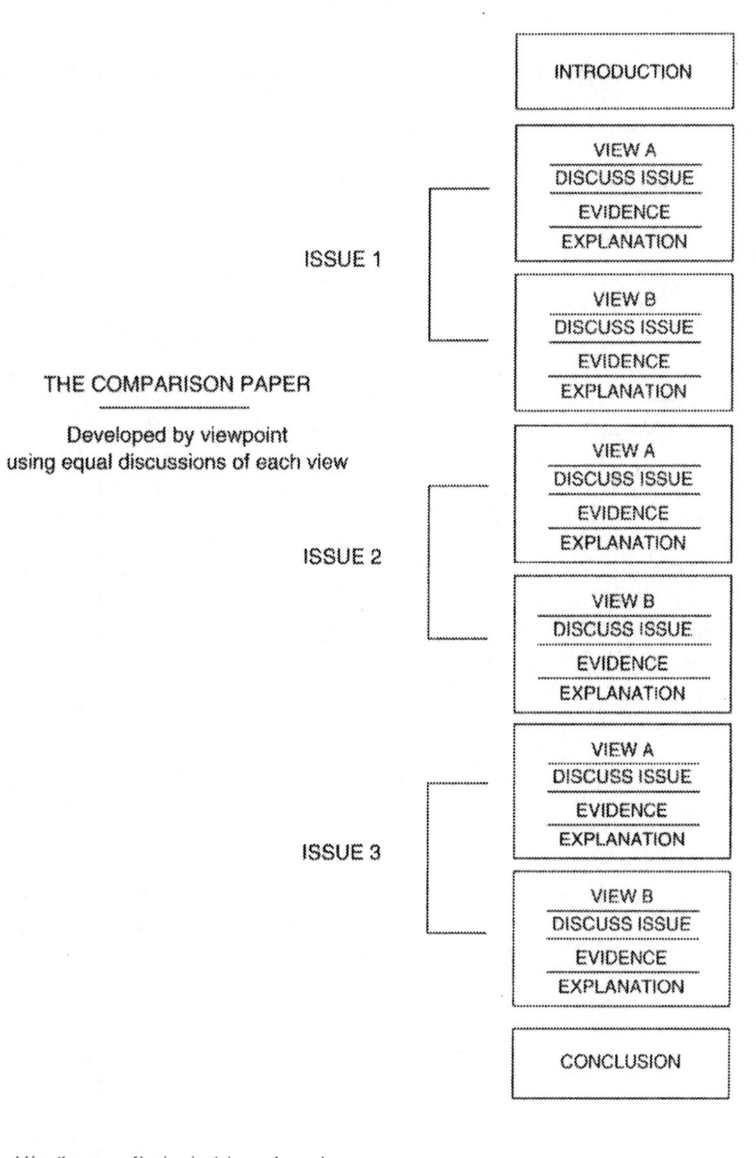

Perhaps ideally there would be diagrams of both schedules under each point of discussion. Realistically the author uses one for each point when needed, and none at all for the general comments stated here.

Tables can play an important role in presentations that develop comparisons. Chapter 6 explored the use of tables for product evaluations and noted that tables are used extensively in such analyses. In one sense, a comparison of two is easier to structure, at least in terms of tables. Because the text is academic, however, the discussion may be quite unlike the product evaluation document. Notice, for example, the following table, in which there are only two columns, but you can sense that the discussion is complicated. There certainly is not a "best buy" approach to this comparison.

Table 2. Major differences between animal-like cells and plantlike cells

Animal-Like Cells	***Plantlike Cells***
Never have a cell wall These cells are *heterotrophic*, that is, they eat other cells or dead and decaying matter.	Always have a cell wall These cells are *autotrophic*, meaning they make their own food. Fungi are the exception in that their cells are plantlike but *heterotrophic*.
Do not contain plastids—these cells do not photosynthesize	Have plastids to photosynthesize or to store nutrition for the plant
Do not contain a central vacuole	Have a central vacuole
Contain centrioles as part of the cytoskeleton	Have no centrioles in the cytoskeleton

This table functions as a summary of the differences between animal and plant cells. It added clarity to a running-text project that was composed in the manner of a term paper. (See Appendix B for locating the table templates on your computer.)

Models 7.A and 7.B

Comments

As noted earlier, two samples are provided for each of the academic project prototypes.

You will notice that all the double-viewpoint models in this chapter are similar in architecture. The first two models clearly demonstrate two popular forms of symmetry.

The critical path project (Sample 7.A) is developed by criteria—a point-for-point discussion of issues.

In the following paper, concerning drafting procedures (Sample 7.B), the project is organized by looking at each subject in turn.

Each of these documents is a comparison of two subjects, but there is no upper limit to the number of subjects. You will notice how balanced the discussions are even though the projects are highly abbreviated.

Model 7.B (1)

Should Today's Draftspersons be Trained in Manual Drafting?
The Practical Limits of CAD Drafting and Manual Drafting

Susan M.

The use of computers is affecting every form of engineering, including architectural engineering. Changes in technological advancements created an enormous impact in the transition from manual labor to the use of more sophisticated machines and computers. The concept of design and drawing documentation has been greatly affected by these changes. This revolution is largely the product of the technological breakthrough created by CAD (Computer-Aided Design/Drafting).

1

The CAD* system is a computer adaptation of the skills of a manual drafter. It is a tool that helps the drafter or designer be more efficient and productive. Although computer-aided drafting is becoming commonplace, and in some cases necessary, in engineering and architectural firms, manual drafting is still valuable to all drafters regardless of their field of interest. In structural engineering and architectural environments, drafting by hand is still relevant in preliminary planning, site developments, sketches, and other situations where the use of CAD is not particularly practical.

2

Manual drafting and design used to be the sole aid in transmitting graphic ideas and concepts in the technical drawing business. Because of higher demands, expensive production cost, and tight deadlines, design professionals sought better ways of producing drawings. Then, specialized drafting equipment and tools emerged to help the designers and draftspersons be more productive and efficient. Unfortunately, these tools made the drawings appear "manufactured" in the sense that they were consistent throughout.

The advantages of the CAD system are quite obvious to any architectural firm. The CAD system allows a drafter to retrieve a detail from the memory bank, display it on the monitor, and make changes for any specific job. The system provides tremendous resources in cost savings and in the production of quality drawings. A big advantage of using computer drafting is speed. Many architectural and engineering firms that use CAD claim the result has been a great increase in production and substantial cost savings.

**CAD refers to computer drafting not to a brand name*

A Basic Comparison Project

The conflicts between the drawing table and the computer screen provide this author with a topic that is more complex than a single-viewpoint project. There are two topics under discussion. Her strategy is to discuss each topic by turn. The project is conveniently developed in halves. A thorough discussion of the benefits of CAD is followed by a discussion of the highlights of manual drafting. This strategy is easy to organize and basically follows the diagram seen here. You will notice that the text could not demonstrate important points without graphic elements to demonstrate her arguments in support of both CAD drafting and manual drafting.

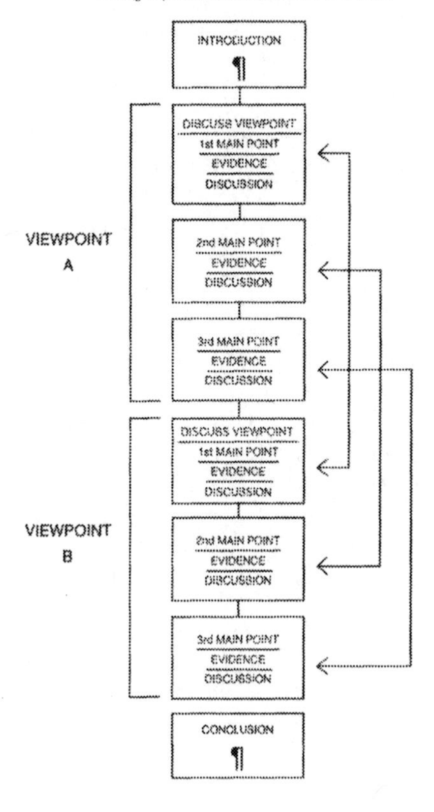

1

The introduction for this project is designed to alert the reader to the current status of the quickly changing drafting industry. Because of the industry shake-up that resulted when CAD was introduced, thousands of drafters had to retool and learn computer skills. (Notice the parenthetical explanation of the acronym in the last sentence.)

2

The text immediately swings into a description of the conflict. The second paragraph opens with reference to CAD, and the third paragraph opens with reference to manual drafting. The two paragraphs set the stage for the discussion by highlighting key issues of concern, such as cost and speed.

Model 7.B (2)

-2-

Computers can also speed up calculations and provide accurate computations for estimates and formula requirements. This makes is easier to find any discrepancies or computation errors that conflict with building and local codes. As we noted, the quality of the originals can be maintained and produced in volume and to specification. Modifications to drawings are easily and quickly generated. The drawings are simply recomposed on the computer when revisions are necessary. Changes will not erase line art, nor will they affect the quality of the drawings produced. There is consistency and uniformity in the drawing and the lettering appearance of the end product. Figure 1 shows a detail view of post-tensioning at slab openings. Notice that the lines are aligned properly, and variations of line weight are apparent.

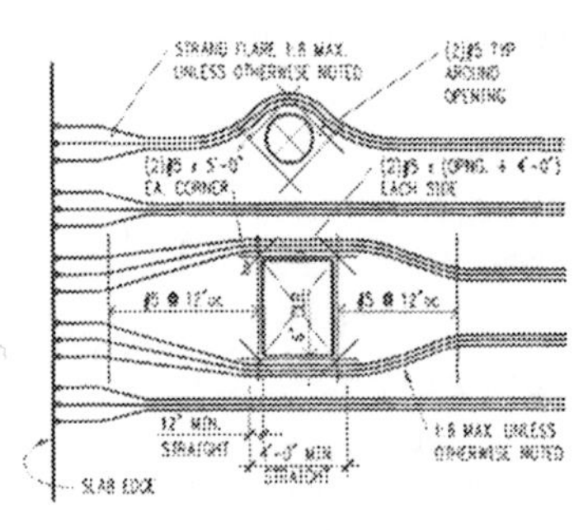

Figure 1
Post-tensioning at Slab Openings (CAD)

Modifications to CAD drawings also take much less time. The duplication of the drawings is as simple as making copies of the originals. Unique to the CAD process of duplication is that many "originals" can be plotted with identical images on all the drawings no matter how many copies are produced.

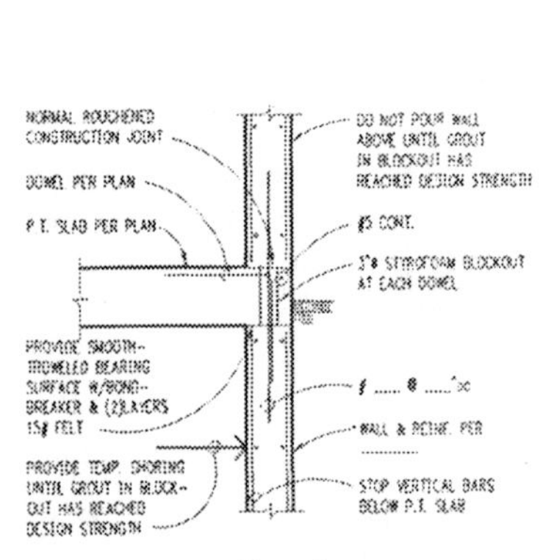

Figure 2
PT Slab Release Joint (CAD)

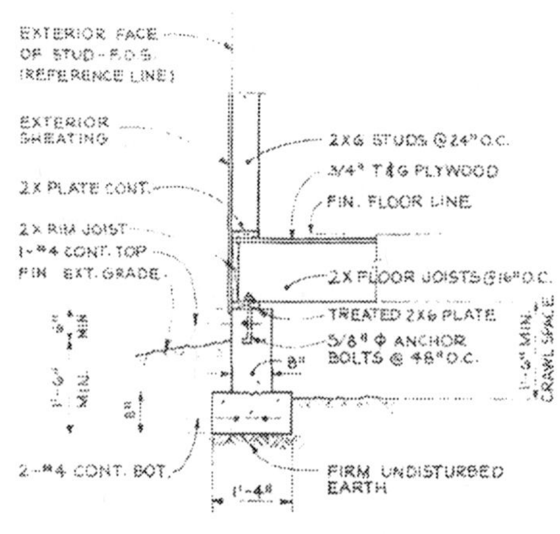

Figure 3
Foundation Section (Manual)

1

The edge that CAD brings to a firm is actually not a simple dollars-and-cents matter, and the text must now explore the issues involved. CAD is introduced in the fourth paragraph at the bottom of the first page. A substantial discussion follows, involving a number of paragraphs.

2

Because this project concerns visual materials, the author has to illustrate the text. The graphics appear as they are needed throughout the discussion.

3

Because the graphics function as samples in the discussion, they are introduced when they are used as evidence. The first illustration demonstrates the precision of CAD.

4

Notice that the side-by-side presentation of CAD and manual drafting is used to demonstrate a key point in the discussion. Since the project is fundamentally a comparison, the side-by-side graphics are an important point of interest.

Model 7.B (3)

There is another side to this story. Figures 2 and 3 show two similar drawings, one done on CAD and one by hand. More company time was spent drafting this section in CAD than manually. Software manufacturers are trying to make the CAD system more user friendly, especially to first-time users, but for the near future, the race to finish drawings by computer and by hand will be a toss up.

1 2 3

In this era of computer-aided design and drafting, manual drafting is becoming less and less important when applying for a job. Most job opportunities require proficiency in CAD. Nonetheless, there are still many job opportunities that call for drafters who have good skills in manual drafting. It is well known that drafters who are accomplished in manual drafting follow a faster, steeper learning curve when using CAD than those who started using the CAD system without any knowledge of manual drafting.

4

Hand-generated drawings are simple and practical. They allow the drafter to change pens or pencils or to darken or lighten lines faster than the CAD system can perform such tasks. Sketch work is also very important in design and drafting. An idea to be developed is much more easily handled when first planned with paper and pencil rather than by setting up the computer to do rough sketches of designs. For example, an isometric view of a drawing done in rough hand sketches allows the drafter to produce a clearer view of a detail on a set of drawings. The sketch in figure 4 of a wall support demonstrates that a sketch is a way to expand ideas and to transfer design proposals from one point to another before drawing the final details.

5

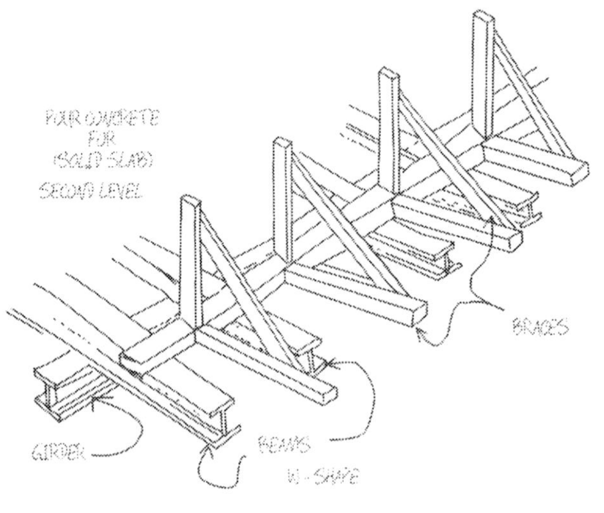

Figure 4
Wall Support Manual)

1 When the text shifts its focus to the second viewpoint, the author opens the discussion with another aspect of the comparison. This tactic is an ideal way to make sure that the readers see the connections and the relevance of the discussion.

2 The transition to manual drafting begins with the side-by-side graphics on the preceding page. The author's position is apparent: she sees the value of both drafting methods. She points out, however, that the new dominance of CAD can conceal the very real need for basic drafting skills.

3 Using essentially the same arguments that defend the use of CAD in the office—speed and cost—the author takes the position that speed and cost are the values of manual drafting in the field. The criteria for the discussion are consistent for the analysis of both techniques.

4 Of course, no one in businesses related to structural drafting will overlook the speed and practicality of sketching. The author is a skilled drafter, and she points out that certain views and certain ideas are quickly rendered with pen or pencil.

5 To support her contention, she introduces another example of graphic evidence. Notice that each of the three manual drawings represents a different purpose for a drawing.

Model 7.B (4)

-4-

The computer simply cannot be used for on-site drawing. Sometimes, when problems occur, engineers or engineering representatives visit sites and make sketches for workers that explain how to solve one problem or another. There is no practical way to take a CAD system to a construction site and to draw the site or the details needed for the office. The sketch in figure 5 is an example of a quickly rendered rough drawing made at a construction site. Notice how practical sketches can be and how the ideas can be transferred from the site to the office.

1

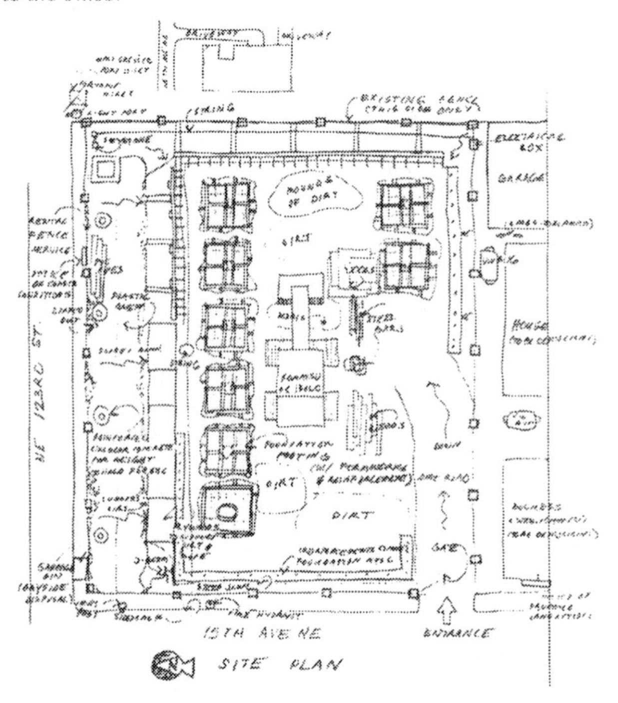

Figure 5.
Construction Site Development

2

3

Hand drawings remain an irreplaceable skill in the field. Besides, pencil or ink sketches also reflect the art of the draftsperson. CAD drawings have no personality and look exactly like any other drawings from a CAD program even if font files are used.

The marketplace for CAD drafters sends out mixed signals. Positions at three different skill levels are available: the manual drafter, the CAD drafter, and the multipurpose drafter. Given the advantages of CAD, the manual drafter with no computer skills has a limited market. On the other hand, the CAD drafter who lacks experience at the drawing board will never be the match of the drafters who know pens and pencils as well as computers and software.

4

5

1 A laptop CAD system is not a practical reality in the near future. Drawing pencils, on the other hand, are as portable as any tool will ever be. The author argues that in the field, manual drawing is the most practical strategy. She could also have noted that, given the rough and dusty conditions, even the construction office trailer is no place for expensive computers.

2 A large figure dominates the final page. When illustrations are finely detailed, they cannot be scaled below the size of conveniently readable print or visible graphic details. As a result, you will occasionally see figures as large as a full page. In the original text of this project, the site development drawing was, in fact, a full-page graphic.

3 This site plan functions as a status report. It was sketched on location, and there was no reason to waste valuable company time to render the drawing on a CAD system. This graphic is excellent evidence of the continued value of manual drafting.

4 The project ends on a practical note. The introduction explained the market trend. Because the body of the text explains the benefits of these drawing methods in terms of skills and applications, the author sums up by looking at the other side of the marketplace: the job market.

5 Is the project evaluative? Yes, the document assumes a very specific position in the final paragraph. Since both CAD and manual drafting are valuable to an architectural firm, Susan does not pick one. In her judgment, the ideal is to develop *both* skills so that a drafter can meet the various demands of work in the field and in the office.

Comparisons

Project Game Plan

- Approach the comparison document as an issue of geometric balance.
- Decide on the number of comparables. Two is the easiest number to control.
- Determine important points of discussion for each item in the comparison.
- Seek points of interest that are shared by the subjects under comparison.
- Allow for unique features that lack points of comparison.
- Prioritize the points of comparison. They become the criteria for your analysis.
- Construct the text by one of two methods:
  - ✔ Develop the text by discussing, in turn, each product, theory, model, or other subject.
  - ✔ As an alternative, discuss each criterion with regard to the subjects that are being compared.

Neutrality and Decision Making

During an election season you will often hear politicians complain about the "equal time" issue of media exposure. It is a popular notion in politics to assume that, all things being equal, fairness means that politician Smith should get the same media time as politician Jones. But all things are not equal or fair. The two comparison models I have been explaining provide a very clear balance between two agents, whether products or policies or theories. Although you could use these highly balanced models to make evaluations or to offer recommendations, the models are better for a neutral perspective that requires only the facts and not your opinion. Investigators are often called on for research and findings, but they may not be asked for proposals.

However, it may be your intention to tilt a discussion in favor of a specific position. If so, you need to use an unbalanced design for the project; you are not under the constraints of an equal-time policy. You need only to explain the opposition's position or, let's say, identify a product's standards, before you take the opposite view for the remainder of the paper. You can take the lion's share of space for your preference regarding a viewpoint or product. Always state the opposite position first so the "first" viewpoint is clear to the reader, but you can be brief.

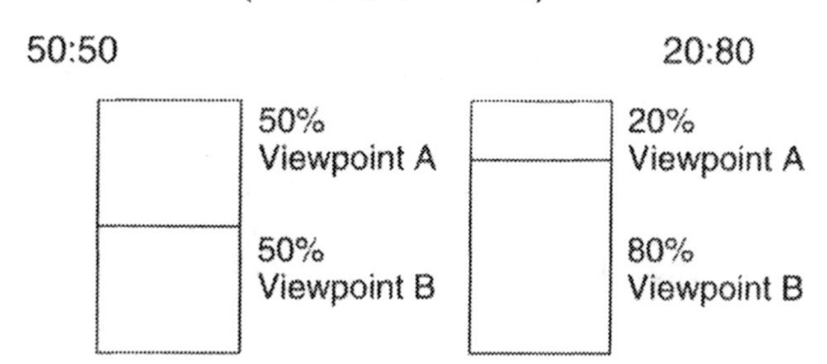

The ratios in the preceding figure indicate likely proportions for the logic structure for each mode. The diagram on page 278 shows the structural concept of the 20:80 model. It is the model of choice if you are a decision maker and you plan to emphasize the better of two points of discussion by giving the superior perspective much more document time.

As the diagram illustrates, the organization of the paper is indicated at the outset in a brief discussion of the viewpoint that is of less concern. The position is developed briefly but the brevity is relative to the length of a project. A five-page document might devote a page to the initial viewpoint, but an extended document might find a five-page presentation to be an appropriate length for the initial position. When the author moves on and begins to develop his or her position at length, the organizational pattern is usually based on the structure of the opening discussion.

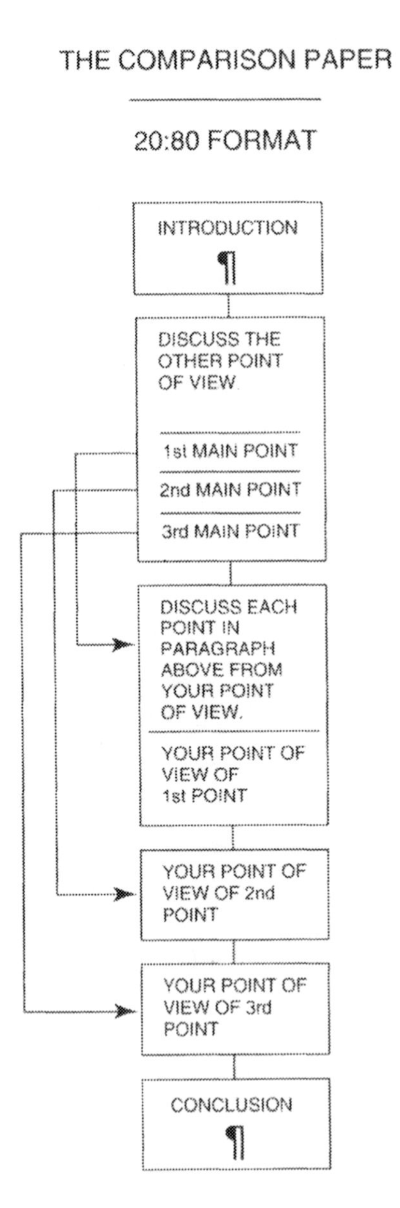

Suppose you are put in charge of a properties development committee where you work. You have to decide whether the business should move to a new location. Currently, you are located near retail markets. However, a proposed location that is near your major wholesale markets is the front-runner as you see it, and the committee wants a recommendation in your report. Most of the document will explain the positive move to a new location. Open the document with a brief discussion of staying at the current address (status quo). Discuss moving costs, the conveniences of locations, trade and commerce, and other issues that the committee identified. Then provide far more space to explaining the superiority of the proposed relocation.

Watch for symmetry problems in using any double-viewpoint analysis. Obviously, there is a complex issue in this relocation scenario because of the existing location. In other words, it is in place; there is no moving cost. *Moving* is the issue, and the issue is lopsided. Do not omit important elements that are lopsided. An ideal balance is not always attainable.

To see the 20:80 technique in practical applications, we could examine comparison models, problem and solution models, or risk analysis. The next discussion concerns problem-and-solution analyses, and you will see models that are handled in the 20:80 style so that you can see the practical value of the design. As long as the project is divisible by two (or more), the decision to allocate more or less attention to any particular areas of concern is up to the author's judgment. Fundamentally, selecting your focus of attention is a type of magnification that allows you to dedicate attention to selective points of interest.

Sample 7.B

Diana E.

Technical Writing

2-21-200X

Outline for Project 2: A Problem-Solution Analysis

Audience: This document is intended to be used by a nonspecialist reader who is interested in the health issues related to diagnostic X-ray and the way in which these difficulties have been overcome through the use of diagnostic ultrasound.

Overcoming the Limitations of Diagnostic X-Ray

I. Introduction

 A. What is X-ray?

  1. What is it used for?

  2. How is it used?

 B. What is ultrasound?

  1. What is it used for?

  2. How is it used?

II. The problems associated with X-ray applications

 A. Can cause cancer and leukemia if patient is overexposed

 B. Can cause genetic damage when reproductive organs are exposed

 C. Can cause poor health in subsequent years

 D. Can kill living cells in human body

 E. Involve risk to the operator

III. Ultrasound offers solutions to the hazards of X-ray

 A. Noninvasive, with no known overexposure risk

 B. Can identify some internal organs that X-rays cannot and without risk

 C. Can be used to *treat* diseases, i.e. clogged arteries

 D. Not harmful

 E. No risk to operator

 F. Can also detect flow of blood that X-rays cannot detect

IV. Conclusion

 A. Summary of problems

 B. Summary of the solutions provided by ultrasound technology

 C. Tips to prevent unnecessary diagnostic screenings

The Problem-Solution Model

Once you see how easy it is to control these double-viewpoint structures, you can readily adapt them to your purposes. Outright comparisons of the sort you looked at in the discussion of balanced comparisons are "lateral," or side by side, so to speak. The same structure will work for the "if-then" logic of an analysis that is intended to focus on problems and solutions. The difference is simply that there is no longer a side-by-side comparison. Rather, the two "halves" are the problem and the solution, and they are often sequential. One section is presumed to precede the other section in time in a problem-solution analysis (although this is not *always* the case).

The actual layouts of the problem-solution projects are much the same as in the models of analysis of comparisons. Here, the project similarly consists of "two" concerns that must be discussed: the problem and the solution (see Sample 7.B). Let's look at *proportion,* however, as a new concern. A substantial body of literature is heavily focused on either problems or solutions, but not both. Although both the problem and the solution are present, the project, whether a scientific paper or a corporate report, will concern itself primarily with one. Why? The answer is probably quite simple. You have heard jokes about solving nonexistent problems, usually poked at bureaucrats, but not at scientists or engineers. Problems precede the solutions—or should.

Entire bodies of literature are directed at problems per se. First, the parameters of a problem must be understood. But these parameters can be judgmental or bureaucratic or otherwise political at times, which will distort the solution. Many times engineers and scientists have been the center of considerable attention, and the issues have clearly been problem-and-solution dilemmas.

The OPEC crisis of the seventies resulted in a clamor for energy analysis. Detroit took the "problem" to be a gasoline issue, and the government, under persuasion, joined the auto industry in a very skewed view of the energy crisis. If investment and research did not concern oil or automotive engineering, it did not count. Never mind that the sun shines every day. Never mind that the wind blows and the earth bubbles up consumable heat. Never mind that there are renewable oils. Never mind that domestic population growth is a correlative of consumption. Only one issue was allowed in the parameter as a problem: a barrel of petroleum. In particular, an empty one.

The outline concerning special ultrasound applications begins with definitions. The body of the outline concerns health hazards associated with X-rays and the low-risk solution of using ultrasound.

Model 7.C (1)

STORM RUNOFF MANAGEMENT

For the Concerned Citizen

A Descriptive Analysis of Growing Problems Connected with Contaminated Runoff

Joseph M.

May 1, 200X

Instructor: David R.

EGR 231

The AIDs epidemic also became quite political, partly because of the necessary order of events in scientific investigation. First, the problem of the autoimmunity had to be studied. Medical professionals did indeed throw every available drug in their arsenal into the ring, but to no avail. Once it became clear that controls were going to depend on new compounds, the research parameters were partly defined. But, *first,* the problem—the virus—had to be understood. For ten years or more the focus concerned the *problem,* because that had to come first. Regrettably, as in the days of polio, or the earlier days of smallpox, we endure tremendous calamity until we *understand* the problem.

On the other hand, if you are working at a corporation that suddenly begins experiencing the "sick structure" syndrome, there may be little need to dwell on the problem. Everyone is having headaches. The supervisor on the first floor observes that her staff had them first, back in July. By September the sick leave statistics are doubling, and the CEO is troubled by the fact that it all seems to be headache-related. The environment has experienced two major changes that were encouraged by a peppy new vice-president. All the fluorescent ceiling lighting on all six floors was upgraded to an energy-saving type in June, and the first floor lobby was carpeted—in July. Everybody assumes that either the mastic or the lighting is the problem. The mastic odor is circulating throughout the building because of the air intakes of the HVAC system, and the lumen levels of the lighting are much too high. If you have to write up recommendations to the CEO, you do not need much of a "problem analysis" at this point. The problem was even reported in the local newspaper. The document will be primarily a discussion of *solutions,* which will include carpeting changes, balanced lighting, and data to show that costs have now doubled (so that the CEO fires the new VP).

It is clear from the outset that this author will deal with problems. The expression "growing problems" is a particularly interesting trigger for a document written in the public interest.

Model 7.C (2)

If one were to ask any group of average citizens what sources of pollution alarmed them the most, many would indicate sources of extreme toxicity such as the military or the chemical industries. Nuclear waste and PCBs for example, attract the most attention due to their dangerously potent nature. Government agencies and private organizations keep a constant vigil over those who handle such dangerous materials. The concern that waterborne pollutants can be an environmental hazard does not raise much alarm.

Now that industrial waste is somewhat in check, and sanitary water systems are being upgraded nationwide, a concern for runoff water management should be seriously considered a priority. According to the EPA/NOAA document *Proposed Development and Approved Guidance—State Coastal Nonpoint Pollution Control Programs,* "nonpoint source pollution has become the largest single source preventing the attainment of water quality standards rationwide" (EPA 1991b). "Nonpoint source pollution" includes septic tank leaching, atmospheric deposition, and agricultural and urban runoff. Atmospheric deposition and septic tank leakage can be direct polluters or borne in runoff water. Agricultural runoff includes livestock waste; erosion from agricultural lands, stream beds and roadways; as well as pesticides, fertilizer and herbicides. Urban runoff includes anything normally deposited on a street or dumped in a storm drain that can be waterborne (see Chart A): sediment, nutrients, metals, bacteria, organic matter, oil, and grease.

Pollutant Contaminations in Urban Runoff			
Contaminant Class	Contaminant	Reported Concentrations	Reference
Nutrients	Nitrogen (mg/l)	5.6–7.1	EPA 1983
	Phosphorus	0.4–05	EPA 1983
Aesthetics	Oil and grease (mg/l)	4.1–15.3	Stenstrom 1984
Suspended solids	TSS (mg/l)	71–1.194	SCCWRP 1990
		141–224	SCCWRP 1990
Metals	Cadmium (μg/l)	3.3–4.2	SCCWRP 1990
	Chromium (μg/l)	11–43	SCCWRP 1990
	Copper (μg/l)	17–138	SCCWRP 1990
	Lead (μg/l)	167–204	EPA 1983
Pathogens	Fecal coliforms (MNP/100 ml)	1,000–21000	EPA 1983

Chart A. Common Pollutants Found in Urban Runoff (Courtesy of National Research Council)

For the problem-solution structure, you could use the same diagram I constructed for the comparison paper, but observe that in comparisons, the subject material is often divided into equal parts for the purpose of comparison. The 50:50 ratio maybe less appropriate for your purposes here. In terms of the project architecture I have been suggesting, the document will have one of the following structures so that it is focused primarily on either problems or solutions.

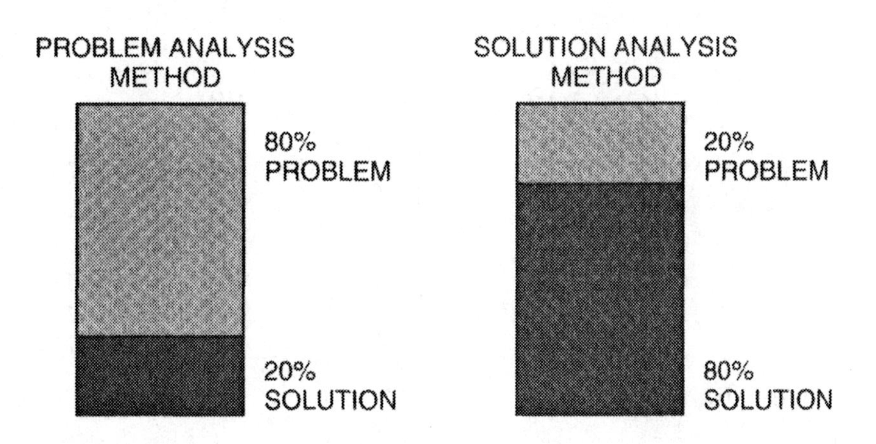

This discussion is interesting in that it highlights the opposition to solutions to the runoff problem. Put in a public context, engineering topics take on additional dimensions—such as opposition—that call for discussion as part of the problem.

Model 7.C (3)

Aside from actual contamination from pollution, there are other liabilities from insufficient storm-water management. The abundance of hard (nonpermeable) surfaces restricts groundwater recharge. Our supply of usable freshwater is dwindling (see Chart B). Water that is sorely needed is deposited where it is needed the least. Even if the runoff water were pure, estuarine waters are overloaded with freshwater. In Puget Sound, plankton, which are essential to the food chain, live in the topmost layer of water.

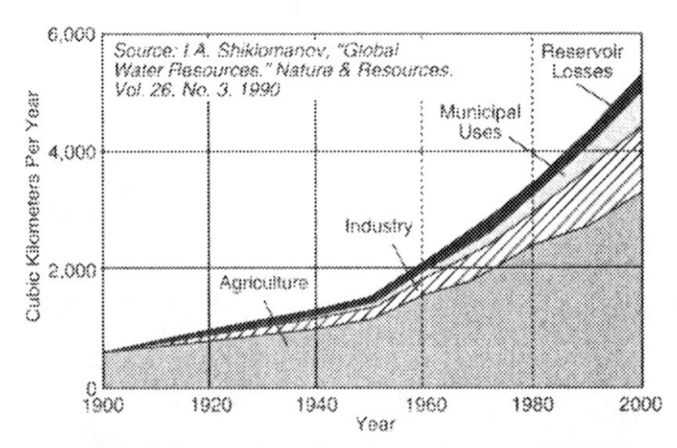

Chart B. Graph Showing Fresh Water Demand Increases Worldwide (Courtesy: I. A Shiklomanov)

A film of freshwater exists because, having a lower density than seawater, freshwater floats. The plankton prospers with a freshwater film of 0.13" or less. In recent years, Puget Sound has averaged 0.25" or more in its freshwater layer, which has caused a depletion of plankton, with obvious effects on the ecosystem. The management of runoff is a simple and affordable option to a wasteful lack of planning; otherwise, the damage to the food chain will continue to go unchecked.

The opposition to storm-water management consists of real-estate developers, the timber industry, housing contractors, industrial polluters, and the businesses that support them. In King County, developers are required to provide runoff treatment on new housing developments. These measures add an obvious expense to tract development, in addition to a sacrifice of valuable land needed to accommodate the treatment facility. Developers are also required to deploy erosion control measures during construction. One does not need an MBA to understand the expenses involved: approximately 10–20% or more in excess of other regions without similar requirements.

If your focus is the "problem," the text will either open or close with some discussion of possible future solutions; otherwise, the concern is only problems, problems, and more problems. The structure is easily plotted. How much space should be devoted to the problem or the solution is strictly a quantitative issue that can be represented in the simple terms you see in the following diagram.

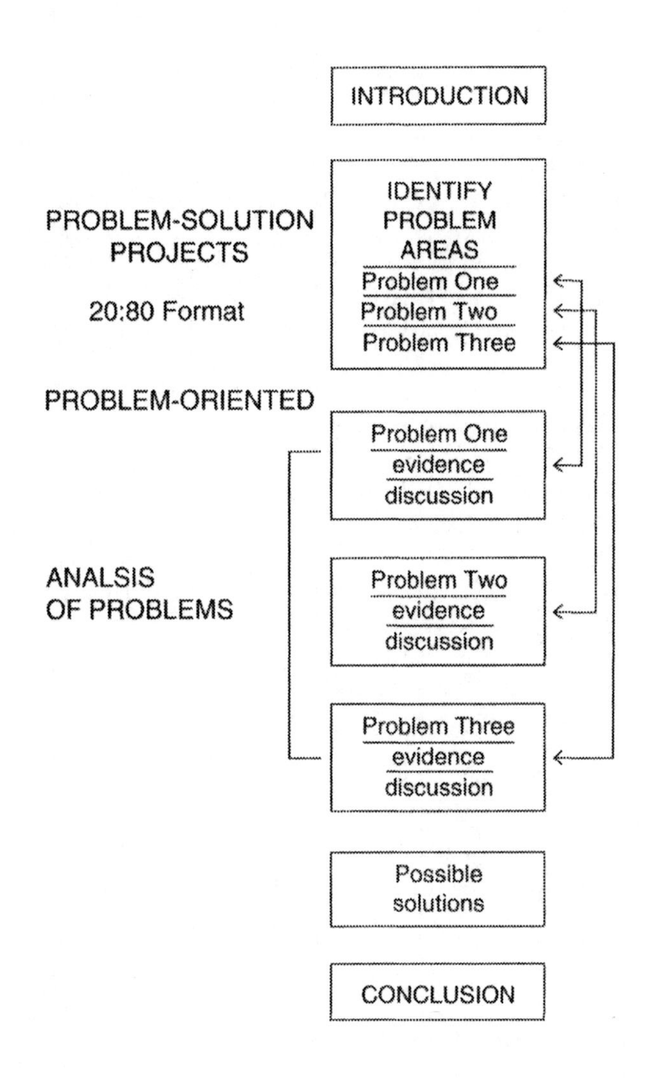

This project concerns rain runoff in new housing areas, and the primary focus is the problem. Charts offer vivid evidence for problem-solution issues because they vividly show changes over a period of time.

Model 7.C (4)

9

There are many solutions to our runoff problems. Fortunately, the technology is available. At new housing developments, storm water is collected in catch basins from streets, parking lots, and residences; the water is then channeled into detention ponds. Here, the sediment settles, and water seeps into the aquifer. Storm systems require an overflow, since mass volumes may flow in a very short time. The overflow may be directed into another pond and/or a bioswale. A bioswale is a flat-bottomed trench planted with native swamp grasses. The grasses absorb nutrients and filter the remaining sediment; at this point, the water should be fit to channel into a natural wetland, where more seepage into the aquifer occurs. Finally, the water may reach a river, lake, or saltwater system. This process, although very effective, is limited to suburban areas where space is available for the ponds and bioswales.

In urban areas the ARS (aquifer recharge system) is the best solution. Proven during use in the Middle East, an ARS filters and returns water to the aquifer by way of an injection well. This system is costly and calls for high maintenance, but it is very sound ecologically. The ARS is the wave of the future. It is hoped that its implementation will result from forethought and not the desperate need due to a crisis. Detention and recharge systems are structural controls; they either reduce the amount of runoff or provide treatment. These controls are usually designed to accommodate the highest storm volume probable over a one-year period. The second major dimension of runoff control involves pollutant controls. Water volume is one dilemma. Water pollution is a broader issue, and a broad-based management policy is critical.

If pollution is reduced at the source, problems with filtration and elimination are lessened considerably. Possible source reduction measures include the following practical measures:

- elimination of illicit connections
- reduction of chemical applications
- mitigation of illegal dumping
- erosion control
- coverage of chemical storage areas
- street sweeping and catch basin cleaning
- prevention and containment of spills.

Adequate solutions are well within our grasp. The facts are evident: if proper action is taken at all levels, we can enjoy an abundance of clean surface- and groundwater. The price is small compared with the dividends.

In the next diagram, the goal is to briefly touch on the problem and devote the text to solutions. The space devoted to the solution is most of the text.

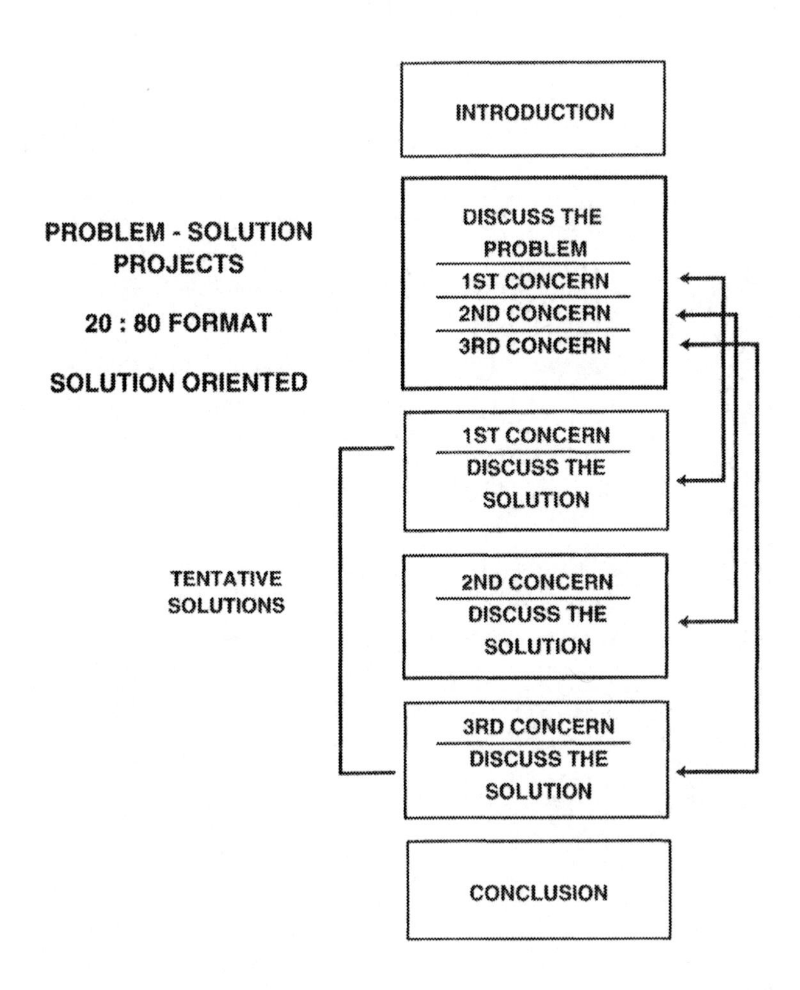

Both of the preceding diagrams are reflected in the two models that accompany this discussion. The storm runoff project is problem oriented. It follows the concept suggested in the first diagram on p. 287. The commuter traffic discussion that follows employs the concept suggested in the second diagram above. The first project concerns problems, and the second project concerns solutions.

The project is extensive and primarily dedicated to problems. Although a page concludes the discussion with solutions, the list in the last paragraph indicates that a great deal more could be discussed.

Sample 7.C

Problem/Solution

Chris W.
EGR 231
12 May 200X

Network Standardization using the OSI Model

Audience: Network professionals and networking systems administrators.
Objective: Discuss the reasons for network standardization and define the layers involved.

I. Introduction

II. History

Briefly review what networks were like before standards were implemented and the **problems** companies had transferring information with each other.

III. Why have Standards?

Discuss reasons why organizations were formed to provide a **solution** of universal protocols in a network.

IV. Defining the Seven Layers of the OSI Model

A. Application

The application layer defines the protocols to be used between the application programs.

B. Presentation

The presentation layer takes care of data type conversion.

C. Session

The session layer establishes and terminates connections and arranges sessions in logical parts.

D. Transport

The transport layer takes care of data transfer, ensuring the integrity of data if desired by the upper layers.

E. Network

The network layer defines how information from the transport layer is sent over networks and how different hosts are addressed.

F. Link

The data link layer defines how the network layer packets are transmitted as bits.

G. Physical

The physical layer defines electrical signaling on the transmission.

V. Conclusion

Models 7.C and 7.D

Comments

Two projects accompany the discussion of problem-solution documents. Both the civil engineering project (Model 7.C) and the transit engineering analysis (Model 7.D) are 20:80 papers.

The author of the storm runoff project surveys an engineering *problem* as his primary concern, and he identifies the solutions briefly at the end of the document. His focus is reader awareness of the pollution potential of urban tract development, which is the problem under discussion.

The following paper briefly discusses an urban traffic problem and promptly moves on to the real issue: the *solution* to the problem. Both of these projects are designed as problem-solving analyses, but they obviously take opposite approaches. The dominant focus of the first concerns problems. The dominant focus of the second concerns solutions.

Contrast the outline on the left and the outline on page 280 and you will see the distinction between a neatly balanced problem-solution discussion (50:50) and one that is designed to explore either the problems or the solutions as a primary focus of attention.

Model 7.D (1)

Commuting: Near-Term Problems for Long-Term Solutions
Hans B.

The Seattle metropolitan area experienced rapid growth in the nineties. It is estimated that 1000 people a week were moving into the city and its suburbs. Growth was incidental in the downtown area but enormous in the outlying districts, where growth was not properly anticipated in the original highway projects. The freeways and arterials were designed and built in the sixties. We have outgrown the design, and the result is large traffic jams. Two popular modifications have been proposed as solutions to the traffic problem: HOV (high-occupancy vehicle) lanes and a light-rail system. Each option has its advantages and disadvantages, and any practical solution will probably consist of a mix of the two alternatives.

The construction of new HOV lanes will be a long and trying process for the commuter. Everybody is aware of traffic delays, lanes that get narrower each year, and detours that go on for months. There is less and less space for the construction road worker, also; in the last year, two workers were killed on the job. Construction will have a high profile and cause much aggravation, but at least tax dollars will be visibly spent. The hardest part about modifying an existing road is that you must keep it functional during the process.

1

Light-rail construction will not clog the freeways with detours, lane narrowing, or rubberneckers. Most construction will be out of sight, and people will hardly be aware that it is happening. On the other hand, neighborhoods will be divided with a long and impassable fence: the new railway. If tunneling or elevation is extensively used to eliminate barriers, then costs will go up ten times, but neighborhoods will not be divided. Some other unique problems will arise from the wide turning radius and grades the light rail will need for commuter trains.

2

In the local area, bridges are the most difficult part of the design problem for any transportation improvements. Simple overpasses are easy to upgrade or construct, but major bridges, such as the ship canal bridge or the floating bridges, require special engineering. Seattle is a challenge for the bridge builder because of the variety of waterways, hills, and large lakes. Earthquakes are a big concern also, so the extensive use of bridges has to be minimized. New technology has made our bridges safer, but we still need to upgrade our older ones as part of an improved transportation package.

3

Problem-and-Solution Analysis 20:80 Ratio/Solution Oriented

The diagram seen here is clearly the opposite of the preceding project. This text concerns commuting, and, assuming that the problems are quite obvious to all concerned, it devotes attention to the solutions. In effect, it is a 20:80 paper that reverses the pattern. It initiates the document with a brief look at today's traffic difficulties. The text then turns to a much larger focus on solutions and the issues involved in improving urban transit. In the case of this text, Hans will be looking at two specific solutions to the problem of traffic flow. Notice that the introduction quickly identifies the prospects for relieving traffic congestion. In some sense, the author is outlining the paper by simply citing HOV lanes and the light-rail as options.

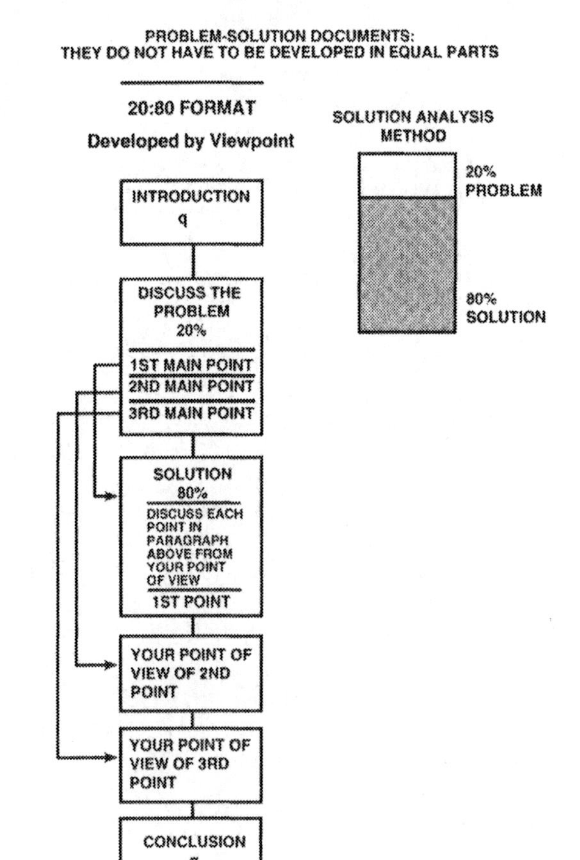

1 HOV lanes are identified as the first solution to traffic congestion. Consistent with the introduction, the first body paragraph moves the issue to the HOV lane concept. Construction feasibility is the first consideration of the paper.

2 The third paragraph shows a shift to light rail. This paragraph concerns the construction problems associated with light rail, so the project is developing by issue. The author will move back and forth from HOV to light rail in a series of short discussions.

3 The last paragraph on the first page is specific to Seattle. Much of the text would apply to any major American city, but because this Northwest city has bodies of water that are difficult to bridge, this unique feature calls for attention.

Model 7.D (2)

- 2 -

Neighborhood impact will be less of a problem for the HOV lanes because very little real estate will be needed for widening the freeway. The present freeways are established, and highway amendments will meet less public resistance than cutting a swath of land for the new rail line. Improvements to the freeway system, such as earthquake proofing, can be done at the same time the new lanes are added. Environmental impact can be minimized with the installation of sound barriers or "lids," such as we see downtown.

1

Environmental impact of another sort is an important consideration when undertaking light-rail construction. Light rail uses electricity, and the trains do not emit any carbon dioxide or smog. Energy efficiency definitely favors light rail and will limit our dependence on oil products. Sound pollution is another concern, and light rail is also quieter than freeways. However, the sentiment "Not in my back yard" is very prevalent, and these objections will be a major force against any rail construction, even though light rail will benefit the community.

With an automobile, the commuter leaves for work when desired and arrives at any desired destination. Convenience and not having to rely on or wait for others are the reasons most commuters drive alone on the freeway. A fundamental part of American culture is the car and the freedom it represents. Overcoming this concept will be the biggest obstacle to the light-rail system or any other mass-transit approach to today's traffic problems.

2

HOV lanes encourage carpooling and reduce the number of cars on the road. Carpooling is more personal than rail travel, since the commuter travels in the company of a coworker, and they obviously have a mutual destination. Most carpool commuters do not drive far out of their way to pick up other members, which matters in terms of convenience. Carpools are somewhat flexible for the individuals, although vacations, sickness, and car troubles can make the carpool slightly less reliable than mass transit.

3

Of course, most commuters are fiercely independent. Light rail will limit their mobility by making them conform to timetables and specific destinations. However, most jobs have set hours and job locations that coincide with these limitations, at least in the downtown area. Downtown buildings have a high density of offices and a huge working population that will benefit from light rail. (Industrial areas will most likely be passed over and they may not benefit from the rail service because of the low job density.)

4

5

1 As you turn to the second page, again note that the author has chosen the technique of using a pair of matched paragraphs to hold the project together. The second focus shifts from feasibility to the issue of public acceptance and public resistance. A paragraph is dedicated to public reaction to each mode of transportation.

2 The third paragraph functions in a way that is similar to the paragraph concerning bridges. The author stops the comparison briefly to look at a specific overall consideration. In this case, he looks at the psychological barriers that challenge highway engineers.

3 Having defined the issue of psychological barriers, the author sets the stage to look at HOV lanes and light rail in terms of the perceptions of the commuter. The solutions may move the cars, but he realizes someone must first be able to move the drivers.

4 Again observe that the discussion evolves in pairs of paragraphs. The two paragraphs address the issue of driver independence and try to suggest solutions to this major point of resistance to reform.

5 Industrial zones pose a special problem that is often overlooked. Because these zones may be found in outlying districts and because the volume may warrant less attention, the author points out that this is a problem that may not be resolved by the solutions under discussion. This kind of openness reflects a strict objectivity that is characteristic of technical documentation.

Model 7.D (3)

- 3 -

Traffic is only one of the problems that results from the existing commuter habits. All commuter cars need to be parked when not in use, which is another problem. The downtown area has a high job density and an equal density of motor vehicles. Parking is limited in the downtown area and costs several hundred dollars a month. If we could invent the Jetsons' solution with a "fold-it-up-in-a-briefcase car," all our problems would be solved. The intent of HOV is to reduce the number of cars that need to be parked in the downtown area. Special areas are set aside for carpool parking, often in choice locations, which helps create an incentive for carpooling.

Light rail moves the parking problem from the city to the suburbs. Large lots have to be constructed next to the stations or else designers risk the wrath of the local residents whose streets become overrun with parked vehicles. These lots, in turn, must be built next to local arterials for easy access to the trains. Mixing personal transportation with mass transit provides the benefit of door-to-door commuting and the opportunity to pick up groceries on the way home. The present Park-and-Ride lots fill daily, which indicates the success of this idea.

Speed of commuting is the goal of any transportation improvement plan. HOV lanes let carpools travel faster than the other traffic. Some of the best advertising for carpools is seeing busses and vans fly past us while we are stopped in traffic. Of course, HOV lanes will eventually become popular enough to slow down their effectiveness. Adding more and more lanes has a limited value because, eventually, we will not be able to get enough people on and off the freeway. The arterials create limitations.

Light rail will never slow down from excessive use. The trains will get longer and more frequent to adjust for the varying volumes of people. After the Los Angeles quake, the Metro Link ridership rose from 12,000 to 30,000 daily riders throughout the city. Regular freeways could never adjust that quickly. The limiting factor in light rail is getting people into and out of the stations in a safe and efficient way.

Not everybody will be satisfied with the decisions made for the future of local transportation. Money will be spent in great quantities for any approach we take to solve this growing traffic problem. The hope is that a compromise will be made that serves the majority of the commuters. In any event, the residents of the greater Seattle metropolitan area will have to adjust their commuting habits in the near future, even with the best of solutions available to us.

1

2

3

4

1

The third page of the text moves on to additional considerations. A major problem is not transit but storage. Where are the cars to be parked? Notice the clever transition sentence that begins, "Traffic is only one of the many problems. . . ."

2

The second paragraph is structured to connect directly to the preceding paragraph. Both solutions have parking volume problems, and the author uses the topic sentence to explain that even the best engineering options simply move around the location of parking lots filled with the same problem: automobiles.

3

Notice that this project assumes that the problem of traffic congestion is a "given." The solutions are, therefore, the focus of the paper. If you sense that the solutions identify a *new* set of problems, you are seeing exactly the point of the analysis that is developed here. Any solution is going to depend on compromise, since the solutions involve vast changes with respect to local economic and social patterns.

4

Speed and adaptability are the final considerations of the discussion. Again note the tight consistency. First the author looks at HOV lanes; then he looks at the light rail system.

5

The conclusion to this project is not of the "summary" variety. I think this brief overview is designed to address the headaches of the changeover phase. If you review the project carefully, you realize that it is a solutions project, but it focuses on the transition difficulties.

6

Because the project is an analysis of *two* alternative *solutions* problems, I have analyzed this project as a comparison paper based on two topics. The paper is, at the same time, a sample of the 20:80 approach to a discussion of solutions. It *also* uses the plus-and-minus strategy we will discuss in the next section. Obviously, the tactics under discussion in Chapter 7 mix well and adapt to an author's specific needs.

Sample 7.D

Ernie J.
Paper #3 Outline

Atrial Fibrillation

Audience: This paper is designed for counselors in the medical profession, particularly those of XYZ Hospital.

Objective: The main focus is to inform counselors of the nature of atrial fibrillation.

I. Atrial fibrillation (AF) affects about 4% of the population over the age of 60 and 10% of the population over the age of 80.
   A. AF has now become one of two "emerging epidemics" in cardiovascular care.
   B. With all the research and the computerized mapping techniques, the process AF takes in the heart has been mapped out, which makes treatments more likely.

II. AF is a disorder of heart rate and rhythm in which the atria are stimulated to contract in a very rapid and or disorganized manner.
   A. Research has shown that there are at least six different locations in the left and right atria where relatively large circular waves can occur.
   B. There are three types of "chronic" AF.
      1. In paroxysmal AF, the episodes are self-terminating and last less than 48 hours.
      2. Persistent AF continues indefinitely, but can be converted to sinus rhythm.
      3. Permanent AF is the most dangerous in that the conversion to sinus rhythm is either impossible or reverts rapidly back to AF.

[Sections III and IV omitted here. DR]

V. A host of treatements are available.
   A. One way of treatment for AF is to use medication to reverse the effects.
   B. Another way to treat AF is cardioversion.
   C. There are a couple of different surgical procedures that can be taken to help the patient with AF.

Problem-Solution Analysis

Project Game Plan

- Structure the paper in two parts.
- Discuss the problem. Subdivide the problem into relevant issues that conform to your needs. Rank the issues and develop them.
- Using the organization of issues established in the problem analysis, develop a second discussion concerning the solution or solutions.
- Develop the two parts—problem and solution—as equal halves if appropriate.
- If appropriate, develop the problem briefly and devote the bulk of the document to the solutions.
- If appropriate, develop the problem at length and abbreviate the discussion of solutions.
- Where multiple solutions exist, make the document either descriptive (neutral) or evaluative (indicates merits of each or indicates the best solution), as appropriate.

The outline on the left is one-half of a problem-solution analysis that placed heavy emphasis on the problem of atrial fibrillation. Two missing sections concerning stroke (part of the problem) were omitted.

Sample 7.E

Nick P.
April 200X

**The Role of Programming Languages
for the Internet and Network Era**

Audience: This paper targets those with some understanding of more traditional computer languages who seek an understanding of the benefits of a network-secure programming language.

Objective: To explain the benefits of the new Java programming languages in the current age of networking and the Internet. The text is organized by environments where the strengths and weaknesses of the languages are apparent.

I. **Introduction**
Introduction to programming languages
- What they are
- What they do
- How they work
- How they play a role in network applications

II. **Environments where traditional languages remain the choice application**

A. The disadvantages of Java
- low compile time
- Does not have a lot of control
- Cannot handle as many applications

B. The benefits of older programming languages
- Fast run time
- Precompiled
- Provide more control and used in many applications

III. **Environments where Java is superior to older languages**

A. The disadvantages of a traditional programming language
- Does not work well with networks
- Must compile to fit many platforms
- Language is large
  - a. 1 bug per 55 lines
  - b. Manual memory management
  - c. Crashing can be a problem

B. The benefits of Java
- Works well with networks
  - a. Compiler smart
  - b. Superior security
  - c. Platform independent
- Easy to program
  - a. Few bugs
  - b. Manages memory
  - c. Small programming dictionary
  - d. Not much crashing

IV. **Conclusion**

The Java language is the language for today's computer systems.

Risk Analysis

If one of these academic models were to serve as a bottom-line analysis, it would likely be a discussion of the advantages and disadvantages of a product or process or strategy of some sort. Think of this type of document as a risk analysis. Documents that analyze risks and rewards are designed to evaluate, of course, so they can be quite decisive—or quite guarded. Here again there is a choice of architecture. If an author has no opinion, or has been asked for none, a document that handles this sort of neutral analysis can be structured in a 50:50 model, where 50% of the text favors the topic of concern, and 50% indicates limitations. On the other hand, this analysis is another one that can be shifted around to take one point of view primarily, even though both must be represented in order to properly describe the topic under consideration. For example, what are the benefits and risks of medical waste incineration? What are the benefits and risks of a manned flight to Mars? What are the benefits and risks of moving your lab to the sixth floor? These topics could be handled with either a neutral or a very persuasive perspective.

Given the subject matter, the question is whether or not to be judgmental, since the 50:50 model is the most likely to be neutral. The 20:80 model may be just as useful if there will be a hands-down decision. We have already analyzed the diagrams for two-sided analyses. There is a specific pattern of discussion, however, if you take a position in a pros-and-cons discussion. Always present the position you are *not* supporting, but do so early in the paper. Since your perspective as the author may be either for or against a position or product in a pros-and-cons discussion, this means that your paper can go in one of two directions: either you touch on the advantages and extensively discuss the disadvantages, or else you briefly survey the disadvantages and then develop extensive support in favor of the advantages. The point is to structure the document so that your view dominates the presentation after a brief introduction of the other viewpoint.

Do not lose sight of fact that your judgment may *not* be called on. The 20:80 model is highly evaluative. If your task is simply to report in a neutral fashion, you would always use the 50:50 style. Then it hardly matters whether the advantages or the disadvantages are presented first. For example, an employee of a corporation explained that her position at work was described as a "document analyst," which turned out to be a writing job that was purely reportage. She read all day and wrote up reports and never had permission to offer an opinion. Her task was to sift through vast numbers of engineering documents and present the pros and cons of each to an engineering team. Nothing more was asked of her.

An analysis of the pros and cons of a topic is often doubled so that the risk analysis becomes a comparison as well. Here the values and limitations of old and new program languages are the two issues of concern.

Model 7.E (1)

COMPUTER MEMORY:

THE ADVANTAGES AND DISADVANTAGES OF STATIC MEMORIES

Audience: Second-year students in electronics who want to know more about computer memories and the utility and limitations of static and dynamic RAMs.

Objective: To explain the nature of computer memory by way of a comparison.

Ansar K.

EGR 231 10–11 a.m.

May 10, 200X

Instructor David R.

In terms of practical applications the advantages-disadvantages model is often coupled with the comparison model, as you will note in the outline proposal, Sample 7.E. Two products or theories of applications may be the focus, but the criteria for analysis are essentially two: the strengths and weaknesses of whatever is being compared. In other words, I noted that an author must establish criteria to judge matters that are intended to be contrasted. The criteria can be many or few, but when they are few, they often are the two basic considerations: favorable and unfavorable characteristics. The most fundamental comparison is to simply ask the question, What are the advantages or disadvantages of this product or these products? Notice the following table concerning a software application.

Of course, a subject such as irradiated foods or pesticides is more recognizably an analysis of risk than, for example the purchase of a software application. The risk can be modest: a minor expense or inadequate performance. Nonetheless, the concept of the pros and cons method of analysis is a practical tool because it becomes a value analysis. In addition the analysis determines superior choice, which is the lesser risk if there is more than one subject under discussion.

Table 1. XYZ Software

THE PLUS SIDE	***THE MINUS SIDE***
English sentences are easy to write.	Does not help with what data is needed Data must be determined by user.
Can actually create tables versus merely giving instructions on how to do so	In order to query, data must first be entered, with no input masks, to make entry easier.
Supports multiple DOS and Windows databases	Cannot launch database directly from program
Makes understanding one-to-many relationships easy	User-friendly RDBMS might be almost as fast.
Inexpensive	FlowMap more limited than Access' Diagram
Five methods of validation good for a range of learning styles	Dictionary does not know nouns
Good for beginners or those having difficulty with relationships	FlowMap limited to three tables visible at a time Editing must be done in DBMS anyway, so some effort is wasted. Limited usefulness for intermediate or advanced user

Risk analysis is often apparent in document titles that declare the intention to analyze some product, process, or object with respect to two fundamental values: utility and limitations.

Model 7.E (2)

There are basically two types of memories: ROM (read-only memory) and RAM (random-access memory.) For RAM, in which the data is permanently stored, the data can be altered by special methods, but there is no "write operation". In ROM, data are permanently stored and can be read from the memory at any time, but there is no write operation, because specified data are either manufactured into the device or programmed into the device by the user. The data cannot be altered. ROM is nonvolatile memory; that is, the stored data are retained when power is removed from the device. Examples of ROM applications are look-up tables, conversions, and programmed instructions.

RAM is a read/write random-access memory. In RAM, if the power is lost, the data is lost. Figure 1 shows the concepts of read and write operations. The term *random access* means that all locations in the memory are equally accessible and they do not require sequential access. RAM semiconductors are manufactured with bipolar or MOS technologies. The bipolar section has only SRAM (static random access memory). but the MOS section has both SRAM and DRAM (dynamic random access memory.) The following discussion will compare static and dynamic memories and explain the advantages and disadvantages of each.

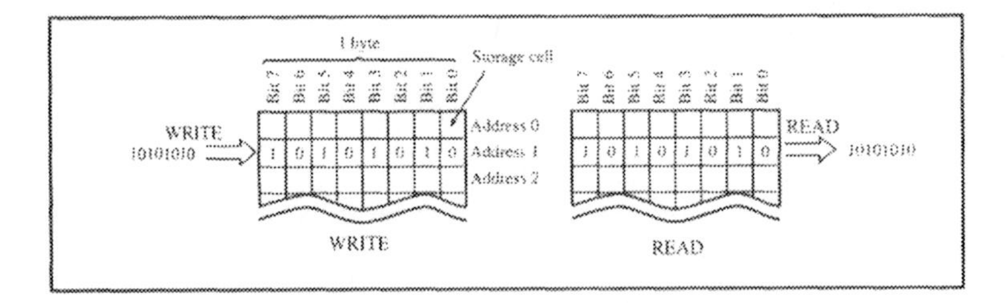

Figure 1. Concept of read, write, and address for generalized memory

It is likely that a table analyzing the "two issues" under discussion will accompany any of the three double-viewpoint documents reviewed in this chapter. The comparison paper can be accompanied by a table based on the criteria developed for the analysis, and the problem-solution model calls for a table similar to a troubleshooting chart. Risk analysis can also be plotted as shown in the preceding table.

Risk can be interpreted in the simplest of comparison terms: pros and cons. (The words *pro* and *con* are awkward and nonspecific. Be specific and refer to strengths, advantages, or other precise terms that fit your discussion.) There are actually many ways in which risk analysis projects can be designed. The following table is from a risk analysis also, and it appears straightforward.

Table 4. Breakdown of possible solutions and their advantages and disadvantages.

Possible Solution	*Advantage*	*Disadvantage*
Defragment System Configuration & TSRs No cost	• No cost • Can speed up hard drive access time	• Limited effectiveness • May take several hours to do
Disable TSRs Configuration & TSRs No cost	• No cost • Can custom tailor to application's needs	• Limited effectiveness • Must be done each time computer boots
Install SIMMs RAM $25–$65 per megabyte	• Possibility of greatly increased performance regardless of CPU type (can prevent CPU stalling) • Relatively inexpensive • Depending on its speed, can be used in subsequent computers	• Does not affect calculations or actual processing speeds
Upgrade CPU CPU $50–$1500	• Increases performance • Cheaper than motherboard or entire system	• Some vendor's machines not chip upgradable (generic clones most likely no problem) • Speed can be limited by existing system's slower components • In other systems, it may not be effective to replace the CPU if other components are due to wear out soon • May have to upgrade BIOS
Upgrade Motherboard CPU $100–$2000	• Increases performance • Comes with upgraded BIOS • Option of local bus	• More expensive than CPU alone • More care needed to choose one
Buy New System CPU $600–$5000*	• Increases performance • Option of local bus • Already configured and set up	• Most costly of all upgrade options • Possibility of vendors cutting corners in quality which reduces performance, in order to be less expensive

These solutions are based from the simplest, least effective to the most extensive and effective. Defragmenting system and Disabling TSRs do not physically alter your system. As a result, they are very limited. Prices were current at time of writing.

The project concerning computer memory is designed as a four-part document. It concerns two subjects—Static RAM and dynamic RAM—from two points of analysis: advantages and disadvantages.

Model 7.E (3)

4

RAM and ROM are addressed in basically the same way. The main difference between the structures of RAMs and ROMs is that RAMs have inputs and read/write control. Figure 3 shows a basic structure of a RAM: a one megabyte with 256* 4 bit organization is used. In read mode (R/W HIGH), four data bits from the selected address appear on the data outputs when the chip select (CS) is LOW. In the write mode (R/W LOW), the four data bits are applied to the data inputs that are stored at the selected address.

During operation five of the eight address lines (A0 through A4) of static RAM are decoded by the low decoder to select one of the 32 rows. Three of the eight address lines (A5 through A7) are decoded by the output column decoder. In the read mode, the output buffers are enabled (the input buffers are disabled), and the four data bits from the selected address appear at the outputs.

The following summary indicates the strengths and weaknesses of static RAM. It is clear that static RAM is good only for smaller systems, for systems that are not expandable, and for systems that do not need more storage capacity.

Advantages of static RAM

- Fully programmable
- Both TTL and MOS technologies are used in its construction.
- The data in/out are bidirectional.
- Consumes less power
- Faster speed

Disadvantages of static RAM

- Very small capacity
- Does not have both row and column decoders
- Costs more than dynamic memory

In fact, the sense of risk in the preceding table is ranked by cost and the analysis is a focus on value— gains versus potential problems— where the basic intent is to identify appropriate choices.

Observe that Model 7.E, which was designed by Ansar to discuss RAM applications, functions in much the same way as the preceding table, but dollar value is of little interest.

To conclude our discussion I should mention the problem of asymmetry. The diagrams you have examined in Chapter 7 are probable solutions to probable problems for writers. As such they are models—geometric generalizations. There is a great deal of truth that will never fit these structures. Logic is often difficult to structure. It can be elusive and can challenge our writing ability.

Perhaps the protestations of special interest groups are a typical mismatch. The protection of endangered owls in Washington and Arizona, and many other endangered animals, has resulted in enormous confrontations between progress-and-growth interests and habitat preservation interests. The conflicting parties are not speaking the same language, so there will be arguments that have no direct counterarguments that are related and that speak in the same terms. Growth is motivated by one set of values. Conservation is motivated by a very different set of values. In this case few of the elements of discussion will "match."

Be aware that the first comparison illustration on page 265 (of equal subjects and shared criteria) is idealized, and it is a "programming" necessity you might say. In other words, the geometry *is* likely to exist, and it certainly *is* convenient for symmetrical analysis. Further, the geometry allows us to structure a neat and tidy technical document that reflects our organizational skills. Always be aware, however, of skewing the balances so that the geometry is forced to emerge. We are now told that Freud had just a bit of compulsion along these lines. He skewed his psychiatric case histories to conform to his theoretical needs. I could call this unprofessional, of course, but it is more fun to think of it as simply retentive. Everything had to fit neatly into place.

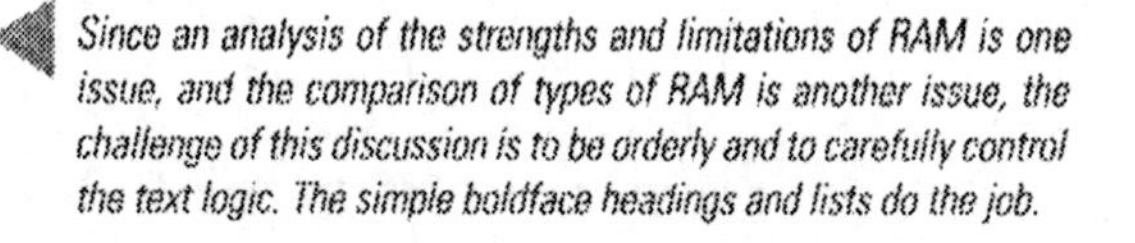

Since an analysis of the strengths and limitations of RAM is one issue, and the comparison of types of RAM is another issue, the challenge of this discussion is to be orderly and to carefully control the text logic. The simple boldface headings and lists do the job.

Model 7.E (4)

7

The dynamic RAM cell stores a bit of information as a charge. Dynamic RAM uses the gate substrate capacitances of the MOS transistor as an elementary memory cell. This results in a much higher density. The MOS transistors act as switches. A low on the R/W line (write mode) enables the tristate input buffer and disables the output buffer. For 1 to be written into the cell, the *D*in line must be HIGH, and the transistor must be turned on by the HIGH on the ROW line. The transistor acts as a closed switch connecting the capacitor to the BIT line. This connection allows the capacitor to charge to the positive voltage, as shown in Figure 5a. When a 0 is to be stored, a LOW is applied to the *D*in line. If the capacitor is storing a 0, it remains uncharged, or if it is storing a 1, it discharges as indicated in Figure 5 b.

Our analysis of dynamic RAM indicates its superiority for better, more reliable, and faster work. Dynamic memory would be the best choice for a computer system.

Advantages of dynamic RAM

- The individual cells are simpler than flip-flops, require less area on the chip, and have lower power consumption.
- Memory cost is much lower than for static RAM.
- A single chip now contains up to 256K memory; however, 16K chips are more widely used.

Disadvantages of dynamic RAM

- Since the charge leaks from the capacitors after every millisecond, the memory cells need to be refreshed because the capacitors cannot hold a charge over a long period of time, and they will lose the stored data.
- The refresh process requires additional memory circuitry and complicates the operation.
- Slower speed.

If you see elements of concern that do not "match" the need for balance, discuss them anyway. The neat geometry is a convenience. On the other hand, the subject matter must, in the end, dictate organization. If the criteria cannot be matched, you should not dismiss the phenomenon. There would be no black holes if scientists did not allow for exceptions to the logic of Einstein's physics. If considerations are one-sided, simply identify their uniqueness and add them to your list of concerns when you develop a comparison structure.

Notice that the pattern of boldface and lists is repeated. This pattern gives the author control of the subject matter and provides the reader with a clear overview of the topics discussed in the paragraphs.

Sample 7.F

Comparison of Zip Drives with CD-RW External Drives

Audience: This article is for people who want to add a larger capacity removable stoarge to their computers.

Objective: To provide a comparison between external Omega Zip Drives and external CD-RW drives using costs, ease of use and performance as criteria.

Advantages:

Zip Drives	CD-RW Drives
Low initial cost per unit	Initial cost is higher for the drive
Lots of Zip Drives in use, especially the 100 MB Drives	Becoming more common but still is not widespread use
Storage life is approximately 10 years	Storage life of disk is considered to be approximately 30 years
Connections: USB Port, Parallel Port, and SCSI Port	Connections: USB Port, Parallel Port, and SCSI Port
Easily installed	Easily installed
	Not affected by magnetic fields
Electro-magnetic.technology	Optical and laser technology
	Fast access
	Can share with newer CD-ROM drives
	Disk is capable of more abuse than the Zip Disk
	No mechanical connections during playing to malfunction

Disadvantages:

Zip Drives	CD-RW Drives
May not be able to share with another computer if it does not have a Zip Drive	Older CD-ROM drives may not be able to read
	Sensitive to light and heat
The gate to the Zip disk as well as the connection to the drive motor could fail	
Cost of new disks from $10-$15* depending on the volume of disks bought.	
Size of larger drive capacity is smaller than capacity of CD-RW. 250MB versus 650MD	

**At the time of this writing.*

Models 7.E and 7.F

Comments

A discussion that concerns advantages and disadvantages can be seen as a comparison. It is a very basic balance in which the measures on the scales of scrutiny include only two conditions: risks and benefits.

If we have two or more comparable items—products, theories, plants, pathogens—and we look at the pros and cons of them, we have a comparison of products measured by another comparison (the two criteria).

In the discussion paper concerning computer memory (Sample 7.E) two types of RAM are analyzed. There are advantages and disadvantages to each. The memory systems are the subject of comparison. The author uses the advantages and disadvantages as very basic criteria.

The following software review (Sample 7.F) takes a similar approach in an analysis of a number of desirable features of design software. In this case, there is only one item under discussion. The document moves from feature to feature in a discussion of the pluses and minuses of the application. The application comes up short in desirable characteristics, and the project is clearly evaluative.

Comparisons are easily constructed in tables where the organization is prehaps more visually apparent than in an outline. This author based his text on the table rather than developing an outline.

Model 7.F (1)

Una Corp's Database Designer Software:

Software Update

Eric

Una Corp of Akron, Ohio, has released Database Designer, a database organization tool that uses English sentences to plan the information structure of a database. The premise behind it is that users can use plain English to create different tables of data and see how they all fit together. Una Corp suggests that the use of this product can increase productivity by as much as 300% because writing English sentences is quicker than designing and redesigning database tables. We are going to look at the demo version of this product from the standpoint of an intermediate Access user as a benchmark. The program's organization, usability, and functionality have advantages for the novice but may not be of any particular value for the designer.

1

The program itself is straightforward. It is DOS-based, which, to Windows DBMS users, lacks aesthetic value and does not allow launching directly into the database. Actions are explained well in the pop-up dialog boxes that appear almost annoyingly after nearly everything in the design stages.

2

Database Designer is organized into three major development segments: sentence writing, sentence examination (by the program), and validation. After validation, a menu allows easy choice of what RDBMS to create. There are no complaints about the program's sequence of design.

The idea that easy-to-use English sentences can run the program is clearly demonstrated. With Database Designer, a simple sentence such as "Each customer is identified by a number, and has a name, address, city, state, zip, phone, fax, and balance" is easily understood to be a table called "Customers" with "name," "address," and "balance" as fields. All levels of user will find it simple to write sentences describing their database structure.

The only catch is that the user has to be able to know and identify parts of speech (at least nouns and verbs). Identification of the words is the trick UCD uses to automatically create a table model. Many important verbs are understood automatically. This should be an easy exercise for most people.

3

Risk Analysis of a Single Topic

This software analysis is based on an advantages-and-disadvantages model. As a method of analysis, this approach is often one of weighing cost values, whereby the investment is weighed by balancing benefits against risks. The outline introduction (note last sentence of the paragraph) illustrates the author's structural approach and indicates a way to go about a pro-and-con analysis. Note that this is *not* a schematic plan as such. Rather, a pro-and-con strategy simply sets up two overriding criteria: thumbs up (A) and thumbs down (B). The questions are straightforward: What are the strengths? What are the weaknesses? In this case, the author is largely critical of the product.

THE COMPARISON PAPER

50:50 FORMAT
Developed by issues using equal discussions of each issue

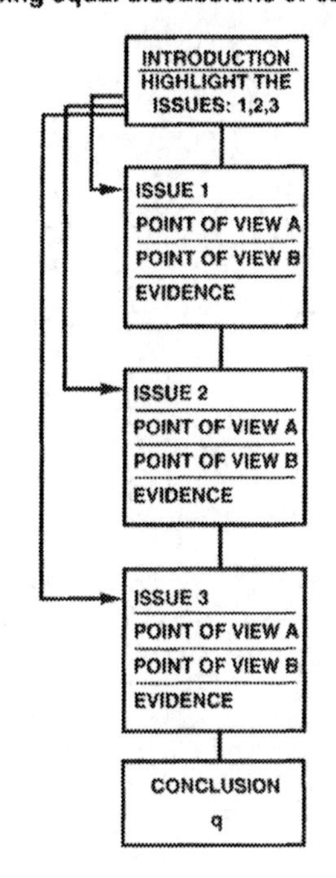

1 The introduction explains that this project will examine a software application. Another popular application will be used as a benchmark, although the project is *not* a comparison project as such. The last sentence of the paragraph is the thesis; the document is clearly evaluative.

2 The next two paragraphs indicate the author's focus on a *single* product, so In one respect this project is unlike the comparison of static RAM and dynamic RAM (Sample 7E). The difference is the number of items under scrutiny. The model remains the same in that both projects are based on the risk analysis concept; both examine the advantages and disadvantages of products.

3 If you look at each paragraph in terms of subject matter, you will note that the project is systematic. The paragraphs move from DOS features to the sentence-based issue of the sentence database, and so on. The pluses and minuses emerge promptly.

Model 7.F (2)

- 2 -

After automatically doing a little work shuffling to more computer-understandable sentence structures, the program gives the user five methods of checking the tables, fields, and joins. This offers a good variety of error checking for a variety of learning styles. It is easy enough to understand (but not too useful in light of the easy-to-use features of Access or the ease of Table Wizard).

1

Error checking is a point of departure for the experienced user. Written validation is good for a verbal summary of each table in the database. Most experienced users already know this information (that is why they set it up that way), but beginners, or those who like to know the big picture, could benefit from it.

2

3

The graphic validation clearly shows the relationships of each individual table and makes it easier to comprehend one-to-one and one-to-many relationships. Of course, Access users who understand how relationships work can get the same information by looking at the join lines in their database diagram.

4

Table graphic generation capabilities is where Una Corp's product is supposed to save time. After entering sentences and changing after validation as necessary, Database Designer can actually create the database .MDB file and its included tables. The problem is that features like input masks and default values must be added by hand anyway in Access, so is time really saved? Most developers end up changing default field size and other properties, so why edit two different times in two different programs?

Function is where the trouble really starts. Since one of the keys to effective database design is careful planning, experienced users already know what data are needed and know how they are interrelated. Database Designer does not cover any principles for effective database design, which is the critical, time-consuming state. Also, since fields usually need tweaking for easier input and validation rules, working with tables in two places can be wasteful.

And the verdict? For Access users, at least experienced ones, Una Corp's Database Designer is not that much of a help. It is fun to use only to see what it does, but that is about it. Access is so user-friendly and powerful that this new DOS-interfaced data organizer really doesn't stand a chance. To be fair, though, Una Corp may be the ticket for those just beginning database design.

5

6

1	On the second page the order discussion continues. As planned, the body of the text began with a discussion of organization and moves on to usability. Nonetheless, the movement is complicated, and you have to follow along carefully as each paragraph shifts your attention to a new consideration.
2	The popular style that Eric uses can be pinpointed in vocabulary choices. The author has a good eye for magazine writing. He refers to the computer as "shuffling" in the top paragraph. In the next paragraph, he uses the popular expression "the big picture." The tone is casual and not particularly critical. It remains neutral with respect to the ups and downs of the software package.
3	The error-checking paragraph is a turning point. The software has been examined in terms of its features until now. You will notice a shift of focus. The following paragraphs examine more features, but from the perspective of the experienced user and the novice.
4	The author is always evaluative and honest, but he is also cautious. He has a tendency to see the good in the bad because of the distinction between types of users.
5	As you could anticipate, the conclusion is hardly a mixed review. The software has its problems. Recalling his benchmark, Access, the author simply declares that the Una Corp package is not what is should be.
6	In the final analysis it is interesting to note that although the text reads rather evenly in its balance between compliments and criticism, in the end the author's position is rather firm. What ultimately swings the decision in this case is the absence of important database design features available in other software.

Risk Analysis

Project Game Plan

- A study of pros and cons will balance well-known forces:
  - ✔ advantages and disadvantages
  - ✔ benefits and drawbacks
  - ✔ pluses and minuses
  - ✔ perks and risks
- The unique feature of a plus-and-minus analysis is that only two basic criteria emerge.
- An overall list of criteria is used to establish the measures that are important.
- These criteria are separated into positive and negative issues or concerns and these lists are then prioritized.
- Tables are often used to clearly reflect the two groups of criteria.
- The project can be constructed in various ways: the pros followed by the cons, the cons followed by the pros, or a discussion of each significant issue in terms of both pros and cons.
- The document can be descriptive (neutral) or evaluative (judgmental).
  - ✔ If the project is descriptive, the positive and negative elements of the analysis will approximately balance each other in the text.
  - ✔ If the project is evaluative, the paper will tilt heavily in the direction of the author's interests, which may focus on the pluses or the minuses.

Activities Academic Projects

If you are interested in academic transfer programs, you need to look closely at the three prototypes that are discussed in this chapter. If your engineering technology program is strictly a two-year focus, it might be interesting to consider these projects even though the industrial applications are more likely to be your focus for work-related writing activities. Each of the following projects should include a table (see Appendix B).

Develop a comparison. *You could compare two of anything from Mac and IBM computers to chemotherapy and radiation treatments to copper and plastic plumbing to faucet water and bottled water. If you want to be more ambitious, think beyond devices and materials and look at two theories, processes, procedures, locations, situations, installations, methods, and so on. For example, you could compare a Mac and an IBM-PC, but if you know enough about the subject you can compare the MAC-OS and the Windows OS instead.*

Use the Comparison Summary (p. 276) as your guide. Develop a project of 1500 words or more.

Develop a problem-solution analysis. *There are three ways to design this document: you can focus primarily on a problem, on a solution, or on a problem and solution.*

If you do not have a technical topic in mind, this project is easily removed from a technical setting and can be used to examine social problems, political problems, personal problem, and so on. Since your focus should be your engineering tech programs, this is an opportunity to look at the social, political, and personal issues that might relate to your program. Safety, security, privacy, and other matters link technologies to a larger social context. Consider the Internet for example. Discuss problems. Discuss solutions. Consider Internet security problems, Internet privacy problems, and so on.

Develop a project of 1500 words or more and use the Problem-Solution Analysis Summary (p. 299) as your guide.

Develop a risk analysis. *Select a single topic that interests you and study its advantages and disadvantages. This project might provide you with the opportunity to write an analysis of personal interest to you or of interest to your supervisor at work. For example, should you subscribe to an Internet service provider? Should your office add two more CAD stations? You could also look at fellow students as your reading audience and address interests of concern to them—perhaps in your program. Should they handle their own upgrades to save money on computer needs?*

Use the Risk Evaluations Summary on p. 316 as your guide, and develop a 1500-word project.

Share a Project

Work with three other members of the class to discuss and edit the third set of technical documents. You can use the same groups that you used previously, or you might develop new groups. New groups might be appropriate for those students who are thinking about a four-year transfer program.

- *Before holding the meeting, review the fifth editing checklist for papers that involve comparison, problem-solution analysis, and risk analysis (see pp. 502–504).*
- *On the day the projects are due, have each member of the editing group explain his or her project in terms of objectives, intended audience, and project development.*
- *Hand the papers around for a critical reading and editing.*
- *Have each member edit the texts for both writing errors and technical errors.*
- *Have each member write a one-paragraph critique at the end of each paper.*

PART III

Technical Presentation Features

Work in Progress

8. Building Structure for Readers

Everyone has his or her own writing style, and I do whatever works for me. I approach writing in pieces: the graphics, the tables, the parts of the text, and the appendices. An outline is a big help in this piecemeal method. You could think of an outline as a road map; it will guide you and suggest where to insert text, graphics, and tables. An outline is not set in stone; you can change its flow more easily than you can change an essay document by simply rearranging its sections. I shape these pieces and organize them in terms of readers.

Documents should be written to the level of the intended audience, and the parts have to fit the reader's needs also. I feel documents should be simple and easy to understand. The documents I have written have been used by a variety of audience levels such as the operator, the supervisor, the engineer, and management levels. I have to address each group differently.

Technology is one issue for a writer. English is another. And foreign languages are also important in global industries. I mentioned earlier that Genus is multicultural, and this is a plus if a company plans to compete in today's international economy. This was proven when the company received an order for parts that were to be installed on vehicles sold to an Arab country. Genus was able to get advice from a couple of our engineers who are fluent in Arabic. This saved a great deal of time in interpreting the customer documents, designing and creating the parts, and supplying parts that the customer could easily understand.

The following list is an example of a process change from the project I was working on. The text explains how labels are to be made and printed. The intended audience for this material is the assembly operator who is familiar with our network computer system.

1. *Using the network computer in your area, double-click on the Reflections icon to access the MAMMAN system.*
2. *At the "Command (MG, 5)?" prompt type "BS U 395" for "Label Production" and press "Enter." The BS is needed to enter into the Barcode Scan function.*
3. *At the "Printer Type" prompt type in "2" and press "Enter." The code 2 is set up for the LaserJet Plus printer in your area.*
4. *At the "Label Form?" prompt type "ORG" and press "Enter." This will identify that you are the originator of the form you are going to create and print.*
5. *At the "How Many Copies of Each Label (1)?" prompt type "1" or the quantity of labels you need and press "Enter." Generally only one label is needed to identify the bin that is to hold the parts.*
6. *At the "Part Number" prompt type in the part number you need on the labels, for example "12345-678" and press "Enter."*
7. *At the "Description" prompt type in what the part is, for example "meter" and press "Enter."*

S. B.

Writers and Their Readers

Two dominant forces shape writing of any kind: the reader and the writer. Technical documentation is no exception, although authors of technical documents may tend to place the emphasis on the subject matter of their projects. The subject is actually driven by the objectives of the author and the needs of the readership. As a result, you need to take a close look at these two "controls." If you have a clear reader profile in mind and a clear objective or objectives, these controls will greatly assist the development of your projects. At times a readership cannot be anticipated with clarity, but if you have a precise target group of readers in mind, the project can be defined by the needs or expectations of the intended audience. The significance of the readers cannot be ignored since readers influence the intent of the author's objectives.

Samples 8.A

SOFTOUCH ASSEMBLY, TESTING, AND UPGRADE INSTRUCTIONS

Audience: This document is primarily written for use by the repair technician whose educational background usually consists of two years of trade/technical school or work experience. It will also be used by the production technician who also will have a similar educational/training background but will be involved only in the testing of the SofTouch (Industrial Personnel Computer) units and not in the repair.

Objectives: This document is written for a company that is ISO 9001 certified. It will be used to guide the repair technician on the assembly, testing, repair, and upgrading of the SofTouch. This document, as written, covers only the testing portion. The other topics will be added later. Also, this layout is not typical of our normal ISO 9001 Work Instructions because I want to explore the use of a paperless document, so I chose this format to ease the conversion to a Microsoft Help File (*.HLP) or a Hypertext File (*.HTML).

CREATING A BASIC 3-D WIREFRAME BOX IN CADKEY

Audience: People with no prior CADKEY experience but with some knowledge of AutoCAD and DOS. An understanding of the hard drives available on a Pentium-based computer similar to those in the NSCC CAD lab is necessary, as is familiarity with subdirectory creation and file manipulation commands. AutoCAD experience beyond first quarter is helpful but not necessary.

Objective: To Construct a basic line, stretch it in the x-axis to create a box, and then to stretch it in the z-axis to create a three-dimensional cube. The sequence of required actions will show the logical processes involved in CADKEY.

You will notice in the samples of this section the care and attention that must be taken to construct a proper understanding of the intended readers of a technical project. Each time you develop a presentation, write down the definition of the target audience and the objectives of your mission. These observations will be very useful.

In previous chapters I frequently noted that technical writing is partly shaped by subject matter, meaning that the subject will dictate certain features of the document. This may be a matter of formatting—for example, a set of instructions may make the best sense in an outline of some sort—or it may mean operating at the symbolic language level—for example, having to use C++ or Visual Basic statements to illustrate programming tasks of a special project. There is, however, another significant "driver" in tech writing that dictates the mission of the writer: the readers. Most writing is, to some measure, written in anticipation of a reader. Even private medical records anticipate a possible audience. Those in the medical profession now keep clinical transcriptions of their daily work, largely as legal protection because of insurance practices and court litigation in their professional world.

Any document must be accessible to its intended reader. If the document cannot be understood, the author is back to square one. In technical documents, the problem is that the subject is often complicated, so it is problem enough to try to articulate the content. The material may also be difficult for a reader to understand, so it does not help if the writing is part of the problem. If an author is not thorough enough, or if he or she fails to define terms or explain theoretical concepts, then the author is at fault. Also, there is the increasing disparity between the technological world that generates products for public use and the public's ability to understand either the products or their applications. So, for example, a manufacturer of electronic instruments must be certain that the product-support apparatus—namely, the written documents such as manuals for instructions in operation, maintenance, and troubleshooting—is in place but also that the documents are *understandable*. Much of the significant hard-copy output of a corporation is directed at the proper utilization of whatever the company does or sells. Clarity is an investment.

The challenge is to reach a wide assortment of readers. Recall our discussion of the electric typewriter shipment. A new electric typewriter will probably be technologically refined to the point at which it is essentially a word processor without a screen. It is too sophisticated to be approached as a simple typewriter, so the corporation that developed it will need to have all types of support material to market and maintain the product. There will have to be simple—"user-friendly"—manuals for the operator, marketing packages for the sales teams, and maintenance instructions for the care-and-repair service sectors. The array of documentation can be large.

Notice that the issue is the product, but notice that the *readers* do not share the same concerns. There are at least three widely different readers: the operator needs instructions for use, the sales reps need instructions for marketing, and the technicians need instructions

If you have a clear perception of your reader and your mission as a writer, you will have better control of the way in which you present technical material.

Samples 8.B

WOA-TXI User's Guide

Audience:

The intention of this guide is to instruct the user in how to equip an IBM or IBM-compatible personal computer and Watson/VIS system with the WOA-TXI application. The purpose of this application is for use in marketing TXI long-distance telephone service.

A prior knowledge of the Watson/VIS system is assumed, as well as a working knowledge of the MS-DOS operating system and PC compatibles (including installing internal expansion boards), and an understanding of computer telecommunications concepts.

Although an overview Watson/VIS installation is included, this document is not intended as a replacement for, or supplement to, the Watson installation guide or user's guide.

HOW TO USE AND OPERATE AN ANALOG MULTIMETER

AUDIENCE: Anybody who possesses a basic knowledge of the concepts of current flow, AC and DC voltage, resistance, and basic semiconductor operation.

OBJECTIVE: To explain the use and operation of an analog multimeter in making measurements of current flow, AC and DC voltage, and resistance.

for repairs. Also, the operator is not likely to have any technical background and is very unlike the technician in this respect. The sales team probably will learn a good deal about the machine, but they do not type for a living and know little about secretarial and office practices—and they cannot repair the digital circuit boards either.

The dilemma that scientific and engineering firms face everyday is how to reach out and communicate technical material. The problem is that when the communication concerns technical matters, the task is never as easy as a telephone call. Consider the 800 or other toll-free numbers I noted earlier. Computer and software manufacturers often provide 800 numbers for anyone who owns one of their products and needs to call for help. The computer industry more or less started this practice, but it probably was not intended as clever marketing. Yes, the 800 number says "we care," but the truth is that it is more of a *necessity* to help users. First, the manufacturer wants second-time-around trade and product loyalty. But, also, the hard-copy manuals are often confusing, misleading, or even wrong. The electronics may also contain glitches. The 800 numbers help hide the effect of zingers.

For a large corporation, the 800 number defuses customers and debugs the systems. Unfortunately, these clever aftermarketing tactics are not strategies that are available to most authors who have to do a little tech writing. Authors do not have the opportunity or the luxury of waiting until the smoke clears. It is now or never. They have to try to meet the reader's needs on the first attempt, or at least the second if authors have a chance to do a rewrite.

Profiles

The easiest approach to understanding the issue of "audience" is to imagine the problem in some other setting rather than to see it as a dilemma in technical communication. You should not overlook the fact that the merchandising of technical products depends greatly on outreach for sales success. The strategies for product merchandising are persuasive devices for targeting an audience. Engineers and engineering techs have a technology to communicate. This situation is not very different from having a product to sell. Technology and production—and they usually come together in some percentage of product-related copy—call for helping an audience understand what you have to say.

For example, if you want to sell a product, you probably need to understand your potential market clientele in the following specific terms:

Income

Education

The audience profile can be complex or simple. The situation depends partly on the identification of skills that are necessary to understand the document.

Samples 8.C

HOW TO DETERMINE THE LINEARITY OF A DIFFERENTIAL PRESSURE DETECTOR'S OUTPUT

Audience: This publication is intended for use by newly trained calibration technicians (USN) and shipyard personnel involved in the calibration and maintenance of high-pressure and temperature-differential pressure detectors and associated equipment.

Objective: The objective of this publication is to provide the technician with a clear and concise general procedure as a guide to the numerous differential pressure detectors and their associated technical manuals. WARNING: This is not a replacement for the applicable technical-manual-approved procedure but a guide to the procedural steps called for in the general calibration of differential pressure detectors common to all such detectors.

TITLE: **TROUBLESHOOTING VFH REPEATERS ON THE TRANS-ALASKA OIL PIPELINE COMMUNICATIONS SYSTEM**

AUDIENCE: Alyeska Pipeline Service Company SCADA Technicians

OBJECTIVE: This article is written to present a logical method for troubleshooting the VHF repeaters along the Trans-Alaska oil pipeline. We approach the subject from a field engineering perspective; rapid and accurate trouble diagnosis with a minimal amount of test equipment is emphasized, and repair by module replacement is stressed. A general procedure for troubleshooting is provided, and its results are analyzed. SCADA technicians involved with communications cross-training will find this information useful.

Age

Interests

Employment

Marital status

Gender

There are other variables if the sale of a product depends on such elements as geographic location. The basic features of a marketing profile are simple enough. A great deal of product merchandising depends on such basic information, not only because the product will appeal to some profile or another, but because the ad copy and every other related document will be designed to communicate to that group. Without losing sight of the fundamental elements of the preceding profile, think in terms of the company where you work.

Consider for a moment the amount of internal documentation that exists in a corporation. The quantity will far exceed the number of glossy paper products that are part of the commerce of whatever the business manufactures. The numbers of audiences or readerships in the building can be extensive, and the distinctions among them are often quite marked. From the boiler room to the executive suites on the top floor, the building is tier upon tier of employees, from the janitors through the rank-and-file to the CEO:

	↑	**CEO**
		Financial Officers
		VPs
		Administrative Support
Buyers		**R&D**
Out Services		**Technical Services**
Repairs		**Production**
Sales		**Secretarial Pools**
	↓	**Building Maintenance**

The objective will vary in length. The subject matter will determine the complexity of the objective, as will the length of the document or the number of goals established by the author.

The "floors" of the building are implicit in this list because each employee differs in terms of the *basic* profile identified earlier: income, education, age, interests, employment, marital status, gender, and so on. The first few of these basic conditions clearly affect the situation of any employee in the company hierarchy.

It is fairly apparent that a mountain of paper is being shuffled among all these groups. The information highway is certainly not new; the paper-trail version is long-standing. The point to observe is the basic need to be aware of readers in any way that can affect their understanding of what you have to say. Yes, education is probably the key concern most of the time, but level of interest, type of interest, gender, and other conditions may matter. A technician who tries to explain the repairs she has just made on a secretary's computer may have difficulty with the explanation. The secretary may be uninterested in anything but the simple operational basics and may have no knowledge whatever of what is going on behind the keyboard.

Realistically, you will have at least three potential reader groups. If you are a technician, you can put yourself on a matrix that reflects those above and those below your skill and knowledge level. For example:

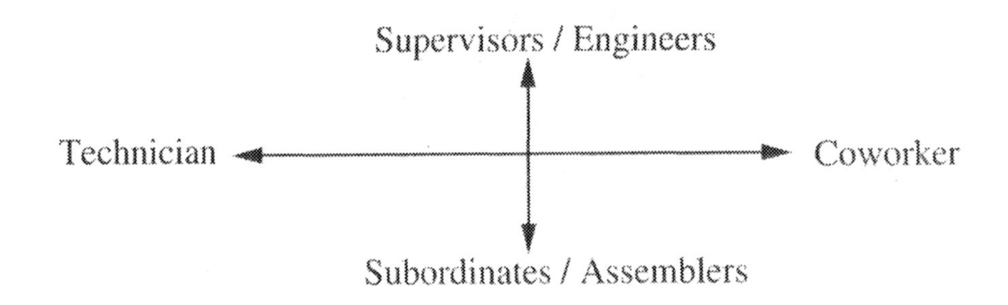

Even these categories reflect a hierarchy that is complicated by important distinctions between, for example, supervisors and engineers. Engineers can be addressed with technical expertise, but supervisors are far less predictable as a group. Some supervisors might have an engineering background; many might not.

The basic question is usually the obvious one: What does the reader know or need to know in order to understand my document? You have heard the expressions "over my head" and "beneath me." These up-down, or over-and-under images are helpful. If you write over the reader's head, the project will be confusing, which is an intellectual response. It may also seem condescending, which can be a psychological response. Any emotional response sets up all sorts of roadblocks. If you are overly simple, people may find the project uninformed or uneducational. The psychological response might be that the project is boring or wasting the reader's time. Obviously, a writer will block communication in any of these situations.

Learning Curves

Generally, your style of writing is somewhat fixed once you identify your audience. This is probably true in most technical writing, with one conspicuous exception. If you examine the writing styles of textbooks and corporate training manuals, you will notice that they are frequently designed to follow a learning curve. The assumption is that you acquire the knowledge of each chapter as you go along and that the knowledge is cumulative. This results in a unique situation in which chapter three is simple but chapter ten is complex. The text moves up the learning curve and expects the reader to be there also.

You know what happens in a math class if you still do not understand chapter four when the instructor is on chapter eight. Writing on a learning curve is a unique undertaking. This text does not stack each chapter on top of the next in a cumulative manner, but most engineering texts certainly do. If you think of the target audience as one that is brought through the reading experience in a learning curve fashion, the following drawing can be used to illustrate the concept. In this case, the reader's position shifts from uninformed to informed because of the education process of the document. Both the document *and* the reader follow the learning curve. Subsequent chapters are more complex than previous chapters. The vocabulary, the mathematics, and a host of elements progress to increasingly complex levels, both of learning and communication.

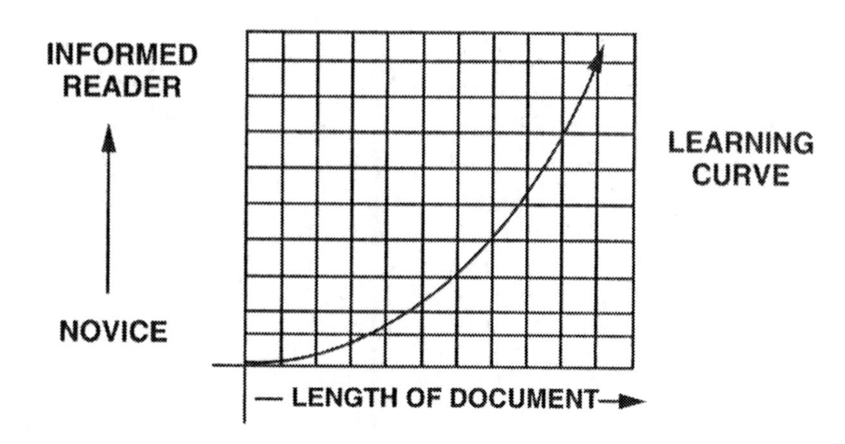

Targeting Your Reader

The solution to the problems of writing for a reader is to target the audience properly. First, it must be defined, as you have noticed in the many sample cover pages throughout this text. You must see the target readership clearly in order to write to it. Second, you must determine how you plan to measure your writing against the reader's needs. You clearly understand how to adapt your language to different audiences in the general conversational environments you encounter at work. In writing you have a similar flexibility, but there are other options you might overlook. Adaptation is not only a matter of vocabulary.

A number of features in any given writing project can be adapted to your reader's needs:

Vocabulary

Definitions

Sentence length

Paragraph length

Extent or depth of material

Manner of explanation

Order, tone, repetition

Headings

Uses and types of graphics

Reading time

Even the first two elements on the list—the most obvious ones—are often poorly managed in technical documents of all kinds. Doctors often overwhelm or underwhelm their patients with too much knowledge or too little respect for the patient's knowledge; CPAs and lawyers know the law but may have difficulty explaining it in English. There are details that can help.

Sentence Length

At the turn of the century there were writers, such as the American novelist Henry James and the Frenchman Marcel Proust, who could write sentences that came close to filling a page. They were fiction writers who loved to embroider the language. Obviously, you

Sample 8.D

You've gotten a clean **compile**[1] on your COBOL program: 0 errors, 0 questionables, 0 warnings. You've **PREP**ped it, and gotten no warning messages. Filled with confidence, you've **RUN** your program, and—oh, no—you see garbage flying across the screen and your heart sinks down to your Adidas. It looks something like this:

*** ERROR 711 ILLEGAL ASCII DIGIT

SOURCE ADDRESS = %003354

SOURCE = 'M'

STATUS = %060701   *P REGISTER* = %002010   INSTRUCTION = CVND

You have a **711** error, and you start to calculate how many meals and how many nights of sleep you will have to skip to get your program in on time.

Relax. A 711 message merely tells you that you've put alphanumeric data (garbage) into a place reserved for numeric data. The error message also tells you where in the program the error happened. All you have to do is to relate what the message says to what you have in your program. It will require a little busy work, some of which may already be done, and some detective work on your part to resolve the problem.

[1]All boldface terms are defined in the glossary.

won't see anything like that in technical writing, but you should consider sentence length as a concern for one obvious reason: the longer the sentence the more complex the content of the sentence becomes. If you equate length with complexity, then the possibility for confusion is going to increase with the length. Also, regard sentences that involve one or more symbolically rendered formulas or calculations as a particularly likely location for confusion. If you simply imagine the three writing "directions" we observed earlier—up, sideways, and down—you can adapt sentence length, and the other criteria, to the readers without difficulty. In terms of writing style, these directions indicate that you can be complex on one hand and quite simple on the other. The middle ground, what might be called the halfway point, is a mixture of simple and complex. This is the level of your coworkers. You might call the coworker level the "simplex" level. When writers write at their own level, they usually fuse simple and complex issues with confidence because they themselves represent the reader's knowledge level.

Paragraph Length

Originally, the paragraphs were much too long in the first two volumes I wrote for the *Wordworks* series. I had to divide hundreds of them in order to make them more readable. I had a tendency to be long-winded because I was trained in a tradition that was academic and in a fashion you might call "scholarly." The results were not particularly friendly and the books were not easy to read. I went back to square one and started with an entirely new set of standards—one of which was the shorter paragraph.

You can write paragraphs that fill a page, but the reader must be highly informed and very interested. On the simple end, you can write three- and four-line paragraphs—but you must realize that there is little depth to such paragraphs. Length again correlates with the complexity of the content. Long paragraphs are ideal for analytical work. Short paragraphs are ideal for instructions, regulations, specifications, and other procedural issues.

Report Length

The third component of the length considerations is the overall weight of your document. The longer the projects become, the less likely it is that they will be read. Put another way, the longer they are, the smaller the audience. Understand that brevity isn't necessarily a blessing. Because the material is content driven, the composition will be as long as it has to be, but there may be plenty of flexibility that will allow you to adapt the overall length to the reader. Again, analytical or research material is likely to be more extensive because of the focus on *why* the findings are as they are. Procedural material can focus on *what* the rules are and may often omit the kind of analysis that usually extends an analytical document.

This sample is targeted at college students. It has a casual tone and a touch of humor. The author was making an effort to approach an old and difficult programming language in a manner that was breezy in style but serious in subject matter.

Sample 8.E

INTRODUCTION

All of us have heard the horror stories about technician-reworked printed circuit boards. We can see them permanently imprinted on many numbers of PCBs. Recently, some serious concerns have been voiced by quality control inspectors regarding the poor quality of technician rework. Although most of the final test technicians have had a soldering class, either in college or through ProPak Control, there is still a decided lack of knowledge in the test area of the techniques and standards involved in the rework of printed circuit boards. The information provided in this pamphlet can be an aid in putting an end to those horror stories. It will give examples, in words and pictures, of the repairs most often made by technicians during troubleshooting. Included also are examples of acceptable and unacceptable soldering and some information on multilayer printed circuit board repair.

Feedback is encouraged so that the information needed will be provided in as complete a manner as possible. Please submit any requests for specific information to your supervisors, who will relay them to the proper authority for an immediate response. The support of quality control areas will be provided to assist us all in upgrading our rework output.

Depth of Material

All three of the length issues are quantitative matters, but there is also a qualitative concern that will override the bean counting. The material must make sense, and the place to determine this issue is on the drawing board. Develop an outline and study the *scale* of the project. Thoroughness is an issue of scale. Discuss the outline. Authors can often predict the intended proportions that a subject calls for if the issue is tightly focused and the discussion of the outline is thorough. At other times, they may have an outline that indicates that the project will run, say, thirty pages. This situation could spell trouble. If this sort of material is explained in ten pages, it will lack thoroughness and read as a hasty or generalized project. Thoroughness will depend on a correct sense of proportion between the content of the material and the length on which you have decided.

The other issue here is the *depth* the reader needs. You can probably assume that theoretical material, for example, is largely the provenance of engineers and not assemblers. The assemblers need to know what to do, but usually they will not need to know why. Engineering technicians will need the theoretical know-how for maintenance but not the advanced level of theory—particularly the math—that engineers may need for design work. Each group can be addressed, in terms of depth, by looking at the "needs" of the group. Of course, education or knowledge level is an important factor here.

Order

The actual order of events in engineering and scientific paperwork is perhaps most often a function of causation. You have a cause and you report the effect. You have an effect and you report the cause. In great measure, much of what you write will be some variation of "if, then" reasoning. Nonetheless, the preference a reader has for certain material and not others can determine the organization of our work. Always consider the interests of the reader when you order the events of your document.

For example, suppose a small construction engineering firm is called in to do a site analysis on a wetland that is to be drained and used for a dozen tract homes. The developers can write their report easily enough along cause-and-effect lines: if they drain the water table in such and such a way, then the site will be of such and such a quality. But the issue of reader preference remains. Who will read it? If the investors read the report, they will want the cost analysis first. The rest is of less concern. If the lawyers read the report, they will be concerned about conformance to federal regulations for wetlands. If the builder reads it, he will be concerned about soil quality, water supplies, septic regulations, and so on.

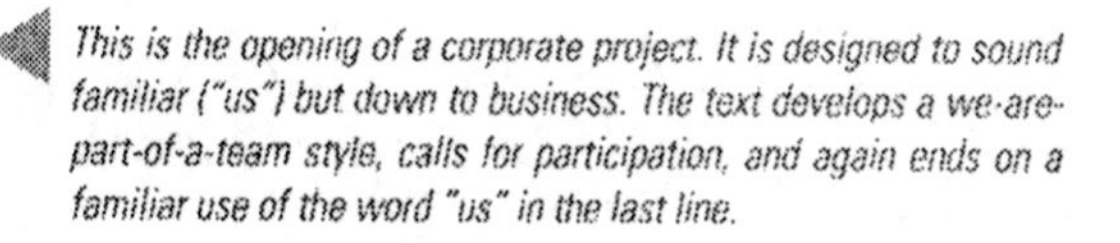
This is the opening of a corporate project. It is designed to sound familiar ("us") but down to business. The text develops a we-are-part-of-a-team style, calls for participation, and again ends on a familiar use of the word "us" in the last line.

Sample 8.F

EXPLANATION:STORAGE DEVICES

Computer storage devices have an interesting recent history. Only since the late 1980s have personal computers boomed in popularity. In the last decade new hardware devices began popping up faster than the average consumer could keep track. We now have superportable storage devices that can hold as much information as many internal hard drives. Up to 2GB (gigabytes) of information can be stored on a small disk that you can carry with you.

You may already be familiar with many common storage devices such as the 5.25≤ or 3.5≤ floppy drive (Figure 1).

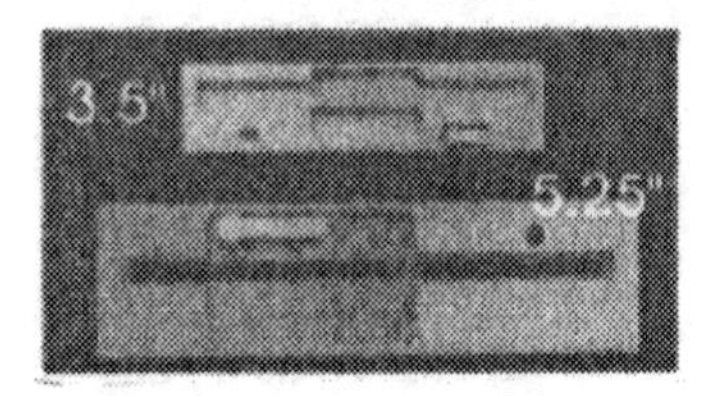

Figure 1. Computer Floppy Drives. (Courtesy of Hardware Central. Copyright 2000, Internet.com Corporation. All rights reserved <http://www.hardwarecentral.com>)

These drives used to be used as the primary hard drive of a computer system. When computers took on a larger and more important role in business, it became extremely important for storage drives to be as trouble free and reliable as possible. Floppy drives were replaced with what we now know as the computer's hard drive. The hard drive is a factory-sealed drive that can store much more information than floppies and has a lower data loss rate because of the factory seal and the inability to use the regular computer hard drive as a mobile drive (Figure 2).

Figure 2. Inside of Typical Hard Drive. (Courtesy of Bason Computer, <http://basoncomputer.com>)

Computer users found the need to transport data from work to home and vice versa. As a student my concern for storage is that I have something portable and compatible with school computer labs. In a pinch I need to be able to take data back and forth from school to home so I can better use my time and resources. It is also important for me to get as much for my money as my computing needs will allow.

The 3.5≤ floppy drives did not have the data capacity that was needed by most users. The industry started working on solutions for portable data storage. Many product concepts were explored at the theoretical level until practical drive options began to appear

In effect, the wetland site report could be the same document for each group with the parts moved around to match various interests. The cost analysis could be nothing but an appendix in the copy that goes to the builder, and the lawyers have little concern about septic tanks. Using an order of importance or order of interest helps structure a document for a reader's needs.

Headings

Organization is an important perception for a reader. A sentence is a unit of organization. A paragraph is a unit of organizational structure. A *page* composed of sentences and paragraphs is yet another framework of organization. Page layout—even at arm's length—can provide a reader with important perceptions concerning the organization of a document if headings are among the author's strategies. Headings insert structural cues into a text and are particularly useful in the running-text format where paragraphs show no apparent visual distinction.

- **Headings divide.** By using headings an author can divide and subdivide a document into units that provide a reader with an understanding of the organization of the overall document. Just as sentences begin with a capital letter and end with a period in a paragraph of, perhaps, ten such distinct units, headings create similar visual distinctions among the paragraphs or larger units of a document. This page, for instance, uses headings to divide the page into understandable units.
- **Headings define.** Unlike the capital letter of a sentence, however, the heading indicates the subject matter of the coming discussion. In this respect, headings are akin to topic outlines; they serve as a topic outline that has been integrated into the text.
- **Headings order.** In addition to dividing and defining units of a document, headings are often numbered. There is no strict rule for numbering headings (or using the alphabet). Perhaps a few headings would not merit numbering, whereas a list of twelve might call for enumeration. The use of an ordering system adds additional clarity to a reader's perception of a document.

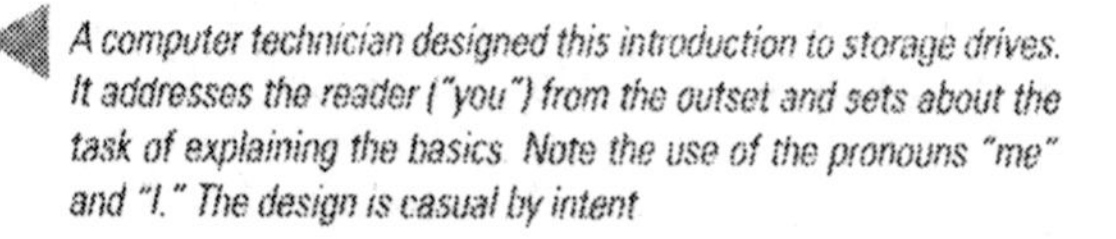

A computer technician designed this introduction to storage drives. It addresses the reader ("you") from the outset and sets about the task of explaining the basics. Note the use of the pronouns "me" and "I." The design is casual by intent

Sample 8.G

The modem (or interface) converts digital signals from the CRT (Site A) into analog signals (for a phone line); this process is called *modulation.* Conversely, the receiving end (Site B) receives analog signals from a phone line, which another modem converts into digital signals for the computer in the process called *demodulation.* (See Figure 1.)

Modern modem technology allows for transmission signals to send and receive in both directions and enables either site to originate a call. Note that this basic communication chain necessitates that the two sites maintain essentially the same conditions.

EQUIPMENT AND INITIAL SETUP

To keep our configuration example of the communication chain relatively uncomplicated, all necessary equipment and hardware for both Site A* and Site B must be available before the actual hookup process begins.

For *each* site, the following equipment is either required or, at least, helpful:

1. Site A: a CRT (or similar input/output device); Site B: a computer (typically with an input/output device attached for bidirectional communications)
2. Modem
3. Dedicated phone line
4. Electrical outlets (typically carrying 115 volts for the modem and CRT)
5. Phone cord
6. Electrical extension cord (helpful)
7. Wall jack
8. RS-232C cabling line

** Note that Site A refers to the "home" or the operator's work site. Site B is the remote site, located elsewhere.*

Tone

This may sound like a stylistic matter, and technical writing is usually sufficiently objective that you might not think that tone would be an issue. Tone is the *manner* of explanation. There is some control here that can be adapted to readers. The text you are reading, for example, is casual in manner. It is a little anecdotal. It is structured to reflect a very practical approach to its content. You can, within limits, construct a "tone" no matter how technical or formal the material. Consider the annual reports of corporations, mutual funds, and other organizations in which the annual accounting is an expected part of their document output. In the worst imaginable downturn these glossy items glow with good cheer. The tone is always upbeat, regardless of the losses in the fine print. The colorful pie graphs will salvage what went well for the year. The real disasters will be veiled in data elsewhere in the report. This example is not directly a concern for engineers and technicians, but estimates and proposals are part of the engineering world, and if they sound bleak instead of enthusiastic, they may not do the job that is desired of them.

In a sense, the simple use of "plain speaking" can be a tone, a highly desirable one that we see, for example, in the popular computer and Internet magazines. You need to simplify complexity whenever it is a practical goal in your writing. Plain speaking in technical discussions is one way to explain what is popularly meant by a "user-friendly" document. If a piece of writing is approachable, nonthreatening, or interesting, tone is probably at work because readers are responding to the author and not just his or her vocabulary.

Read the full-page samples and you will sense the way in which the writers adapted to readers. You can hardly miss the tones. Also, notice the obvious decisions the authors made regarding vocabulary, sentence length, and paragraph length. The samples are presented in an approximate order from casual (Sample 8.D) to very formal (Sample 8.H) and very technical (Sample 8.I).

Uses and Types of Graphics

Graphics speak. The illustrations you use must abide by the standards of the discussion here. One of the most common problems that engineers and technicians encounter is the need to simplify illustrations when they are dealing with the public. The graphics must target the readers. Legends and callouts, for example, have to be shortened or enlarged depending on the audience. An assembly drawing may need many callouts to identify component parts for a rookie assembler but nearly none for the senior technician. Schematics will serve their purpose only if the reader knows how to read them. The same is true of construction drawings, tables, charts, graphs, and a host of other graphics.

This sample is technical and addresses a peer-group readership that is familiar with the subject matter and can understand the technology.

Sample 8.H

For each bridge, the element with the lowest C/D ratio was used to determine the vulnerability factor, since failure of any one of the elements would be sufficient to cause failure of the structure. Although the methodology used in obtaining the C/D ratios varied slightly from engineer to engineer, the more important considerations—consistency and relative rankings—remained the same. The reliability analysis required consideration of the following four major steps:

Step 1 Identify Lateral Seismic Load-Carrying System Components

This step required the identification of the lateral load resisting mechanism based on the data compiled. The type of interaction between the superstructure and substructure was established. Any redundancies present in the lateral system were also identified in this step.

Step 2 Assess Superstructure Seismic Performance

The superstructure was assessed for stability; in particular, the potential for dropping of deck spans was thoroughly examined. The effective as well as adverse features of deck restrainer devices and seat extenders at expansion hinges and abutments were thoroughly evaluated.

Step 3 Assess Substructure Seismic Performance

The substructure—consisting of columns, foundation elements, and abutments—was evaluated for any attributes of ductile performance. The performance criterion to be met by the substructure components was maintenance of vertical load carrying capacity during and following a major seismic event. The potential for large column support or abutment displacements was thoroughly examined, especially in view of any indications of soft soil conditions or liquefaction potential. The vulnerability of bearings was evaluated in this step.

Step 4 Assess Reliability of Lateral Load Resisting System

The reliability of the seismic lateral load resisting system, in terms of the redundancies present in the lateral structural system, was evaluated based on the results of Steps 1–3. The potential for the formation of possible collapse mechanisms was thoroughly examined.

Life Hazard Factor

The intent of the life hazard factor, L, introduced here is to quantitatively reflect the extent of a life hazard due to a seismic event, in terms of the likelihood of severe personal injury or death. The primary parameters that were considered are formulated to

Reading Time

To some extent the length of time it takes to read a document is a function of everything under discussion in this chapter. I have had only one occasion to time the reading of documents. I wrote a pamphlet on fetal alcohol syndrome for the large family practice of a local area hospital. I sat in the lobby one morning and scrutinized the clientele. I watched them pick up magazines for a minute or two, look at a few pages, and then put the magazines down. Shortly, they would grab the next magazines and repeat the cycle. These magazines were not the usual dentist office selection of *National Geographic* and *Sports Illustrated.* The magazines were quick reads, such as *People, US,* and *Cosmopolitan.* I wrote the pamphlet to match the cycle; it took the clients, on average, two minutes to read the pamphlet. Allowing for low levels of education and poor reading skills, I might double that time, but I felt the time element was a major consideration.

Obviously, time is money, and reading time is a real driver for a lot of us in our work. Quick-reference material is quite valuable. Any device you can add to your writing skills that speeds up the reading of a document is usually appreciated by the reader—but you must never "cut" the content at the expense of thoroughness. You cannot leave out the last page of the instructions. What you can do is put all your design ideas together to maximize ease of communication. The document will then be as "fast" as it is going to be.

Perhaps the popularity of the "telegraphic," or "cryptic," style of technical writing is somehow connected to the issue of reading time or speed. Rather than discuss the cryptic method under a heading concerning writing style, let me simply put forward a few ideas concerning reading time so that you can see whether there is actually any utility in stylistic brevity.

Telegraphic is the word that is often used to describe the method of writing that replaces a sentence such as

> Connect the left speaker cable to the left jack mount on the left speaker

with a shorter version that might read

> Connect left speaker cable to left jack mount on left speaker.

Apart from being awkward English, consider what the author gained here. The truth is that the omitted words are always the *least* time consuming elements of the sentence. They are usually (1) articles of speech that serve very simple logic functions, and (2) they are usually what are called monosyllables—mere bleeps of a voice. To the eye, the time saved is near zero.

In both style and layout the load-carrying sample is very formal. The language uses precise engineering terminology throughout.

Sample 8.I

- The \NWCLIENT directory was created to fulfill this need. Key files copied to the directory included LSL.COM, the MLID for the NIC you selected (such as NE2000.COM for an NE2000 NIC, IPXODI.COM, and VLM.EXE), as well as its related virtual loadable modules. In addition, a NET.CFG file was created that contains the parameters for the MLID, as well as the following section:

```
NETWARE DOS REQUESTER
        FIRST NETWORK DRIVE=F
```

- A file called STARTNET.BAT is created in \NWCLIENT. Its contents are similar to the following lines:

```
CD \NWCLIENT
LSL
NE2000
IPXODI
VLM
CD\
```

If you choose not to have your CONFIG.SYS and AUTOEXEC.BAT files updated automatically, you must do it manually now.

You must now restart your computer. The STARTNET batch file will be executed to load the network programs, finishing with VLM. If everything works correctly, you will be given the command prompt C:>.

The clumsiness of the cryptic method, particularly in the hands of unaccomplished writers, is the *real* problem. The method is *very* clunky, and writers pursue it simply because they think it is a tech writing technique. It should not be. It is for this reason that, in Chapter 9, I will always refer to figures as figures (for example, "see Figure 8") and not as "figs." Replacing a two-syllable word with a one-syllable word means little to the mind's eye in any terms whatsoever—except for the confusion that may result. Avoid the false economy of the telegraphic style.

Repetition

You should also consider the possible use of slowing your writing down. Haste can make waste, especially on a learning curve. Repeating key points of concern can be valuable. In longer texts, summing up material is valuable. Once key information is presented, the information can be used repeatedly without any appearance of repetition. You are reading a text that uses all three tools: repetition, summaries, and reiteration. Notice that I just repeated myself in the last five or ten lines. Usually, repetition is not boring. In technical writing, repetition can help simplify the complexity of the material. For instructional use, repeating key ideas is simply a learning-curve tool to help people understand and retain concepts. Page 323 and 325 use repetition. I let the ideas develop slowly by overlapping the logic of each paragraph with the next paragraph.

It does not take a professional writer to address the conditions under discussion here. You do not really need to use the word "style." Adapting to a reader's needs is quite easy to manage once you are aware of the tools you need to do the job.

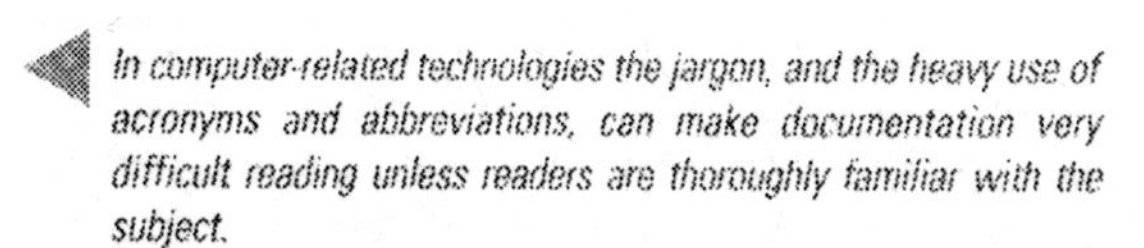

In computer-related technologies the jargon, and the heavy use of acronyms and abbreviations, can make documentation very difficult reading unless readers are thoroughly familiar with the subject.

Vocabulary in Technical Documenation

You may not have noticed that the discussion up to this point omitted the first two items on the list of conditions that address or determine the success of targeting readers. Vocabulary and definitions are by far the two most basic concerns an author must have for reader orientation. If the readers do not understand the vocabulary, the project will not fly. Technical topics usually call for technical vocabularies. The problem is a simple one: who will understand what? Notice the three brief samples in samples 8.J.

Clarity and consistency in vocabulary are desirable goals for any write, I would hope, but the challenge in scientific and technical fields happens to be unique. Consider this detail from a Hewlett-Packard advertisement.

> The new MS ChemStation series makes great mass spectrometry available to every lab.
>
> The HP 5988A GC/MS or LC/MS is your choice if you want all the most useful big system capabilities: EI, positive CI, negative CI, DLI/LC/MS, thermospray, DIP, DCI, FAB, and high mass to 2000 amu.
>
> The MSD and all the HP GC/MS and LC/MS systems are also available combined with the HP1000 RTE-6/VM computer in multi-instrument systems.

Why is it that the engineering fields appear to be so dense? First, the *subject matter* is dense, but that is obvious. Density is a difficulty in content, and so technology is one problem for a writer. The other problem is the *way* the material is presented.

There is a often difficulty in the way writers form or articulate the technical terrain. For example, notice the number of abbreviations and acronyms in the preceding sample. Since the 1940s the popularity of abbreviations and acronyms has greatly increased the conceptual density of scientific writing. A hundred years ago, or even fifty years ago, a document simply said less in the same space. A compression has resulted from the popularity of abbreviations and acronyms; that is, twice as much information occupies half the amount of space. You might take this to mean a doubling of intensity. For a reader it is more like a fourfold increase.

Samples 8.J

The for statement uses three control expressions, separated by semicolons, to control the looping process: the initialize expression, the test expression, and the update expression. The FOR loop is a pretest loop.

The initialize expression is executed once, at the beginning of the loop prior to the execution of any other loop statements. An example of an initialize statement might be "n =0;" or "var = 25". This sets the variable that is incremented by the update statement to an initial value.

HOW TO INPUT DATA AND CALCULATE

1. Press **GoTo** (function key **F5**), type "INPUT" (or "I"), and press **Enter.**
2. To make changes, move the cursor to the location of the change, and type in the change and press **Enter.**

   Example: Move the cursor to column L and down to the value of ZF1. Change the value of ZF1 from 0.4 to 0.5 by typing in ".5" and then pressing **Enter.** Now move the cursor down one to change ZF2 to 0.5 as well by typing in ".5" and then pressing **Enter.**

3. Press **Calculate** (function key **F9**).

All computers use a raster scanning technique that contains a microprocessor, bus, ROM (Read Only Memory), RAM (Random Access Memory), keyboard I/O (Input/Output), mass storage I/O, and video I/O section (see Appendix, Figure 7). To display a character on the screen of a television set, a computer has to have a special device called a character generator ROM to convert ASCII bytes (American Standard Code for Information Interchange) or a 6-bit value character to a tiny dot matrix pattern for displaying on the screen (see Appendix, Figure 8).

What do you get when you take an already dense subject and crunch it? You get a document that is a challenge to read—and a challenge to write. This compression is one feature that gives today's writing its unique character, and this density—both in content and form—calls for writing strategies that can cope. It is not a coincidence that technical writing as a field or specialty emerged in the last fifty years. As writing gained in density, the need for control and clarity became obvious. The first challenge is vocabulary and how to share it.

Another contemporary problem is tech slang or the kind of garble that simply isn't necessary. The editorial policy of a technical or trade magazine is usually guided by the need for accessible writing. From early on in their development, computer magazines called for clear writing. The magazine *Microcomputing* (no longer available) invited its readers to contribute articles, but the would-be authors had to know the rules, one of which follows:

Do write in English—not computerese.

> One thing: Please try hard to use as few buzzwords as possible. Remember that *Microcomputing* is trying to bootstrap newcomers into this field, not scare them away. If you understand your subject, you shouldn't have to be obscure.

It is not necessary to say, "I will output the decimal equivalent" or "I can't seem to access the database." Many educators are frustrated by the popularity of this sort of talk. The French government has even tried to legislate control over computerese. Nontraditional (new) usage is always a frustration for a traditionalist, but what matters is that there is a need to communicate, and slang or any other unnecessary use of a nonstandard vocabulary turns into confusion for all readers. Readers do not need it.

On the other hand, there is a real problem with the enormous vocabularies of the engineering and scientific fields. The fields seldom overlap, and so the vocabularies become separate languages. In addition, these vocabularies continue to grow. In other words, technicians and engineers may share certain basics, such as mathematical systems and the terminology to go with them, but otherwise, the languages of engineers—civil, mechanical, chemical, industrial, aeronautical, nuclear, electrical, mechanical, and so on—are likely to be distinct. As a result, every engineer has to manage every word with care. Your words will be the measure of you. If you have a reader profile and address the target with accuracy, you will achieve your goal of being understood. Furthermore, you will be understood with a minimum effort, which is a goal toward which every writer should strive.

The uppermost sample speaks precisely but without the parenthetical explanations that are in the lower samples. The amount of explaining will depend on the intended readership.

Samples 8.K

Most fasteners used in aircraft structures have been assigned NAS (National Aerospace Standard), MS (Military Standards), AN (Army Navy), or MIL (Military) numbers. However, in some cases, vendor or manufacturer part numbers are used, such as numbers starting with BAC (Boeing Aircraft Company), HS (Hi-Shear), HL (Hi-Lok), CH (Cherry), or RV (Olympic). In other cases, the part numbers used on older drawings may have changed several times over the years. An example of this is the evolutionary name and part number change of the Universal Head Rivet, which has had the following sequence of designations:

AN470 MS20470 NAS1242

It is because of these changes that more than one drawing code may also have been assigned to essentially the same fastener over the years.

The pO_2 electrode is commonly referred to as a silver (Ag)–silver chloride (AgCl) reference electrode. Application of a small amount of voltage to the electrode creates an electrochemical reaction. Oxidation or release of electrons occurs at the negative pole of the electrode (anode), and reduction or plating of electrons occurs at the positive pole (cathode). The reaction creates a potential difference or electromotive force (EMF) that is proportional to the concentration of pO_2 in the blood.

Language is critical because there are three goals that depend on language precision:

Accuracy of subject matter

Accuracy of technical expression

Accuracy in targeting the reader

One tactic in vocabulary selection is word choice. The gauge is your reader. Another tactic is to use definitions when and where you need them.

Definitions

Definitions are frequently inserted into technical projects. There are a number of ways to build them into your documents. You must first decide whether the reader needs a definition, how much of one, and how immediately. For example, if you put a list of terms at the end of a paper, the location is least immediate. However, if most of your readers are likely to understand the concepts you plan to define, then the glossary at the end will be available, but it will be unobtrusive to savvy readers. You can also place definitions in the paper where they occur, or are needed, if most readers will need them. You have various options for defining. The following discussion will explain the methods you see in the samples.

In-text Definitions

You can always stop the text and insert a one-sentence definition into a paragraph. You might think that the practice would make the text choppy, but it usually will not have this effect. If readers do not need the definitions, however, the frequent use of definitions may seem like interruptions. The first illustration in Samples 8.J uses the in-text definition technique:

> The FOR loop is a pretest loop.

Abbreviation Definitions

It is quite common, as you know, to abbreviate concepts or devices or just about anything. If speakers can make some sense out of the abbreviation by sounding it out as a syllable or two, then the abbreviation is particularly popular and likely to stay in the language: sonar, radar, ROM, RAM, byte. Many such acronyms have been adopted into our daily vocabulary. Everyone knows what a laser is, but only engineers know that it means

Both of these samples use parenthetical explanations with considerable success. Notice that, even when used frequently, parenthetical details do not interrupt the text if they are brief.

Sample 8.L

- 4 -

A rated speed is necessary. This can be handled by three methods: a pull rope, a small single-phase electric motor, and a start capacitor. If the mechanical means are used, power to the idler is not applied until after the motor has started spinning and the rope or power to the single-phase motor has been removed. To balance the voltages and currents in the three-phase output, a pair of run capacitors can be used. A disconnect switch is required by most local electrical codes for each piece of equipment. If a plug and receptacle is used to connect power to the equipment, this meets the disconnect requirement.

Overload protection[3] is required for each motor. This can be built in to the motor or provided separately. Check the motor nameplate. If it does not say "built-in overload protection," then it must be supplied separately. Typically, a thermal overload relay and magnetic contactor[4] are used for controlling the motor. The magnetic contactor is a heavy-duty relay for turning motors on and off. It is designed to handle the high starting currents of motors. There are also mechanical (manual) contactors available with thermal overload protection as part of the switch. The two wires carrying the single-phase 220 VAC power are called lines 1 and 2. These are connected to terminals 1 and 2 of the idler motor respectively. The wire coming from the third terminal of the idler motor is called line 3.

[3] Overload protection is the result of a device that operates on excessive current (but not necessarily in short circuit) to cause and maintain the interruption of current flow to the device governed.
[4] A device that repeatedly establishes or interrupts an electric power circuit.

"light amplification by the stimulated emission of radiation." There are thousands of these acronyms, probably equal in number to conventional abbreviations of the less clever type. In writing they are frequently explained, when first cited, by coupling the abbreviated form with the long form (see Samples 8.J and 8.K):

> The RAM (random access memory) was. . . .
>
> He worked for the National Aeronautics and Space Administration (NASA).
>
> It was measured in kilohertz (kHz).

This tactic serves as a type of definition, though clearly operating as a minimal explanation. The term is explained only once, when it is first used. The abbreviation can be inside or outside the parentheses. Abbreviations and acronyms can also be handled in lists (see Sample 8.M).

Footnote Definitions

There is one value to putting definitions at the bottom of a page: they serve the purpose for readers who need them immediately, but they are out of the way of knowledgeable readers (see Sample 8.L). This is a good compromise for a mixed readership and a handy solution for a middle-ground readership that calls for a blend of simple and complex material ("simplex," as I called it earlier). Footnotes are usually numbered consecutively for the length of the project. If you have only one or two terms per page, here and there, the alternative use of symbols will work quite well, but they get cumbersome if you have more than two on a page. The usual sequence is the asterisk (*) and the dagger (†). Other symbols are then called into use, although readers are most likely to be familiar with the repetition of a double asterisk (**) or a double dagger (‡) if you need additional symbols. Numbers may look a little cleaner.

Another value of the footnote definition has to do with length. If a word or concept calls for a substantial explanation, the footnote is a practical approach. The in-text definition should be only a sentence or possibly two. A footnote can easily handle ten or more typed lines because they are single-spaced and because they are often set in smaller type—10-point, for example.

This sample illustrates the use of footnotes to explain highlighted terms of importance. If the notes are removed to the end of the text, the convenience of the footnotes is lost.

Sample 8.M

HTML Is the Language of the World Wide Web

With the rapid outdating of so many programming languages that often required much study to master, many people are reluctant to learn yet another computer language. HTML may be a programming language, but it is a high-level one. This means that it uses English-derived commands and does not make heavy use of complex formulas or even simple mathematics. Its syntax, or rules of code structure and use, is simple and intuitive, making it possibly the easiest programming language to learn. Once the logic of its syntax is understood, mastering HTML merely involves learning its relatively short list of codes and their capabilities. If programming languages were espresso, HTML would be a decaf skinny latté.

Common Acronyms and Abbreviations

HTML	Hypertext Markup Language
WYSIWYG	What You See Is What You Get
WWW	World Wide Web
SGML	Standard Generalized Markup Language
ASCII	American Standard Code for Information Interchange
URL	Uniform or Universal Resource Locator
ISP	Internet Service Provider
HTTP	Hypertext Transfer Protocol
DOS	Disk Operating System
GUI	Graphical User Interface
CSS	Cascading Style Sheets
DHTML	Dynamic Hypertext Markup Language
XML	Extensible Markup Language

Many Web site designers use WYSIWYG (an acronym for "What You See Is What You Get") programs that can be used to create Web pages without the complex

In-text Lists

There is no harm in inserting a list of terms *in* the text of a document at some point (see Sample 8.N). Such a list can be five terms or twenty-five. The format can be an entire page, a boxed insert, or even a sidebar (a narrow column down one side). Magazines use sidebars for many purposes, and the device will not cram a page if the text is printed on facing pages. In magazines the facing pages, in effect, double the visual spread available at any given time, so a sidebar does not encroach on the overall appearance. In a one-sided format of the sort you would create, the boxed insert is the most useful way to list terms, particularly if they will take up only three or four inches of space. Simply draw a box around the group of terms (see p. 352). The informed readers skip the box. The novice stops to learn.

List of Terms

The final device is the glossary (Sample 8.O). The word *glossary* is not used very often in technical writing, and you probably would use one of the more popular titles such as "List of Terms"; even "Definitions" will do. It might also be a list of symbols and their explanations, which is called a *symbology*. The list of terms is usually placed at the end of the project on a separate page. It may run pages at times, and it may be illustrated if, for example, an author is identifying parts of something. The basic layout practices are quite simple. You might boldface the terms, indent the definition, and single-space the text. Double-space between the definitions, which are listed alphabetically—although they can be numbered.

Usually, vocabulary that is to be found in the glossary is identified in the text at first use by italics or boldface or quotes, and the reader is then told that such highlighted terms are defined in the glossary. For example,

> It is a practical convenience to be sure that the O scope has a *timester probe* (see the glossary for explanations of *italicized* terms).

You will find this technique used in Model 5.0(2) on page 190 and Sample 8.D on page 332.

You can also build the list of terms as a group of footnotes that have been removed from the text. I find this to be a more complicated mechanism, but it is popular in academic writing. Essentially, all the footnotes are removed from the page bottoms and put in the back in a section called "Endnotes." If most of the notes are definitions, they probably better serve readers at the bottom of the pages where they are needed. Once they are clumped together in the back, they are not a convenient reference because they are not alphabetical. They are, instead, numbered sequentially.

◀ *As this table shows, an in-text list can be quite substantial and yet take up little space if the author is simply explaining the meaning of abbreviations or acronyms.*

Sample 8.N

-2-

It is important to know how to disassemble your computer, keep track of what plug-in device goes to what plug, and recognize key components if you plan to replace the motherboard. Keep in mind that with IBM-compatible computers, this task is expected to happen. The creation and development of the IBM compatibles has always been intended for integrating.

The designs of these types of computer systems are flexible and upgradable to adapt to many different types of computer components. These systems are designed so that you can do what you are about to do! Before ordering your motherboard, get to know your computer by going through the first two steps of replacing the motherboard: disassemble of the computer case and proceed with the following preparations.

Case The case is the frame that contains all the computer components. It usually is metal and comes in two main parts: the shell, which holds all the parts, and the cover, which is easily taken off.

CD-ROM This data-storage format is just like a read-only disk drive for compact disks (CDs) except that it is full of computer data instead of music.

CPU This stands for "central processing unit." This is the brain chip of the whole system.

ISA An old type of peripheral slot still commonly used. Often referred to as a "16-bit slot."

Motherboard This is the body of your whole system. It allows communication to go on between the CPU and outside devices, peripherals, ports, and keyboard. It does this by using the on-board memory.

PCI This is a newly developed slot. Small but compact with connection points, it is mostly a 32-bit device nowadays.

Peripheral A piece of equipment, such as a printer or a monitor, that can be used with a computer.

Peripheral Card Each outside piece of equipment has a peripheral card that allows for communication with the motherboard and CPU. A video card, sound card, and modem card are all peripheral cards. Usually a cable or cord connects the peripheral with its card, but a card can also be a peripheral in and of itself. Sometimes a motherboard has a built-in peripheral card, such as a controller card for hard and floppy drives.

VLB Also called "VESA local bus." A certain type of peripheral card slot that is not commonly used anymore.

Some rules are made to be broken, and a list of terms could perfectly well be placed in the front of a text. After all, a "list of equipment needed" always goes in the front for perfectly logical reasons. When you think about it, the "list of terms" is the equipment needed to understand the document. A symbology is a case in point. If you have a list of symbols that you have defined, it can be placed in the back as a variation on the concept of the list of definitions. On the other hand, the symbology might be a valuable asset in the front of the document also. Use your audience as your guide. The following symbology, for example, is an appendix, but it could be positioned in the opening pages of the author's project if the symbology will be of value of most readers.

PCB Dimensional Inspection Symbols

UNDERWRITER'S LABORATORY LOGO

PERPENDICULARITY (one feature to be 90 degrees to another feature.)

PARALLELISM (One feature to be parallel to another feature.)

CONCENTRICITY (One hole, bore, c'sink, etc. to be concentric to another.)

FLATNESS (The indicated surface to be flat. Warpage to be within tolerance.)

PROFILE (Surface roughness not to exceed the specified allowable limit.)

COUNTER BORE (The hidden line shows a counter bore on the far side of the hole.)

VIEW DETAIL (Elsewhere on the dwg. is an enlargement of a more detailed view.)

HIDDEN DETAIL (Shows an item on the far side. A standoff or Pem-nut is typical.)

FLAG NOTE (Shown to draw attention to details that are of particular importance.)

Feel free to explain terms in your own words when it is appropriate. You can also use dictionary definitions from technical dictionaries (which should be identified as a source), but you should understand that definitions are conditional, not absolute. Terms and concepts are conditioned by how you used them and how you want the reader to understand them. You may often find that the version in your technical dictionary has to be explained! In the same sense that you target the readers by word choice, you target them by word explanations. You have to decide how extensive the definitions must be and when enough is enough.

A block of definitions (in this case a sidebar) is a particularly popular device in magazines. It can be conveniently used in single-sided projects as well.

Sample 8.0

-12-

GLOSSARY OF TERMS

Bias current—A steady, constant current that presets the operating threshold or operating point of a circuit.

Bias voltage—A steady voltage that presets the operating threshold or operating point of a circuit.

Dynamic conditions—The performance characteristics of a device or circuit under AC signal operating conditions.

Forward bias—Forward voltage or current in a transistor.

Gain—Same as amplification.

Impedance—The total opposition offered by a circuit to the flow of alternating current.

NPN—A bipolar transistor in which the emitter and collector layers' are N-type material. The base is P-type material.

Reverse bias—Reverse voltage or current in a transistor.

Static conditions—an operating characteristic determined from constant rather than fluctuating values. The DC characteristics of transistors.

SYMBOLS

V_{BB}—The base voltage supply for a transistor.

V_C—The collector voltage of a transistor.

V_{CB}—DC collector to base voltage.

V_{CC}—The collector supply voltage of a transistor.

V_{EB}—DC emitter to base voltage.

V_{CE}—DC collector to emitter voltage.

V_{EE}—The emitter supply voltage of a transistor.

The Symbology

The standard symbols that are used in a discipline, such as digital electronics, pose a special problem. Symbols and codes often need to be explained, particularly if there is some likelihood that readers will be unfamiliar with these handy shortcuts. Because technical graphics are highly detailed, symbols and codes are most likely to appear in illustrations, although they also may be important for the written text.

If codes can be put in alphabetical order, design the list accordingly. For symbols, list the symbols on the left and the explanation to the right, or list the symbols laterally and place the explanations below them. Since this arrangement is *not* alphabetical, organize the symbols in the manner that is popularly accepted in your field.

Place the symbology before or after the glossary, or in place of the glossary. If the text is for novices, you might do well to put the symbology in or near the front of the text rather than in the back.

DESIGNATORS:

C	Capacitor	Q	Transistor
CR or D	Diode	R	Resistor
F or X	Fuse	RP	Resistor pack
J	Jack	SW	Switch socket
JP	Jumper	T	Transformer
K	Relay	U	Intergrated Circuit
L	Inductor	P	Plug

SYMBOLS:

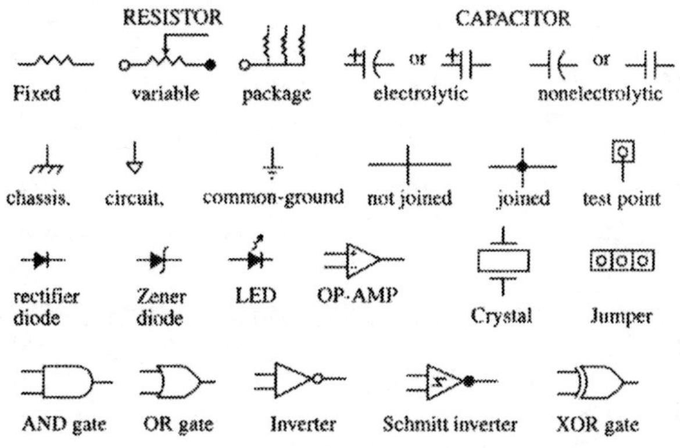

RESISTOR: Fixed, variable, package

CAPACITOR: electrolytic, nonelectrolytic

chassis. circuit. common-ground not joined joined test point

rectifier diode, Zener diode, LED, OP-AMP, Crystal, Jumper

AND gate, OR gate, Inverter, Schmitt inverter, XOR gate

If a vocabulary is quite extensive, the terminology is best removed from the text and explained in a list of terms, abbreviations, or symbols.

Samples 8.P

This mode is used with the **index register.** The first byte is the operational code, and the second one is called the **offset address.** To operate this kind of instruction, the offset address is added to the index register. The result presents the **operand address.**

Another type of compatibility problem is less obvious. A **double-density disk** (360 K, 5.25≤ K or 720 K, 3.5≤) and a **high-density disk** (1.2 M5.25≤ or 1.44 M. 3.5≤) are not compatible. Though you can format as well as read and write to a double-density disk in a high-density drive, you cannot do the reverse.

Hence, it is called a mobile ground. **It is very important that this test circuit is turned off to prevent unimpeded current from rushing into the oscilloscope ground and causing damage.** The X probe is "hot"; it is directly connected to one side of the transformer. It will reject the full AC potential where it touches, which will seek the probe ground.

WARNING

If a substitute fastener is to be used, it must be equal to or better than the original in ALL respects, and it must be approved by Engineering PRIOR to installation.

Boldface

A widely used feature in technical writing is boldface typescript, which is used to signal important considerations. For example, all the terms that are to be defined in one way or another can be boldfaced (or italicized) to highlight the terms (see Sample 8.D). This highlighting can be a first-use boldface, or the terms can be highlighted throughout. The very first occurrence in the document will call for an explanation of the boldfaced word (see the footnote in Sample 8.D). There are several examples of boldfaced terms in the samples on p. 358.

Another prominent use of boldface (and italics to a lesser extent) is to alert readers to the warnings that are frequently needed in technical documents. The usual problems involve potentially costly errors that can damage apparatus, or risks that involve the reader's health and well being. Do not hesitate to use boldface when it serves the purpose of reaching for your reader's attention.

The use of uppercase or italics can serve the same purpose as boldface. The idea is simply to highlight such features as jargon, particularly if the terms are somewhat similar to conventional words. This situation is most apparent in programming and computer sciences.

Observe Sample 8.I on p. 342 and you will notice that the use of caps adds considerable clarity to the author's text. Most of the terms are abbreviations and acronyms that would be capitalized in any case. The intent of using uppercase lettering, italics, or boldface is to add a visual cue. The cue distinguishes the vocabulary and adds clarity to the reader's perception of the text. Notice my heavy use of boldface in Appendix B.

Unfortunately, the samples here do not begin to show how complex the levels of vocabulary or types of jargon can become. Notice the following sample on page 360. The author graduated with a two-year engineering-tech degree in network technology. He was developing a user's manual for the university where he works as a technician. He was very alert to the conventions that would help clarify the text, and at least a few of the typographical devices should be apparent to you (see Sample 8.Q). If you now go back and look at Samples 8.J on p. 346, you will realize that they need to be designed with visual cues similar to what you see here.

The boldface samples point out the emphasis that you can add to today's computer-generated documents. Your instructor might suggest a protocol for boldface and italics that is appropriate in your field.

Sample 8.Q

Conventions

The following typographical conventions are used:

Commands	Commands appear in bold print and should be typed exactly as they appear.
Variables	Variables appear in italic print and should be replaced with an appropriate value when they are typed into the computer. The names of the variables will be as self-explanatory as possible.
`Prompts`	Prompts and other output generated by the computer are displayed in the Courier typeface.
<Command Keys>	Command keys are enclosed with < > and are used to denote pressing a particular key (Enter, Escape etc.).
"Text in quotes"	Bold text in double quotes refers to a manual section heading.
"Text in quotes"	Plain text in double quotes refers to a menu item.
'Single quotes'	Single quotes have no particular meaning in this document and are used only to make the text more readable.

For example:

1. To display the protection bits of a file, at the system prompt . . .

```
watson:/ul/users/sudduth%
. . . type . . .
ls -l file_name <enter>
```

2. The computer will display something like . . .

```
-rwxr-xr-x 1 sudduth 7379 Mar 11 12:04 file_name
```

The author of Sample 8.Q decided to establish a set of conventions for all his documents. They begin with the table of conventions on the left and this explanation for the reader's convenience.

Publishers go a step beyond our visual methods of typeface management and create vivid documents with imaginative alternatives that would not normally occur to us. Sample 8.R is handled in the style of a professional publication. The author wrote the text as a college project in technical writing, but he had the computer skills to embed keys into the running text. He had seen this practice in manuals and set about to design a similar product.

Obviously, the best judge of your success at targeting an audience is the target itself. You will benefit from a trial run of some sort. If someone or some small group will read your work, you can test the readability quickly enough. The readers should, of course, be typical of your planned audience. Ask for their perceptions and make a note of what worked and what did not. Revise accordingly—and be very open minded about criticism.

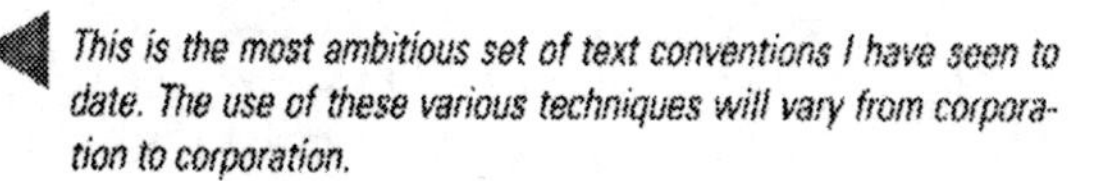

This is the most ambitious set of text conventions I have seen to date. The use of these various techniques will vary from corporation to corporation.

Sample 8.R

4

Generally, main directories such as your different mathematics-based classes will be your main directories located at Home position. Directories can be created in many ways. The procedure that I prefer for creating a directory is as follows:

1. Type in the name of the directory that you want to create by holding down the α key in accordance with typing the buttons that have the letters printed to the bottom right of them. Press the ENTER key.
2. Type in these letters in this order with the α key: CRDIR
3. Press the ENTER key to store the new directory.

Here is an example that you can try. The following line shows you the sequence of keys to press to create a directory named MATH:

Pressing the Alpha key (α) twice locks the button down so that you can type a word with many letters instead of pressing α before every letter you want to type. Pressing ENTER unlocks the Alpha key. The directory that was just created (named MATH) should have appeared in one of the available boxes at the bottom of your screen. If it did not appear and you cannot find it, refer to the italicized writing in the section Storing Variables. When you locate the new directory, press the □ button directly beneath it to open that directory. (It should be empty.) Now you have a place to store your variables.

Samples 8.A through 8.R

Comments

Samples 8.A, 8.B, and 8.C were taken from the title pages of a number of projects, both corporate and academic. Observe the precision of each target description, which is a useful tool for project development.

The full-page samples 8.D through 8.H are indications of style changes that can be made. Notice that they are generally user friendly, but they vary from the casual magazine-article style of the COBOL project to the more formal, but instructive, manner of the modem discussion. The sample concerning bridge structures (8.H) is typical of a project addressed to professional readers in both its formal manner and technical content.

Samples 8.J through 8.O illustrate the various ways in which vocabulary can be defined in a text. The first two sample pages demonstrate in-text definitions of terms and abbreviations. Sample 8.L demonstrates the use of footnotes for definitions.

Samples 8.M and 8.N use in-text lists. These can be handled in sidebars or text boxes or text. Sample 8.O is a glossary from the end of a text.

Sample 8.P demonstrates the practical uses of boldface. Sample 8.Q is an unusually thorough list of type conventions used in a manual.

Summary

- Measure the needs of the readers.
- Determine what level of complexity they will understand.
- Develop the text at the level desired.
  - ✔ Use appropriate vocabulary and appropriate definitions, the two important controls.
  - ✔ Adapt the lengths of sentences, paragraphs, and documents to the readers.
  - ✔ Carefully measure the complexity of math, symbologies, and symbolic graphics (schematics).
- Organize the body of the document in terms of reader interest. The content could be sequential, prioritized, or structured in some other manner.
- Adopt an appropriate tone—usually formal.

Activities Audience Targeting

Develop a brief report (500 to 1000 words) in a memo to your instructor. Use one of the following suggestions for the exercise.

Develop a paragraph that calls for the definition of at least six terms, acronyms, or abbreviations. Construct several versions that use different techniques for handling the definitions. Review your target audience and decide which of the samples is superior to the others.

If you are presently employed at a company where you have access to brochures, sales materials, manuals, and other documents, collect a few and analyze the way in which each uniquely targets an audience. What elements are used to target the intended readers? Can you profile the intended readers?

---

*Look at the advertising copy in several magazines, including one from the popular press (*Sports Illustrated*, for example) and one that is dedicated to a selective readership (*MacWorld*, for example). Can you see elements that are used to target readerships? Can you profile the intended readers?*

If you have an important but impossibly difficult manual that is indispensable, write an improved version that will make a garbled section of the manual readable. You are, in this case, the author and the reader. Profile yourself.

Share a Project

1. Work with three other members of the class to analyze audience profiles. The groups can consist of students in one or more specialty areas.

- *Discuss the students in your programs as readership groups and decide if any of the following aspects of the possible profile are relevant, and discuss.*

Income	**Employment**
Education	**Marital status**
Age	**Gender**
Interests	**Technical training**

- *As a group, propose a profile of the selected students and delegate a member to write the profile at the meeting. Focus at least half of the profile on the precise technical knowledge the target group does or does not have.*
- *Discuss how you think the target group is best addressed in terms of the following considerations:*

Vocabulary	**Extent of material**
Definitions	**Length of material**
Sentence length	**Manner of explanation**
Paragraph length	**Order, tone, repetition**
Reading time	**Uses and types of graphics**

You may review whatever textbooks you have with you as part of the discussion—including the price.

2. Work with two other members of the class to develop brief project samples (500 words) that are designed to target specific audiences. For your audiences, each team member should use one of the basic three prototypes suggested on p. 328.

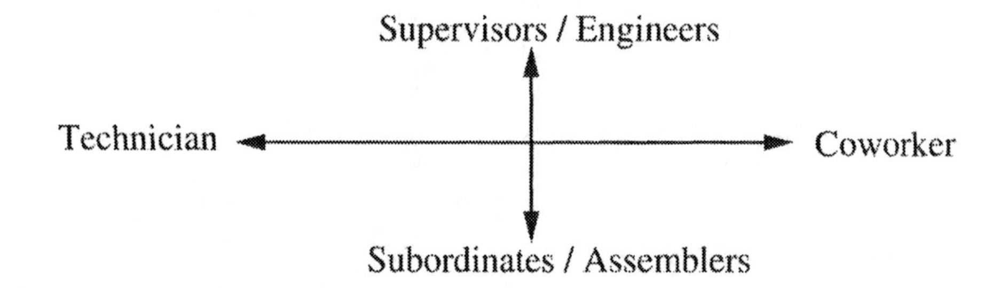

- *Develop a brief profile of the target audience.*
- *Have each member develop a sample text.*
- *Discuss how you think the target group is best addressed in terms of the following considerations:*

Vocabulary	**Extent of material**
Definitions	**Length of material**
Sentence length	**Manner of explanation**
Paragraph length	**Order, tone, repetition**
Reading time	**Uses and types of graphics**

- *Share the outcomes and discuss.*

---

3. A special task group of volunteers could undertake the following interesting project.

It would be informative for members of the class to see various document types from the workplace from those class members who are currently working in a technical environment. Four or five such members could form a group to discuss document design and readership. Each member would present and discuss an assortment of company documents. The discussion would be based on the conditions identified in the first group exercise.

In order for the class to have an opportunity to hear the discussion and see the documents, the special task group would be seated as a panel, at several tables, in front of the larger group, and the samples, once passed among panel members, could be passed to the audience.

For this activity any of the following items could be discussed:

Specifications
Training manuals
Repair manuals
Maintencance manuals
Safety guidelines
Abstracts
Minutes
Bulletins
Business letters
Memos
Promotions
Brochures
Fliers
Proposals
Newsletters
Policy manuals
Lab reports
Site reports
Progress reports
Environmental studies

Work In Progress

9. The Rough Draft

I started on my rough draft once I had my outline, graphics, tables, and red-lined texts. I put the graphics and tables in sequence. This gave me a visual image of the distribution process flow problem and shaped my document. The types of graphics I used ranged from the test setup schematics (shown previously), to flowcharts, photos, and figures. Shown below is a simple example of a flowchart for the changes regarding meter stock handling.

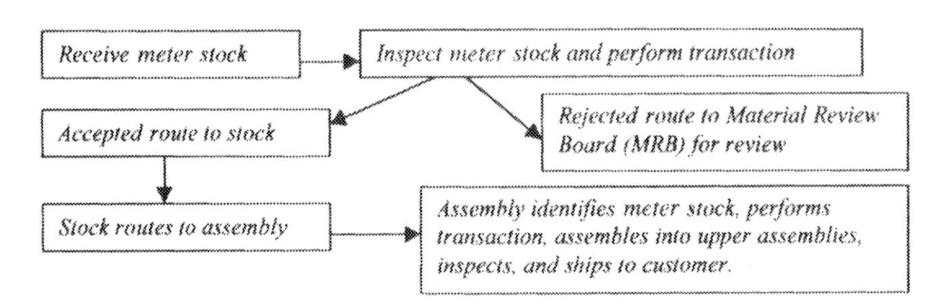

Meter Stock Handling Baseline Flowchart

Since I received graphics from several different departments, it took a day or two to confirm where they fit in the flow. The flowcharts will be used to identify where an action is performed in the distribution process. Inspection and Assembly will use the test setup schematic figures as an aid in testing the parts. Assembly will use the photos as an aid in assembling the parts into upper-level assemblies. The tables will be used in the procedure descriptions and instructions by all departments.

The red-lined texts from different departments took a day or two to assemble so that they would match the graphics and the overall sequence. As I mentioned earlier, the distribution process is used by multiple departments, and each department has variables within it. The trick is to align everything so it will work for all departments. The other trick is to write it out clearly!

Once I had everything in line I used the various types of software to input the changes into the existing documents I had previously saved onto my personal drive. I then saved my changes and printed a copy of these to use as my rough draft. I write "To" on the changed documents when these are submitted for final approval. For now, I write "Rough Draft" on them for identification.

S. B.

Modeling Concepts into Illustrations

Technical writing has provided us with a medium through which we can understand the technologies of the future. It has done so by setting our writing method back about 3000 years. Our first efforts at writing consisted of images. Painting preceded writing, we can be certain. Language probably developed before it dawned on us to either paint or write, but writing was probably pictorial in at least a few of the oldest writing systems, such as Chinese pictographs and Egyptian hieroglyphs. Visual presentations of knowledge are an ever-present need, and writing that emerges from scientific and engineering fields is particularly dependent on graphic systems of communication.

Engineers often use graphics as a modeling activity, a process for constructing working designs of their ideas. Graphic mediums are indispensable as the interim logic of the creative process used in scientific fields. Between the theory and the application, you will find the modeling, the design work, the adaptation of theory to practice. Much of this sort of work is graphic, whether you are looking at DNA chains of ping pong balls or computer simulations of air flow dynamics on a wing design or blueprints for the HVAC system of a new building.* The following drawings were created by an author for a paper on solid modeling with AutoCAD.

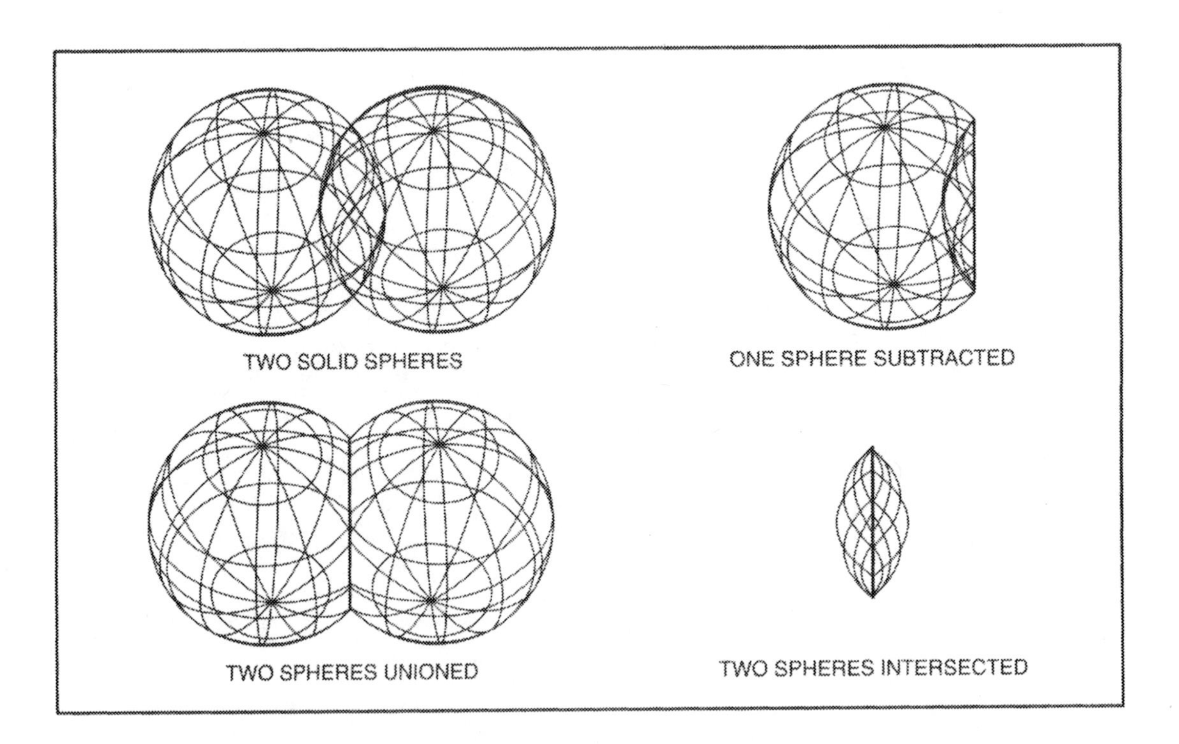

From the practical standpoint, many writers are unfamiliar with the challenge of integrating graphics into a text, but it is a critical concern for good technical documentation. Publishers have always requested (required) that a text be separate from the graphics that an author wants to use to accompany a text. The tech writing or publications department of your corporation may operate on similar standards. From the point of view of publishers, a manuscript is a long way from being finished. Their work begins only when the writer is finished. They prefer to see the manuscript as a text. They want to position the graphics.

* *An excellent analysis of engineering in terms of "visualization" can be found in a study from MIT by Eugene Ferguson:* Engineering and the Mind's Eye.

However, there are drawbacks to separating the text from the graphic elements. If the author is allowed more control of where the graphics are supposed to be positioned, then he or she is in effective control over how they are to be *used.* Graphics are not decorations. They are tools. They are a language also, one of many available to engineers and technicians. In addition, writers do not usually write for a publisher. Most authors usually write for very limited distributions. For example, you may be writing term projects for your professors in an engineering program, or you may be writing for your supervisor at work or for a small committee. The work may be intended for a narrow range of clients. You might also be self-publishing for a business. The project may simply reflect the needs of a small company, and there may be no one to develop a document all the way through to a publishable level of completion except you. The following illustration was one of many three-color drawings from a project concerning optics. It was intended for one reader.

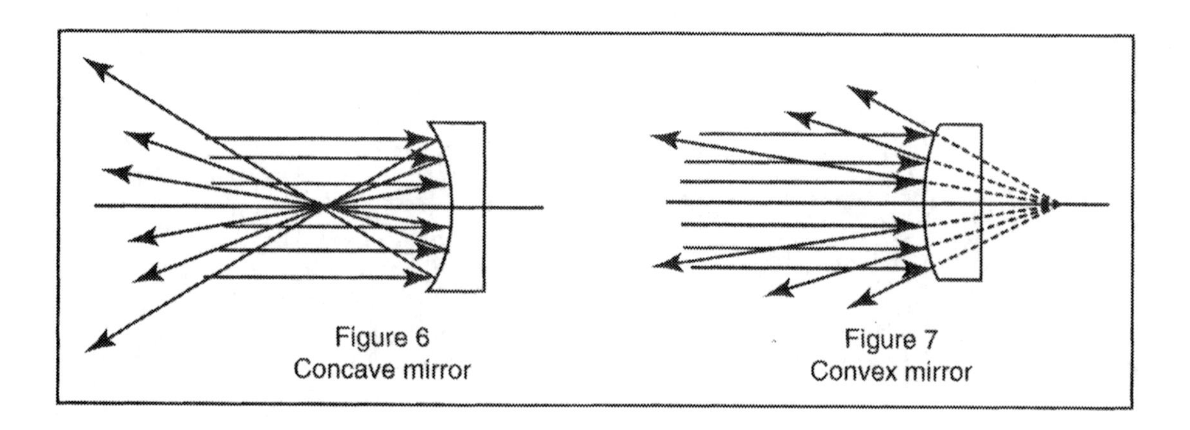
Figure 6
Concave mirror

Figure 7
Convex mirror

Yet another consideration is the amazing progress in word processing equipment in the last ten years. Desktop publishing has reached such a high level of craft that student work now looks camera-ready. Not only do writers have the benefits of word processing features, they also have design assists such as the template that helped build the table in Sample 9.B. There are a number of graph and table templates and similar features in most up-to-date software installations. What used to be thought of as "manuscripts" are now looking like galley copy or dummy copy from a professional designer! Laser printers, ink jets, scanners, dry copiers, and a host of processors and support software have dramatically changed an author's ability to package technical documentation. And the work often appears in color.

Sample 9.A

8

Appendix A

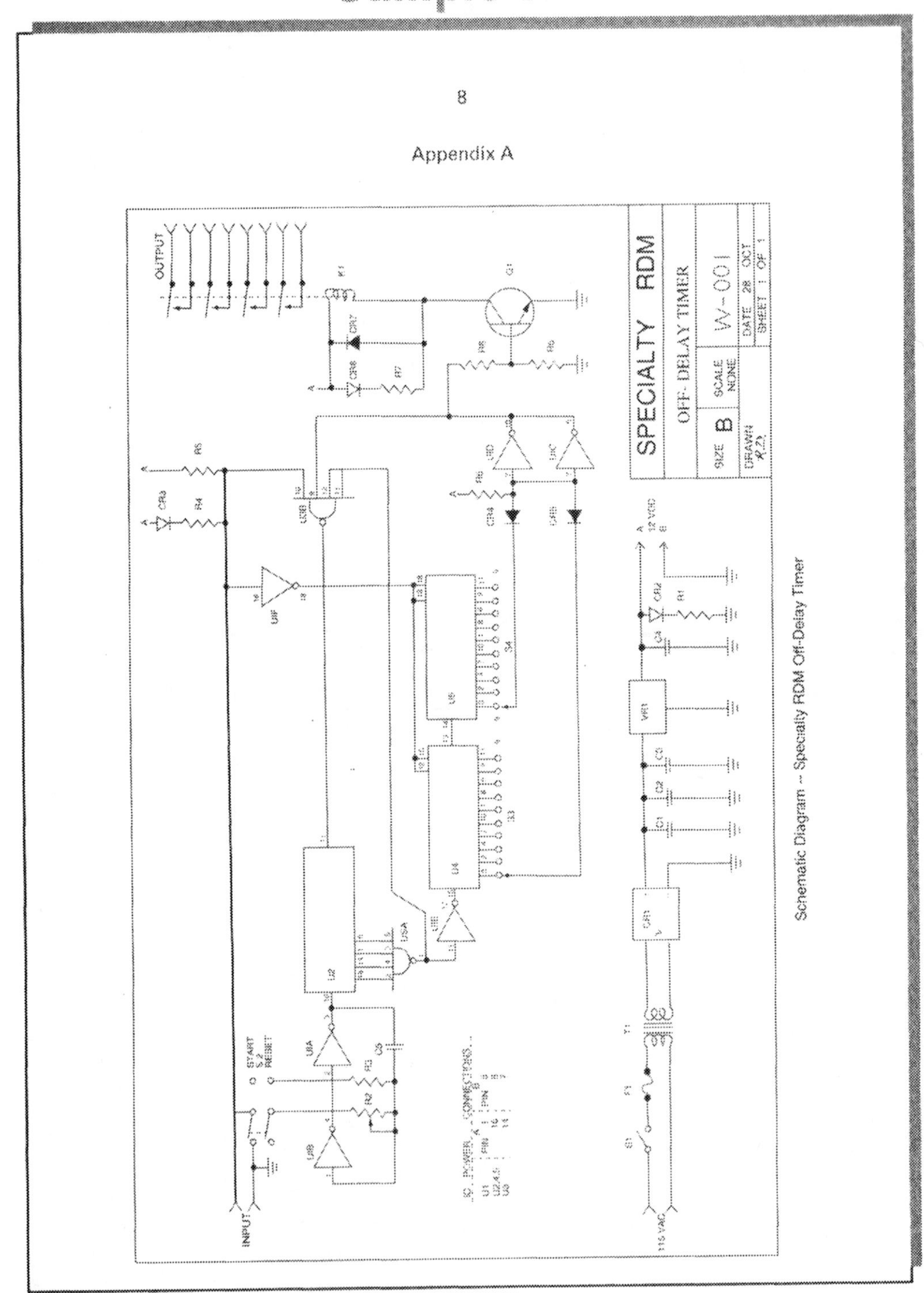

Schematic Diagram -- Specialty RDM Off-Delay Timer

First, consider the support systems. Your projects do not have to be elegant, but if you are using a word processor, you are on your way to a professional-looking product. Laser printing is considered the best available. Ink jets are improving. Graphics can be imported, or you can create pasteups and photocopy the final copy. Pasteups are often superior unless you have access to very high end scanners.

As you know, in conventional printing processes the illustrations are usually translated into a dot matrix for printing. A true black-and-white photograph uses gradations of gray and has no matrix pattern. This has always been a barrier for reproduction with ink. Unlike the emulsions on photographic paper, inks will not register gradations unless they are composed in dots. The ink bleeds, seeps, runs, and otherwise turns into a mess unless it is composed in those little dots you see in the pictures used by newspapers. The visibility of dots per inch (dpi or lpi) is usually very noticeable at around 100. Newspapers will use an even lower resolution because the ink is poor, and the coarse paper is destined to be discarded anyway. An inexpensive scanner operates at around 300 dpi. High-end scanners might be sold at much higher dpi ratings, but 300 dpi will meet your needs.

It is very practical to look seriously at creating a fully integrated text in technical mock-ups because of such equipment. On the other hand, by any measure, scanners are only basic tools; with a video-capture card you have the software to freeze-frame video images and transport them into your system. Or consider the GIF editor, software designed for viewing pictures on your computer screen. Because the images are digitized, you can then print them, preferably with a high-resolution color laser printer. For schematic work, there is the multicolored ink-jet printer. In color photography, images can now be rendered in digital format for alteration in Photoshop or other applications. A relatively recent article I published had a photo that had unsightly power lines; the publisher "removed" them by digital correction.

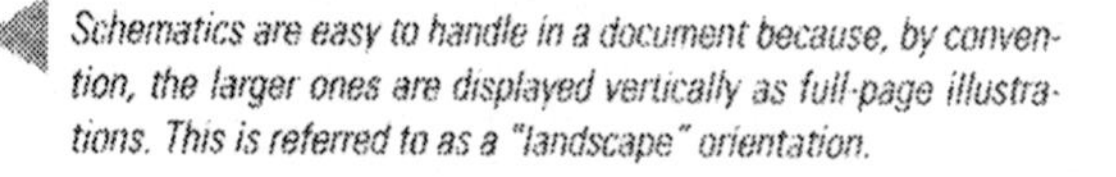
Schematics are easy to handle in a document because, by convention, the larger ones are displayed vertically as full-page illustrations. This is referred to as a "landscape" orientation.

Sample 9.B

- 4 -

Table 2. Regional Transportation Options

	Considerations						***Effects***		
	Costs		***Traffic Impact***		***Public Safety***		***Environmental Impact***		***Vehicle Reduction***
ALTER-NATIVES	Construc-tion	Opera-tional	Construc-tion	Opera-tional	Riders	Traffic	Pollution Reduction	Reduction of Fuel Usage	Single Occupancy
Heavy Rail or "Rapid Transit"	*Negative* Astronomi-cal to tax-payer	*Negative* Very ex-pensive to tax-payer	*Negative* Very signif-icant to ex-isting flow as currently proposed	*N/A* Moderate impact on existing flow	*Positive* If crime is addressed with air-port laws	*Neutral* If access to system is isolated from traffic	*Positive* If ridership significantly increases	*Positive* If ridership significantly increases	*Positive* May de-crease tem-porarily until population increases
Increase Existing Bus Ser-vice	*Negative* Moderate increase to taxpayer	*Negative* Moderate increase to tax-payer and riders	*Negative* Mild to in-significant as currently exists	*Negative* Increase in vehicles will add to congestion without HOV lanes.	*Negative* Weapons and firearms not regu-lated	*Negative* Hazards increase as vehi-cles in-crease.	*Negative* Increase in air and noise pollu-tion	*Neutral* Depends on increase in ridership.	*Neutral* Until travel time and public safety have been addressed
Increase HOV Lanes	*Negative* Significant increase to taxpayer	*Negative* Moderate increase to tax-payer	*Negative* Significant effect on normal flow	*Positive* Insignifi-cant effect on flow	*Neutral* Insignifi-cant	*Neutral* Depends on driver skill	*Negative* Increase in air and noise pollution	*Positive* Mild de-crease in consumption	*Positive* Significant reduction

That was two years ago. Now I can use a digital camera and shoot a "file" and mail a disk that is ready for cosmetic touchups. Then there are the CATIA* systems for creating drawings that, as you know, can then be rotated and viewed from any angle because the computers "visualize"—a 767 wing assembly, let's say—from any angle on command (only $80,000 a "seat," as one engineer explained). For a fraction of the price, you can learn basic CAD for generating drawings, such as the one in Sample 9.C.

Without a computer you may feel that you have been left at the starting gate. However, you do not need all this apparatus. Graphics designers are creative people, and they got along quite nicely without the conveniences. The new breed of electronic assists simply simulate digitally the original method of creating the layout for a graphic insert.

The traditional method of integrating graphics into a text is also the easiest. It may serve your purposes to use the conventional pasteup. Many authors, even those with considerable computer experience, will occasionally use the pasteup, either for convenience or for technical reasons concerning the illustrations. All it takes is a glue stick. The idea is to leave spaces for graphic elements, which are glued into position. The finished product is then copied. Here is the method:

1) **Feel free to use "borrowed" illustrations if you are a college student.** You are under no strict obligation concerning copyrights, partly because you have no intention of publishing the results; however you must credit the source of the illustration. Under the caption, identify the author and the title of the work that was the source for the graphic. If the author, in turn, borrowed the graphic, ignore his source. Credit the source in your hands. For formal academic projects, these sources can be fully identified in bibliographies or footnotes. For corporate use, you will need to seek permission to use someone else's illustrations. A company usually wants to use original graphics. The real convenience of borrowing for college use is that no one expects you to be an artist or a photographer.

* *Computer-Graphics Aided Three-Dimensional Interactive Application.*

Tables often exceed the width of a page and are presented in the landscape position. These are numbered pages of the text.

Sample 9.C

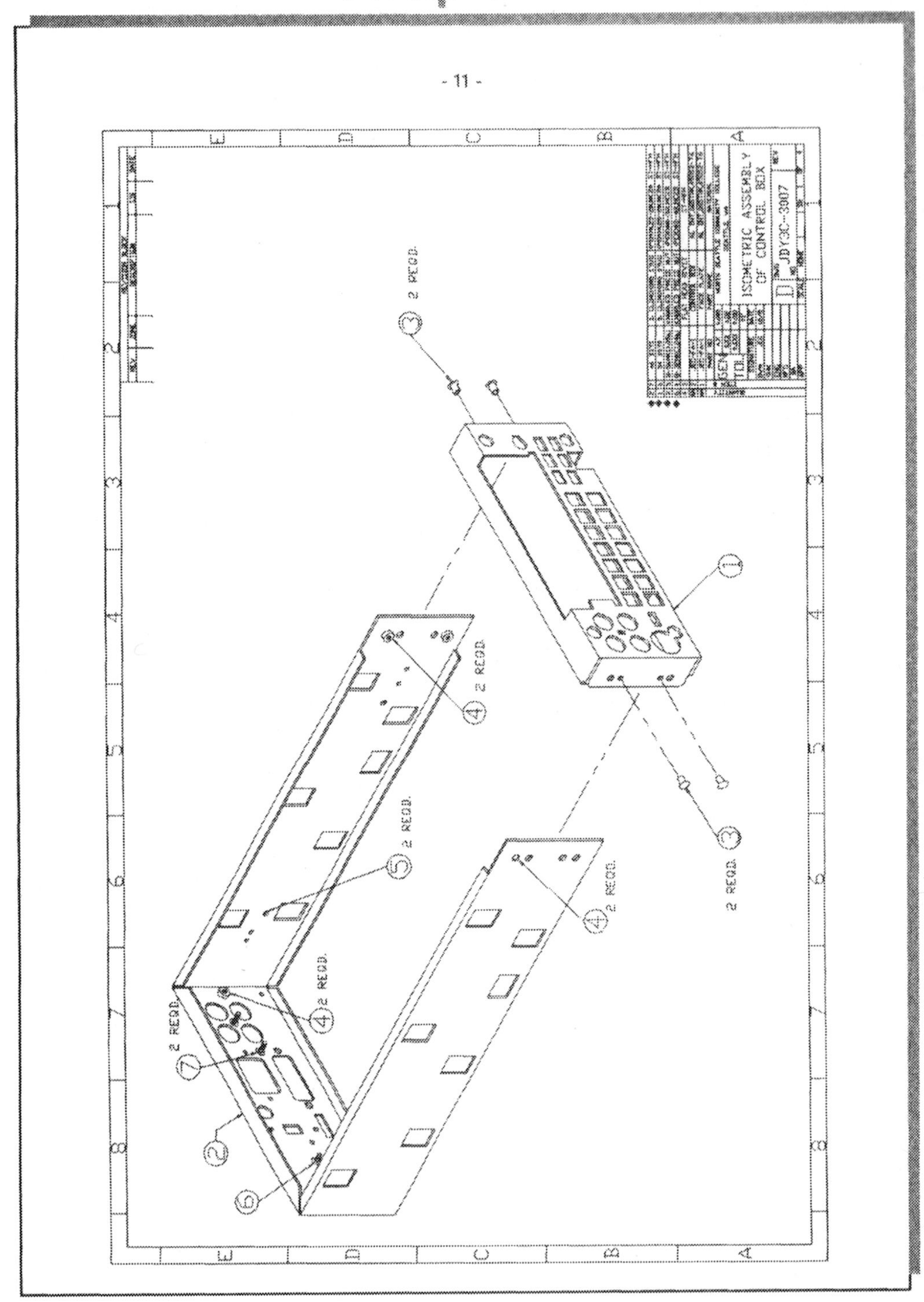

2) **Try to use a good copier if the project is important.** Countertop office copiers are adequate, but remember that you will later be copying the copy. Successive generations of copy always lose quality. By the third generation, for example, my video editing is a mess. Fortunately, there is not a dramatic loss in resolution on dry copy. The higher your standards at the outset, the better the result in the long run. The very high end machines at the local photocopy shop usually work very well.

3) **Color photographs will copy quite accurately in color, but the second generation is iffy.** I have used the pasteup as the final copy when color balance was important.

4) **Do not use too much glue or it will crinkle the page.** Do not use "temporary" glue sticks because they tend to lack adhesion. Use a one-inch stroke from a permanent glue stick. You can tear it loose if necessary. Do not use cellophane tape because your fingerprint may be on the tape, which can leave a smudge on the copy.

5) **The edges of the illustrations used on a pasteup are the curse of the project.** The copier sees the shadows on the edges of your graphic as lines and promptly creates them for you. The best copiers usually do not have this problem. To avoid shadows, be sure the pasteup is flat against the glass platen. Everyone has solutions for dealing with shadow lines. The usual one is to mask the borders of the graphic. The edges can be covered with liquid correction fluid or correction tape. Transparent cellophane tape works as well. The graphic can also be torn on all edges; that is to say, scissors leave the sharp edge that makes the copier see the line of shadow around your picture. If you just tear the picture with your hands, the rough edge casts less of a shadow. This is an old trick photographers use once in a while. You can sometimes even erase light shadow lines.

An assortment of other graphic considerations call for the vertical framing of the landscape layout. The left side is the top, always. If it is reversed it is considered upside-down.

I use pasteups extensively because I do not have handy access to high-end scanners, and I usually use simple line art, as you will have noticed. I use AutoCAD graphics, but importing them was such a headache until two years ago that I photocopied the masters on a reduced scale to fit my work and used conventional pasteups. All the material—text and graphics—was on disk, but it was faster to use a glue stick for the final integration of the parts because of software compatibility problems. The current AutoCAD for Windows (13 or 14) imports quite easily and would now solve my difficulty.

A few other "older" tools that might be of use include templates and transfer lettering. Chartpak, Koh-I-Noor, and Mecanorma are a few of the companies that manufacture these materials. You are probably familiar with the *symbol templates* used for a lot of schematic design work in a variety of engineering fields. Plastic templates are a quick and neat convenience and require no electronic support. *Transfers* are large sheets of decals that are rubbed into position. Graphic designers historically depended on them, and they are handy to have available. The catalogs of transfers illustrate hundreds of fonts and a wide variety of sizes. There are also extensive graphic components, of the clip-art variety, available in transfers. For example, there are sheets of map symbols and of landscape architecture symbols. If you want to create an aerial view of a landscape, you can construct it with the symbols. There are sheets of representational materials, such as brick walls or stonework, sheets of dot matrix patterns, and so on.

All these graphics have been programmed into software applications, but you might want to know about the original resource. The sheets of transfers from a local art and graphic-design supplier may be a lot less expensive than a software application you would seldom use. Also, transfers allow you to create interesting graphic buildups, such as adding aerial drawings to an aerial photograph of, let's say, a five-acre lot development. This can be done with Photoshop or similar software, also, but one method may be preferable to another at different times.

Using transfers is a slow process. Before software speeded up the design work, there were efforts to automate at least the lettering procedure. There are machines such as the Kroy lettering system that will "type" large transfer letters onto transparent tape. There are applications, such as placing callouts on a large photograph, for which such a system is quite appropriate and not easily replaced with standard word processors at any reasonable cost.

A final note on pasteups: It is possible, and practical, to go so far as to paste-up the *written* text of the project, also. In other words, in a worst-case scenario in which you cannot figure out how much space to allow for graphics, or in which you do not have the time or patience to anticipate spacing, there is another way to go. Create the entire text as a printed document and use the printout as part of the paste-up. For example, the first graphic goes on page two after the first paragraph. Cut off the rest of the page of text. Mount the paragraph at the top of a fresh sheet of paper. Mount the graphic below it. Cut some more text into position to finish the page. For a college paper in the range of ten pages, this method does not take long. It has only one minor limitation, and that is that you cannot type along the sides of the graphics. You can place text only above or below the text cutouts, which obviously have to be full width.

To Construct a Graphics Box

1. Make sure your drawing toolbar is displayed. If it is not, go to **View** on the top toolbar, then **Toolbars,** and click on **Drawing.**
2. On the drawing toolbar, click on the square icon that has an **A** in it.
3. Draw a box by clicking on the point where you'd like the box to be and then dragging away from the point until you get the shape and size box you want.
4. To move the text box, click its border and drag the box where you'd like it to go.

Computer buildups are simply another version of a pasteup. Instead of a glue stick, you click on toolbars among the screen icons to find a drawing tool. Follow the instructions and construct a box of any size desired. The box perimeter can be left in place, but it can also be removed without losing the space. The computer will maintain the graphic position even when you delete the border of the figure box, and the box becomes transparent. If you do not have a scanner, you can develop a pasteup with your printout of the text if you have all the graphic boxes in place. Using a copy machine, you can scale your graphics to the size of the boxes you have created and then glue the graphics in position. Make a final a photocopy.

Graphics can be electronically imported in a host of ways. *Microsoft Draw* will embed graphics into *Word* and other programs. These images are considered superior to Microsoft *Paint* output. Claris *Works* is an all-in-one application that includes word processing, graphics, spreadsheets, and other features. IBM and Macintosh scanners are popular tools, and in most cases come with graphic manipulation software to alter your graphics. You can now take images directly off the Internet and import them into *Word*. You can also use a file from a digital camera.

A Graphics Sampler

This section is a sampling of graphics created by engineering and engineering tech students. These graphics will encourage you to experiment with graphic media. The samples illustrate a variety of strategies you can use to develop graphic support from among the broad assortment of graphic types that are conveniences you can use to meet specific needs. Some of the productions are quite simple, and others are sophisticated outputs of computer software. Use whatever works. You should feel free to use existing resources, but all the samples here are originals.

Photographs

I use photographs extensively for magazine articles. Polaroids are quite high in quality and certainly serve for mock-up work or any job short of professional. A photographer works with me on my magazine articles and he takes "roids" by the dozen to isolate and design and balance a picture. Only then does he get out the expensive color transparency film and shoot in 4" × 5" format. The black-and-white Polaroids are crisp, and they copy quite well. For color, I would suggest a 35-mm slide film with high color saturation on the order of Fuji Velvia. If your photography work is close to what you intended, it will serve your purpose, and the pictures will always serve to show a professional photographer or editing staff exactly what you want done.

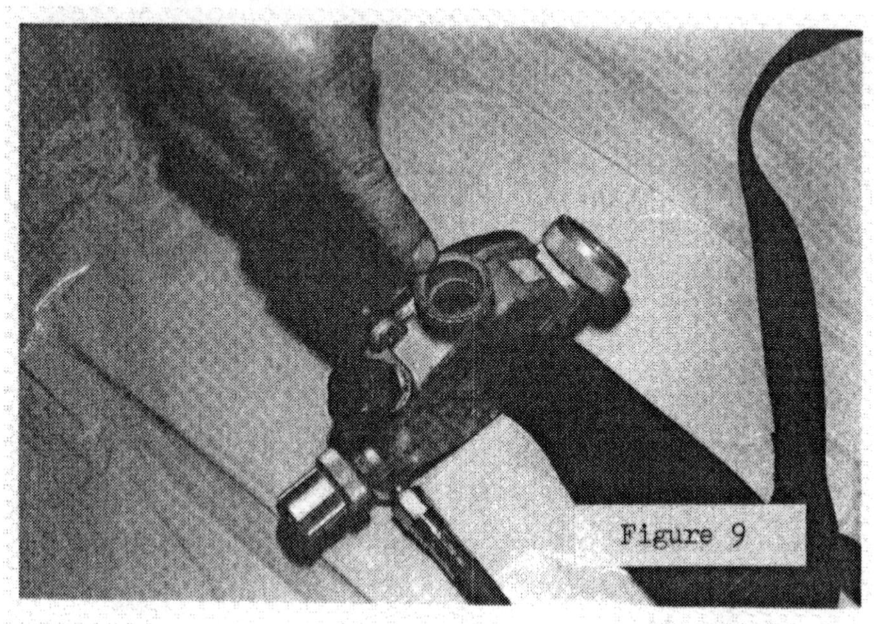

Inspect the air outlet.

Sample 9.D

5

The Documents folder, located in the middle of the Start Menu contains shortcuts to documents or files that have recently been opened. These files can easily be reopened by using the following three-step process.

1) To look in the Documents folder bring the cursor over the menu and click the left mouse button. (See Figure 4.)

2) Put the mouse arrow over the ▸ arrow symbol until the "My Documents" window opens. (See Figure 4.)

3) Listed here are all the files you have opened since the start of your OS. (See Figure 4.)

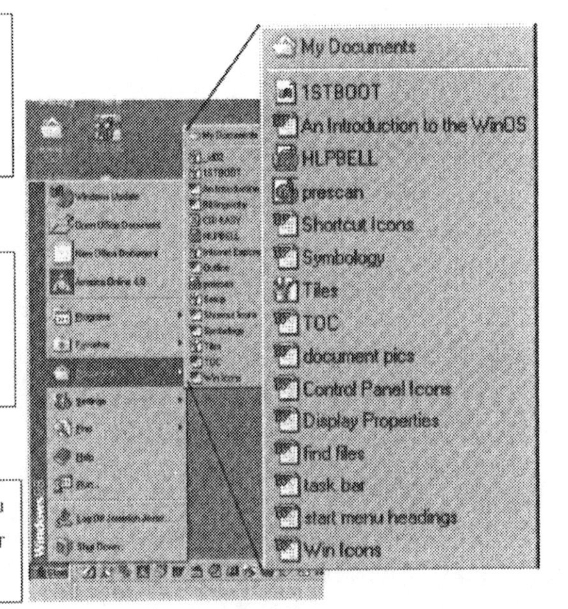

Figure 4. How to Access Documents

Note: The files located in "Documents" are not the actual files but rather the historical blueprint of the most recently active files. If I were to delete the "Win Icons" document illustrated in Figure 4 from its original directory location, the document would still appear in the "Documents" folder because it would not change the fact that I had already opened the file earlier.

(Screen shot reprinted by permission from the Microsoft Corporation)

Screen Captures

In recent years I have seen a growth in the popularity of screen captures. It is difficult to discuss computer applications without the captures. The screen can be divided up and enlarged, piece by piece, so that menus or icon bars or other features can be presented or used in a discussion. Note the screen capture in Sample 9.D. Observe also that you will occasionally see an image that lacks an acknowledgment. If an author uses a number of images—such as screen scans—from the same source, the acknowledgment for *all* of the images was placed below the list of illustrations in the front of the document.

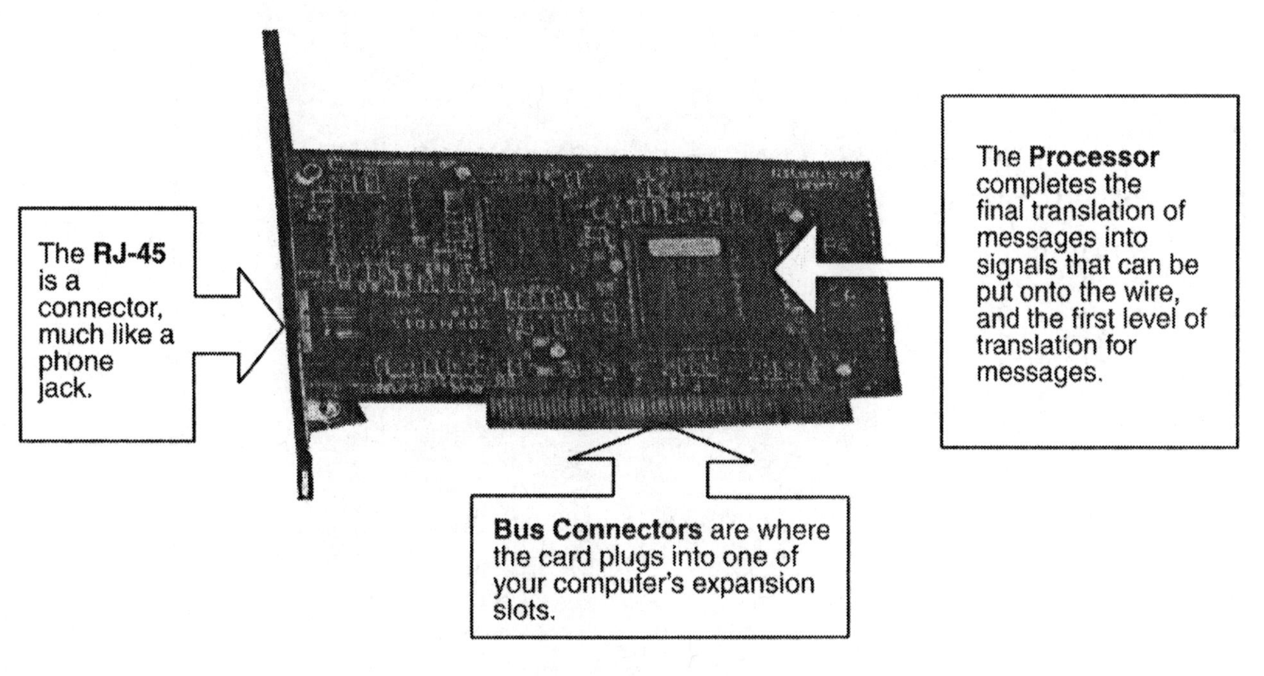
The **RJ-45** is a connector, much like a phone jack.
The **Processor** completes the final translation of messages into signals that can be put onto the wire, and the first level of translation for messages.
Bus Connectors are where the card plugs into one of your computer's expansion slots.

Figure 1. ***A Linksys Etherfast 10/100 LAN Card***
(Photo by Chris McCurdy)

Scans

A scan is basically a digital adaptation of a photograph, which can easily be altered. With Photoshop, arrows and callouts can be added to identify parts of the scan. A high-end scanner will do justice to the original illustration, but quality will be questionable if the equipment is not the best. Even the quality of the paper can seriously affect scan quality once it is printed. The author added callout boxes to the screen scan in the preceding image. There is a full page of text with a scan in Sample 9.E.

Screen captures are an important new breed of illustration that emerged in recent decades. These graphics contribute to the clarity of computer manuals of every type.

Sample 9.E

By manipulating these devices with software, the user can perform tasks such as writing a letter, printing a report, playing a game, or monitoring other machinery.

A. CPU Slot

B. IDE Connection for Hard Drive

C. I/O BUS

D. Memory (RAM) Slots

E. Peripheral Device Connectors

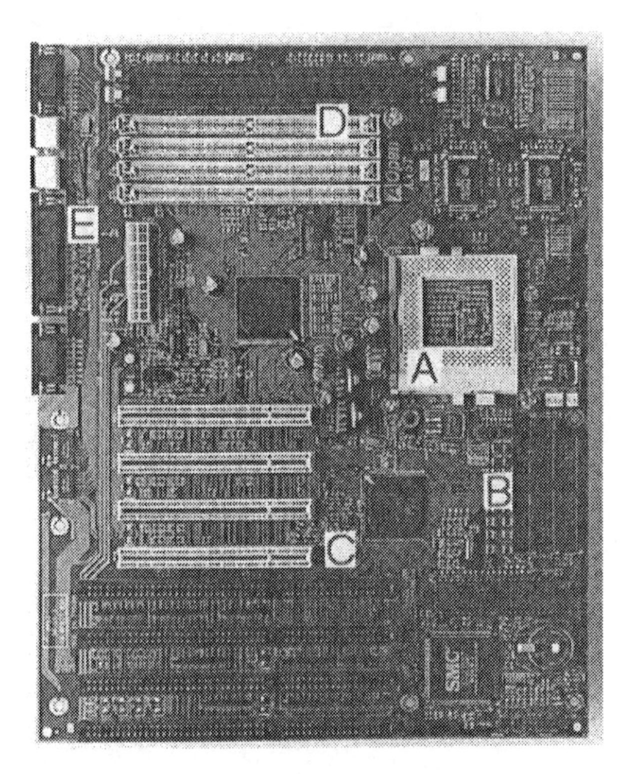

Figure 4. Computer System Motherboard

The Main Memory (RAM)

In every computer system, the CPU needs information to act. The faster it gets the information, the faster it can compute each instruction and reach the results. However, a CPU operates many times faster than either a disk drive or the I/O bus. In order to avoid bottlenecks in the flow, a special type of memory called main memory, or RAM (Random Access Memory), is used to temporarily store information (see Figure 5). On a personal

Freehand Drawings

I seldom run into someone blessed with this skill, but from time to time, a project is turned in with original artwork. Consider tracing. Tracing is a compromise that has been used with some success. Needless to say, in the days before photography, the craft of freehand drawing was the best source available to early researchers, such as the daring voyagers of Chapter 1.

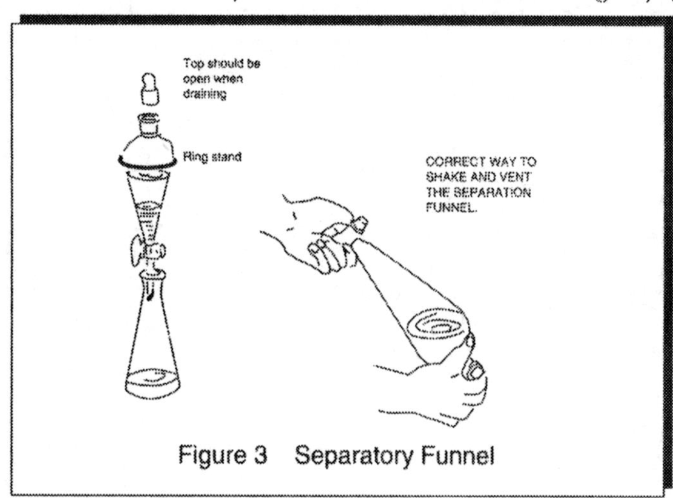

Figure 3 Separatory Funnel

Architectural Drawings

Orthographic, isometric, oblique and perspective views are the stock-in-trade of construction engineers as well as mechanical and civil engineers. Indispensable to engineering, these drawings are critical to conceptualizing ideas, and there is no better evidence that modeling sits dead center between theory and application in engineering. These drawings used to originate as lead on vellum, or ink on Mylar, but increasingly they are computer-generated products.

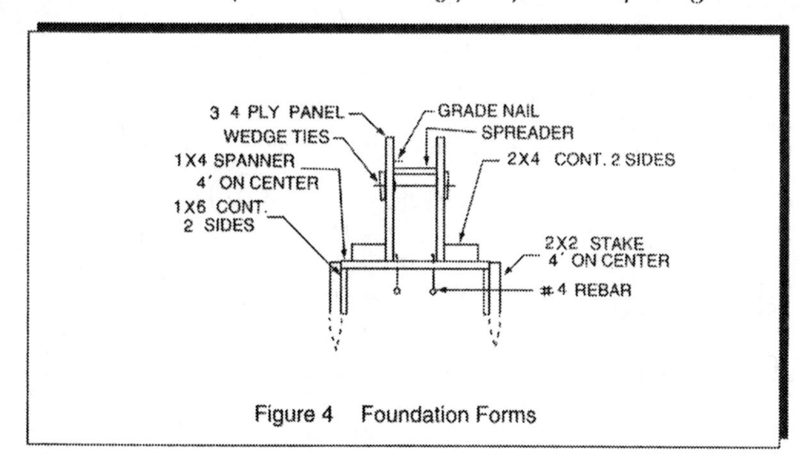

Figure 4 Foundation Forms

This graphic, like most samples in this text, was created by an engineering tech student. This digital photograph was taken by him, scanned in, and trimmed with Photoshop features—all in full color.

Computer-Aided Drawings

Although CAD systems are now used extensively to generate engineering drawings and electromechanical drawings, the utility for the systems is endless. Many of the more visual concepts of *Wordworks* were created on CAD from sketches and then used as pasteups for the textbook manuscripts.

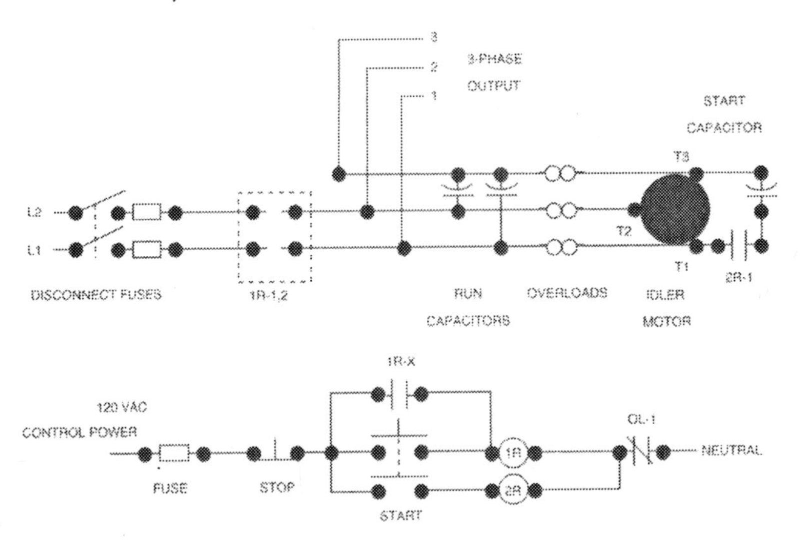

Figure 1. Schematic for the Rotary Converter.

Pictorial Diagrams

Realistic illustrations, such as assembly drawings of electronic circuit boards or components, are useful for technicians and represent a level of drawing that avoids schematics. The following figure was created by using Visio.

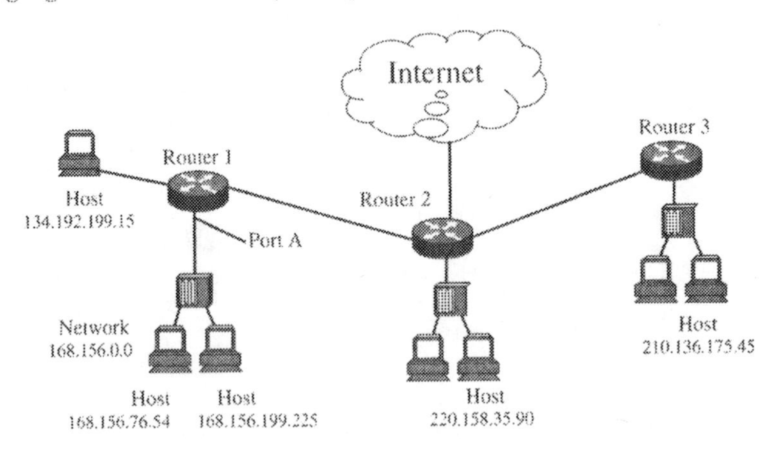

Figure 1. IP host 168.156.76.225 must send information through Router 1 and Router 2 to reach the destination host 220.158.35.90.

Schematic Drawings

There is no better place to see the crossover between language and illustration than in schematic drawings. The symbols have to be understood in order for an observer to "read" the "drawing," so this is a unique medium, certainly one central to tech writing. Here are two conventions for symbologies in electronics.

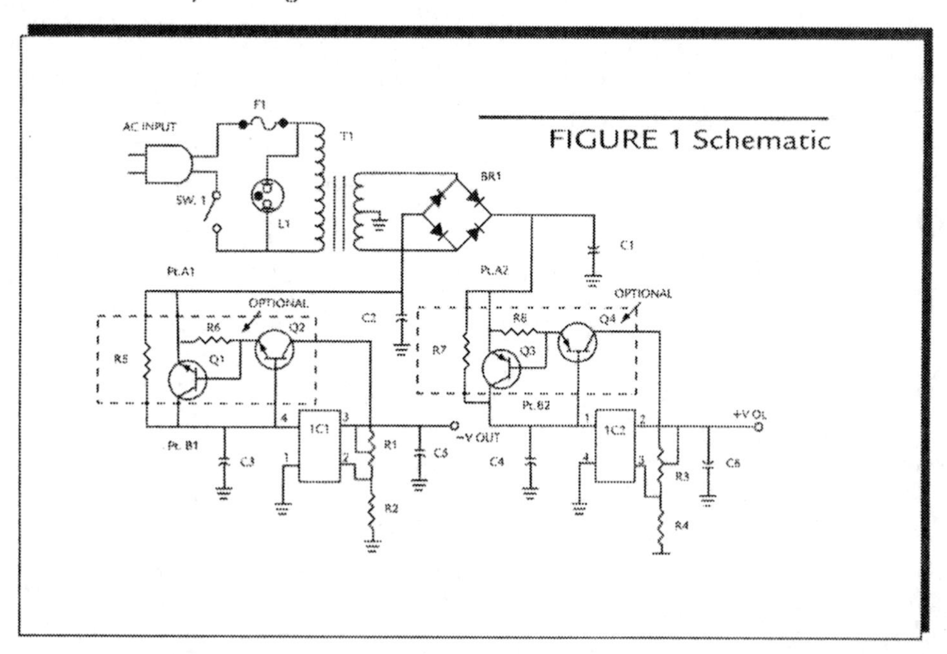

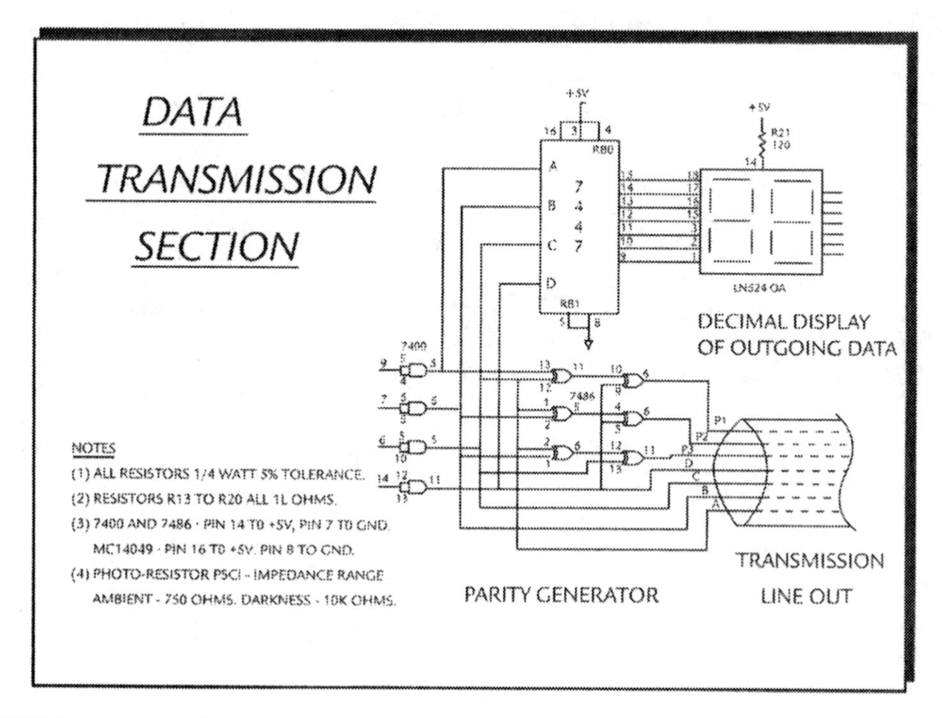

Cutaways

Dear to the heart of anybody who was brought up on *Popular Mechanics* and *Popular Science*, the cutaway is always fascinating to see. "So that is how it works" is the usual response to a vivid cutaway. The idea is to create a realistic view—but from the inside out. The usual approach is to make the subject simply appear to be sliced open. At industrial trade shows, this practice gets carried right out onto the showroom floor—occasionally in huge models that weigh tons and that have been cut open with some very heavy duty equipment, indeed. For publishers, a variation of the cutaway concept is handled with successive sheets of plastic overlays. You have seen them in textbooks. First, here is the skeleton, then the organs, then the muscles, then the epidermis, then the shirt and shoes. I confess a huge fondness for the entire gamut of cutaway concepts, which run from the simple line drawings in the popular *The Way It Works* books to the gifted artwork of David Macaulay in his well-known explorations of architecture. With no pun intended, surgery textbooks use cutaways extensively. The artist Melo Esclavon designed the following drawing. She has a skill writers would envy.

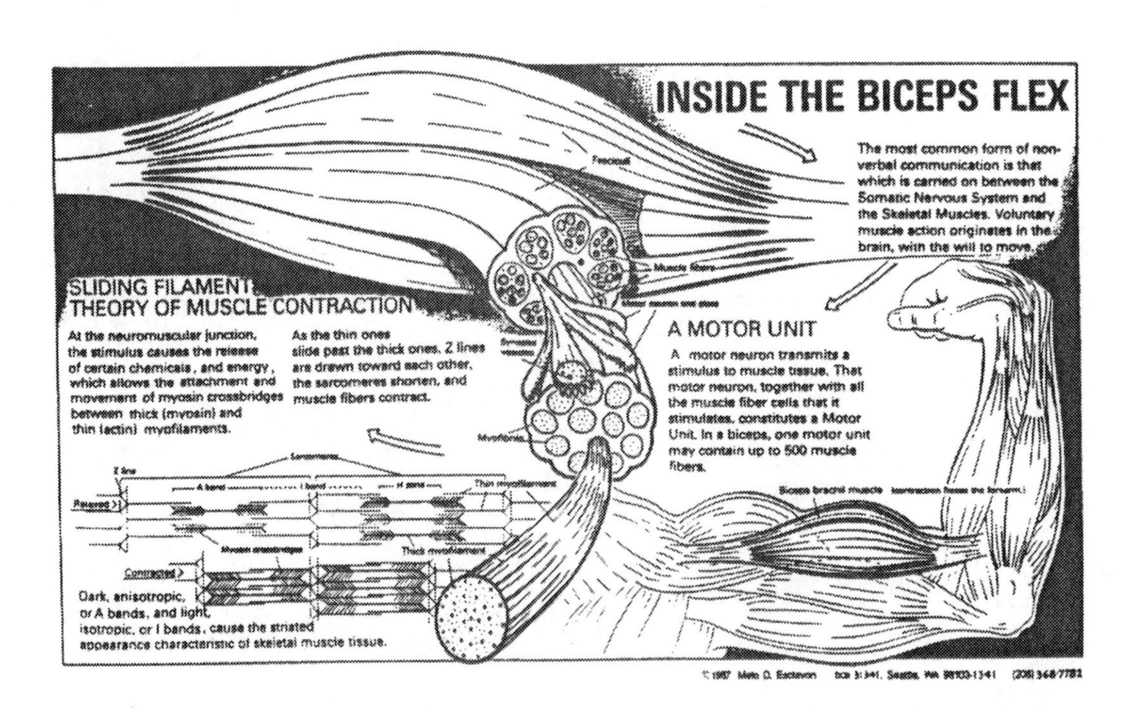

Exploded Views

Here is another favorite of technicians and engineers. The exploded view is another way to visualize assemblies. The difference is in the way the image is presented to the reader. The cutaway shows a device "assembled," so to speak. Exploded views float in the air in an image of ultimate orderliness. They visualize disassembly or assembly with absolute clarity.

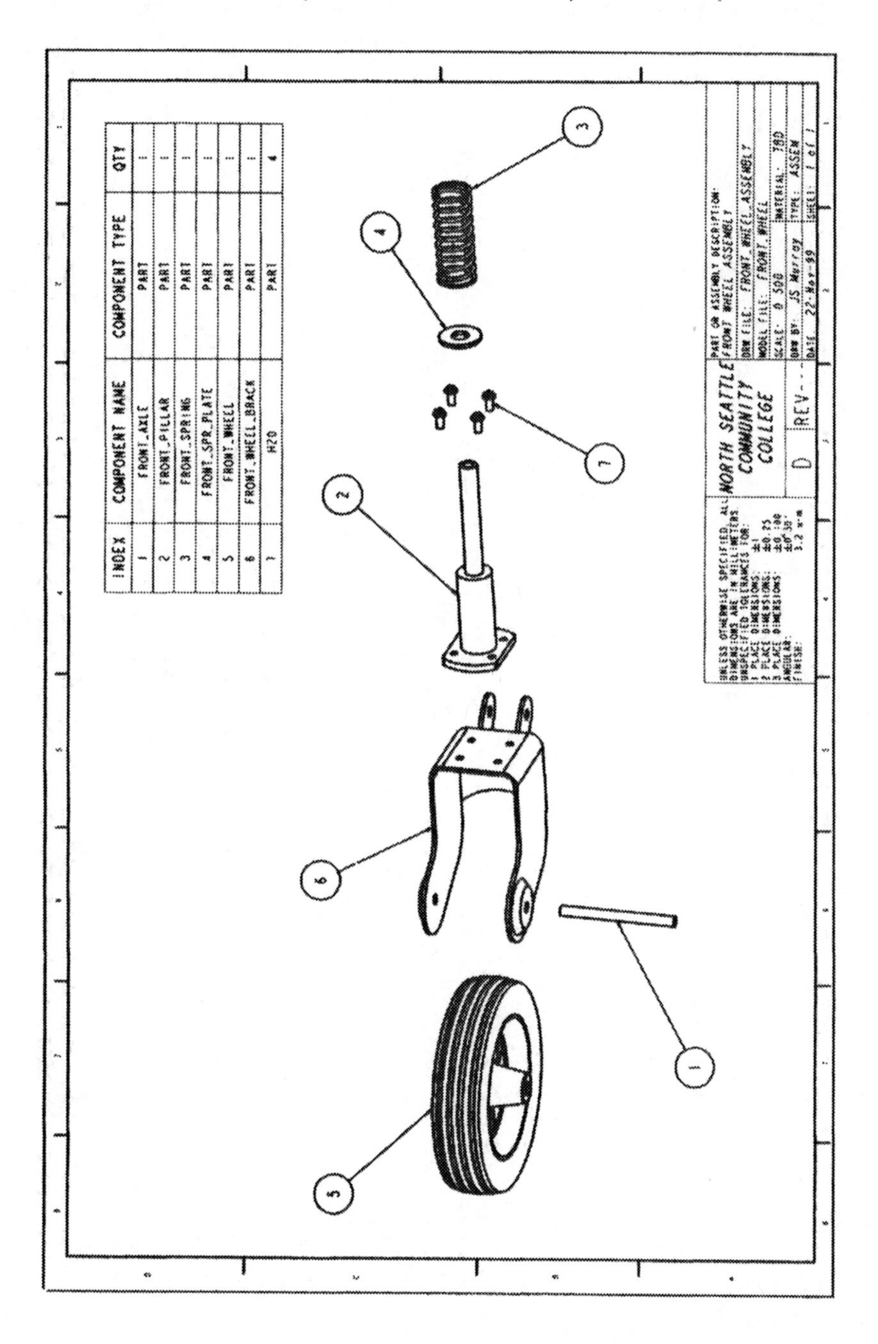

Tables

Like the schematic, the table shares something of two worlds: language and graphics. Tables visualize categories. The grid is not the "measure" of a table as it would be in a chart or graph. The spaces *in* the grid are the point of interest: the categories. The boxes may contain commentary, statistical data, measures, or images. A table might contain only silhouettes of birds in flight, for example, which clearly indicates the graphic potential of tables. Even with text in the boxes, the table remains visual—and it is a quick reference, which is why it is such a practical tool in books of engineering tables. Software is helpful for constructing tables. *Word* can be used to construct tables. *Word 7* and later versions have a large selection of table templates (see Appendix B).

Table 1. Important Characteristics of Selected Digital Data Storage Devices

	Data Access Time (Milliseconds)	*Data Transfer Rate (kbyte/second)*	*Maximum Available Size (mega byte)*	*$ Per Mega byte**	*Pros*	*Cons*
Floppy Drive	Typically above 500 ms	50 kB/s	1.44 MB	$50 for drive $0.80/MB for disks	Poratbility Price	Capacity Speed
Hard Drive	15 ms	3000 kB/s and above	1028 MB and above	$0.85/MB	Speed Capacity	Lack of portability
Magneto-Optical Drive	30 ms	1000 kB/s	128 MB	$500 for drive $1.28/MB for disks	Speed Portability Capacity	Price
Tape Drive	1000 ms to many minutes	50 kB/s	305 MB	$300 for drive $0.08/MB for tapes	Price Portability	Speed
CD ROM	280 ms	300 kB/s	640 MB	$300 for drive $0.12/MB for disks	Capacity Portability	Not writeable

*As of this writing.

If you need many tables in your work, you should consider using Microsoft *Excel* 7.0 (or a more recent version), a package that has the capability of building a variety of tables. *Excel* is available separately or as part of Microsoft *Office*. The program is an excellent tool for engineers.

Charts and Graphs

Engineers may or may not have much use for graphs. Pie charts are the simplest of the group, and they do not commonly appear in engineering documents. The line graph is a flow analysis and often is measured in time. The bar graph measures quantities. A surface chart is a mixture of the bar graph and the line graph. All these devices are often enhanced with graphic tricks—perhaps a graphics designer will overlay a pie graph on a scanned image of a silver dollar. Bar graphs are redesigned to replace the bars with graphics—stacks of oil barrels for oil production charts, for example. The two following samples used conventional software features found on many general programs. *Word* has a graph utility.

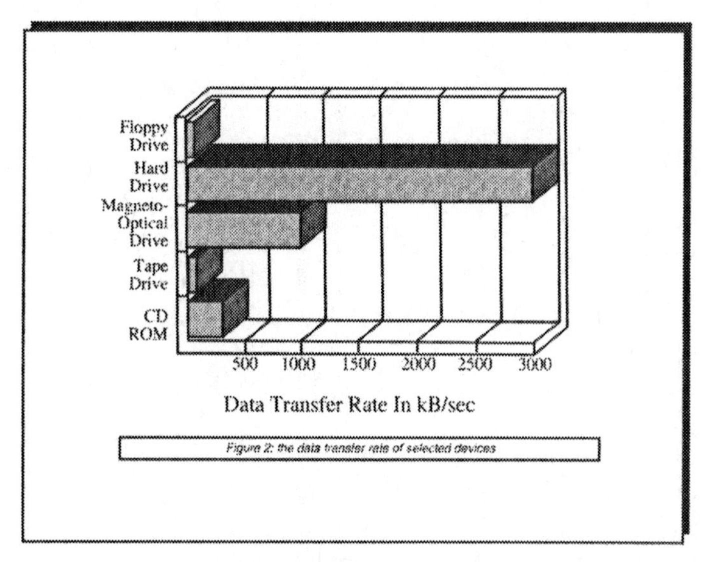

Figure 2: the data transfer rate of selected devices

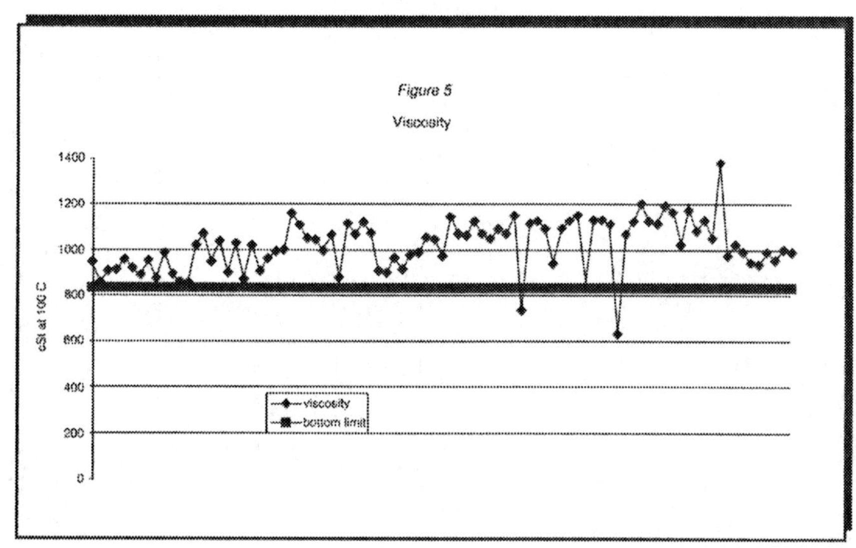

Sample 9.F

- The Job Scheduling for SG040R to generate client appointments cannot run until the SM4006 and SM4007 screens are prepared.
- You should wait until after the 10th day of the current month to get an accurate count of the currently enrolled clients who will receive renewal forms.

The following flowchart begins on the first week of the current quarter:

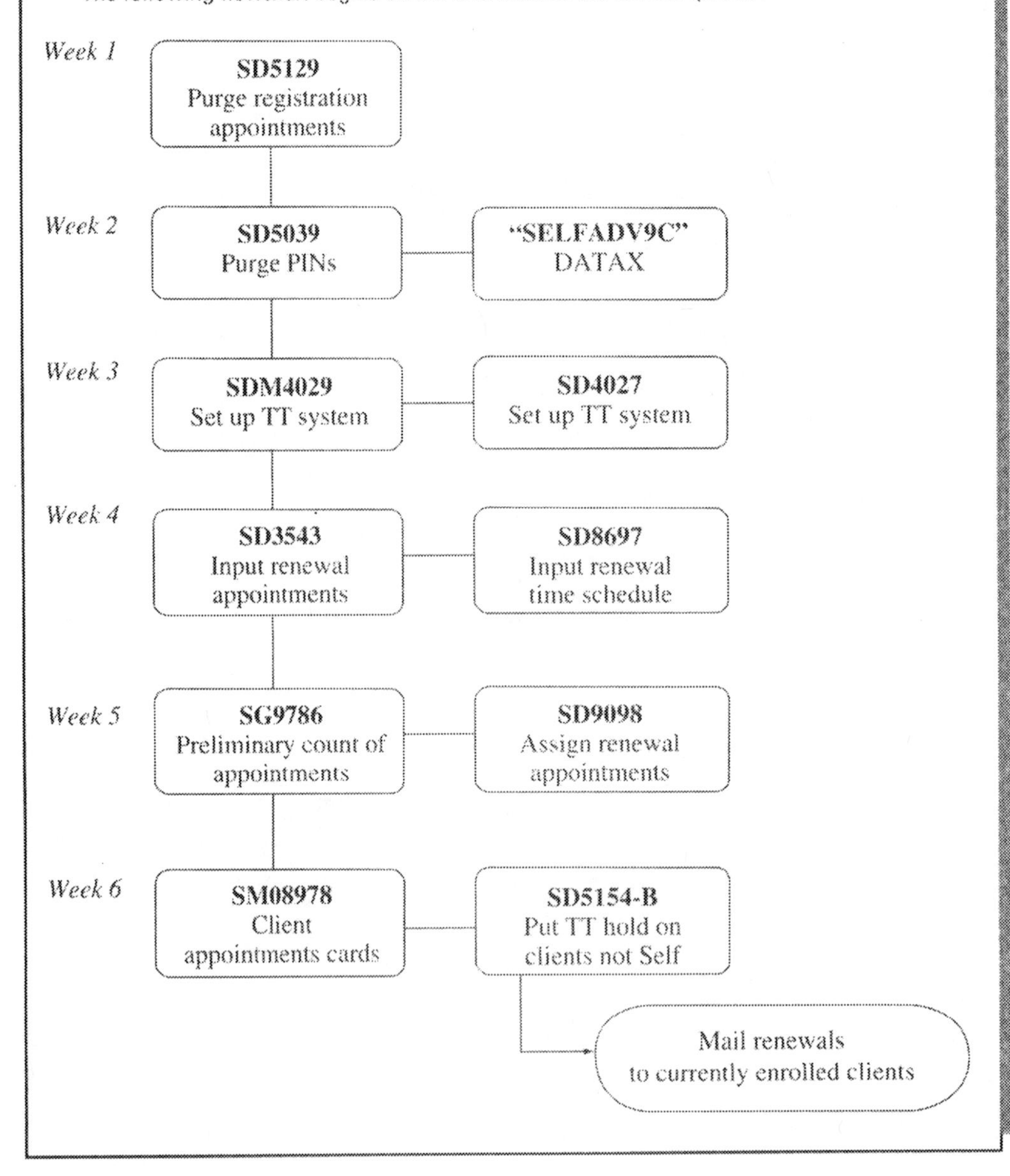

Flowcharts

Of all the chart types, the flowchart may be the most practical tool for engineers and technicians because it functions as a linear path of conceptual issues, usually cause-and-effect relationships. As a result, it is a handy way to graphically present a troubleshooting process, for example. You may need a dedicated software application, such as *FlowCharting PDQ* from Patton and Patton Software for complete flow chart capability, but a simple flowchart can be drawn by hand or created with basic computer features. Flowcharts are often used without designated meanings for the shapes. The simple style appears in Sample 9.F.

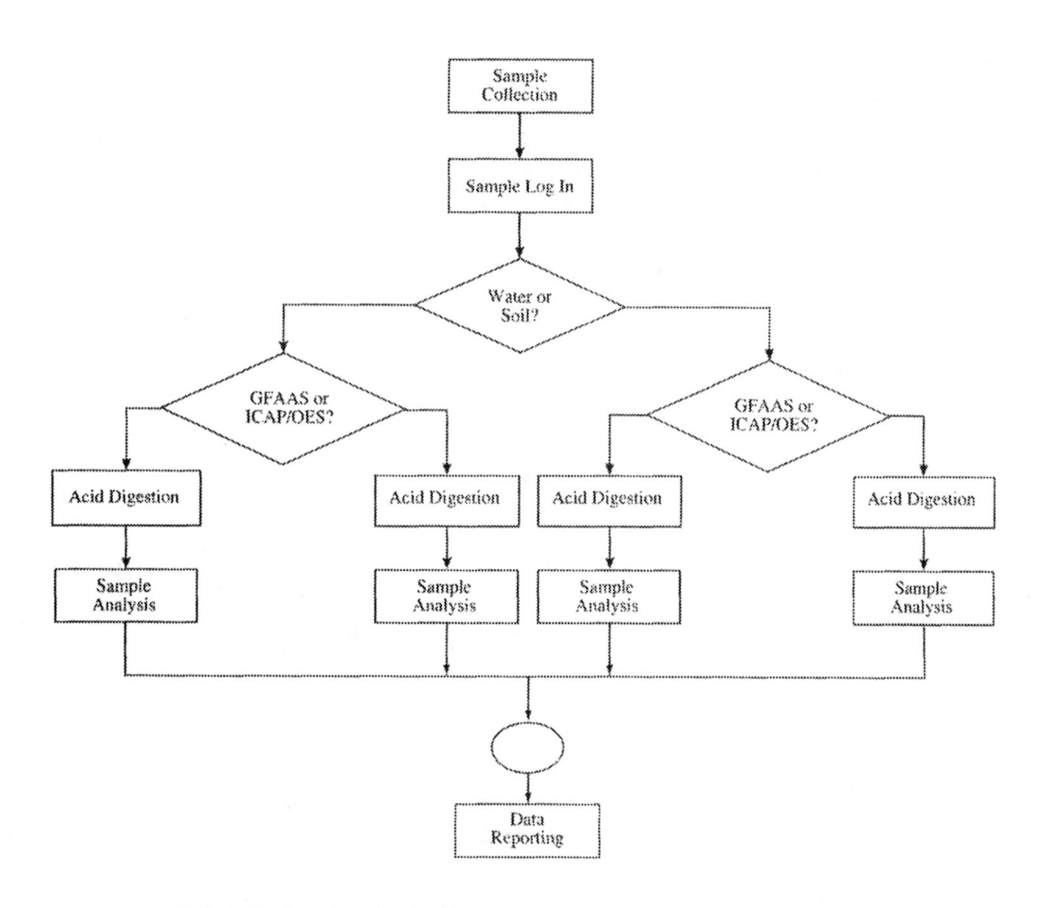

Figure 1. The flow of samples through the laboratory from sample collection to data reporting

Flow charts are of two types. In many engineering documents the specific shape of the chart elements are assigned meanings. Equally popular is the style presented here, in which a shape is used repeatedly with no intended meanings.

Other Options

There are a few additional techniques, such as cartooning, that are used from time to time. When you consider that graphics must speak to the audience as clearly as the written text, you realize that such ideas as cartooning are not entirely out of the question. In the early days of assembly language, MIT published a book titled *The FORTRAN Coloring Book* that used cartoons throughout, and it must have been a blessing to data processing students. Swann and Johnson, with the help of a fictional Professor E. McSquared, published a major textbook for calculus, entirely in comic-book-style story blocks (see pp. 434-435).

Consider visualizations of any kind that you think are appropriate for your work. Even the humble block diagram, such as the following one—or the more complex variation in the subsequent one—is often a workhorse for clarity in any document in which it is used.

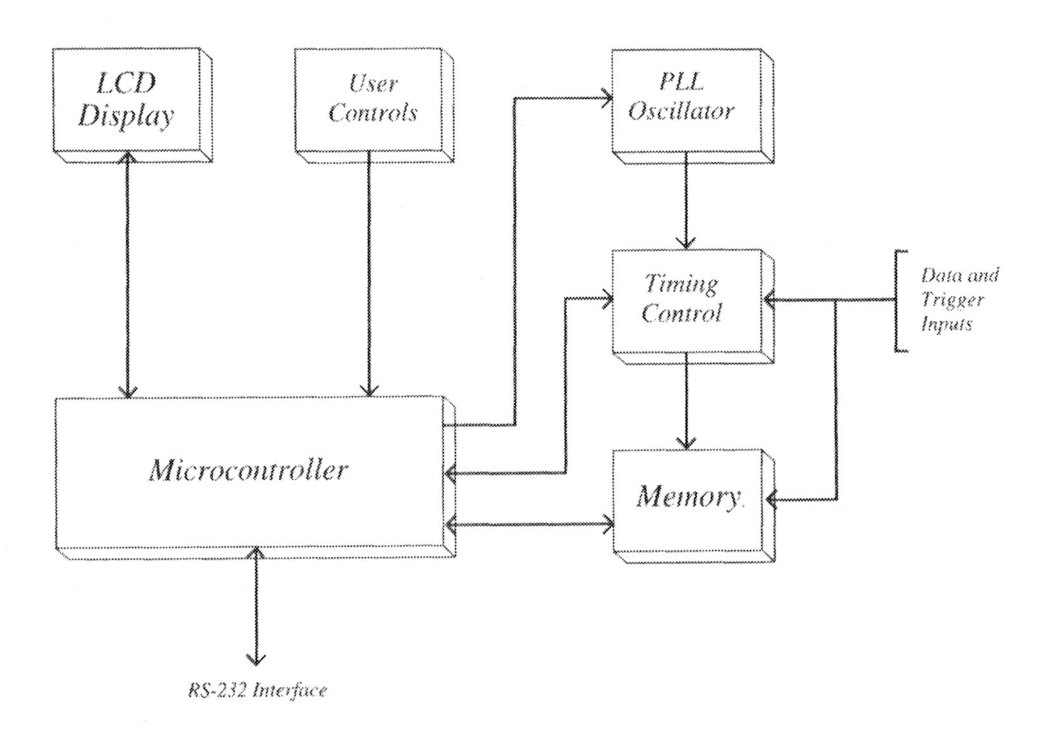

Figure 2 - System Block Diagram

Computer applications have enlarged our understanding and use of block diagrams. Using visual features from a program such as Visio, an author can add graphic elements to enhance the diagram. The end result is a vivid pictorial diagram that otherwise would be impossible to construct without artistic skills. This sample from network technology demonstrates the program, which has gained considerable popularity in network installation engineering.

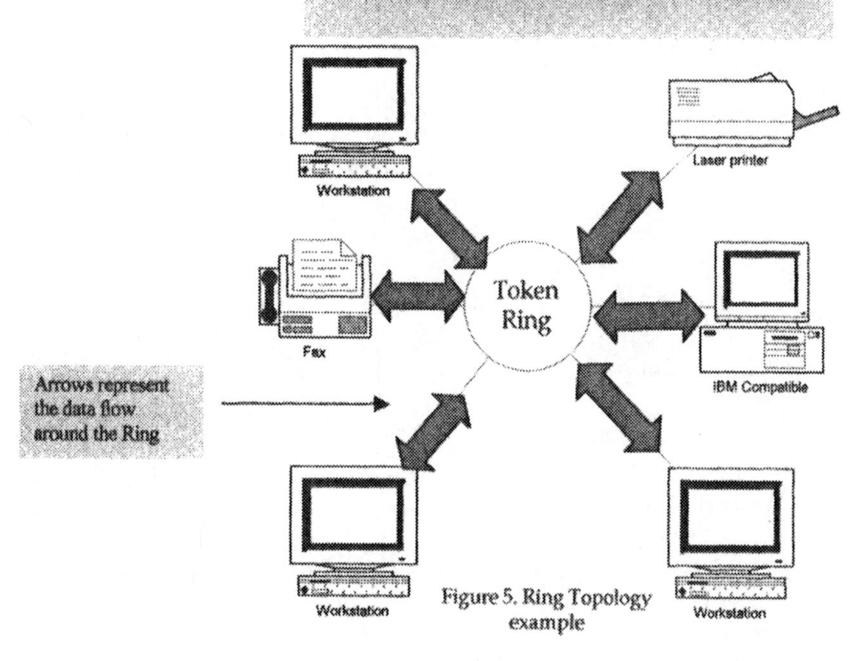

Figure 5. Ring Topology example

Computer modeling has come of age also. CAD systems can now construct much more than the engineering drawings for which they were originally intended. The following three-dimensional rendering simulates the product that was detailed in a set of drawings.

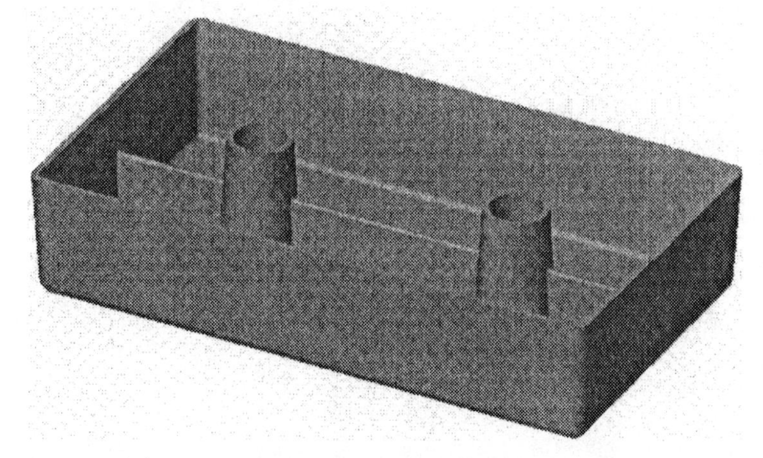

Your only limitation in developing visualizations is the temptation to use the *real* thing. Obviously a real sample is superior to a simulation, but you do not need a five-ton mastodon to make the point, even if it did still exist. Graphics as models are convenient substitutes. We can leave the rattlesnake in his case. A cutaway drawing of his venom sacs will do just fine.

The Graphics

Comments

Although the graphic "types" that are illustrated in this chapter are probably self-explanatory, it should be noted that these devices are vehicles of communication. They may or may not be effective for a given discipline.

For example, the typical schematic used in electrical engineering is used by few other engineers. Digital gate schematics are used for very specialized applications (see p. 452). Even the simple text boxes or "windows" used by software engineers and software writers are used in very few engineering disciplines. On the other end of the spectrum, photographs and representational drawings are the most likely candidates for the widest possible readership.

You will select the graphic media that conform to the needs of your engineering field. A few types will noticeably dominate any given engineering area. Within that range of graphics you can simplify or complicate the graphic content to conform to the subject matter of the text and the needs of your readers.

Summary

- Graphics represent a modeling process.
- The models simulate important features of engineering fields.
- Photographs and representational drawings reach the widest audiences.
- Pictorial diagrams, cutaways, and exploded views are practical tools for the analysis of parts.
- CAD productions, architectural drawings, and schematics are symbolic and usually address knowledgeable readers.
- Tables and graphs vary in complexity and readability. They can be adapted to audience needs.
- Tables and graphs have utility in a wide variety of engineering fields.
- Learning-curve documents use graphics that progress from simple to complex.
- The block diagram is considered the most basic and the simplest form of most schematic drawing systems.

Activities Graphic Tools

Design or analyze the graphics that are appropriate for the following projects.

Develop an extended definition of a term or process or concept that involves a graphic. Link a paragraph of text to the graphic in such a way as to create a satisfactory definition that uses both text and your illustration.

Build a flowchart that represents some process that is relevant to your program of studies or your work. If you cannot develop a procedural flowchart, you might develop an organizational flowchart of the company where you work. Use a computer to construct the chart.

Develop a line graph, bar graph, or pie graph to explain some feature of your technical specialty. Alternatively, if you wish, develop something of personal interest, perhaps a graph of rising housing costs in your area. Use a computer to construct the graph and add appropriate labels.

Surf a few useful sites on the Web (see the Writer's Handbook, Chapter 10). Copy a variety of graphics and charts and tables of importance in your technical specialty. Briefly explain the prevalent graphic types. Use the explanation as a cover memo and submit the package to your instructor.

Develop a brief report (500 words) in a memo to your instructor. Use one of the following suggestions for the exercise.

- *Visit the periodical section of the campus library. Select a number of different technical magazines from different disciplines. Examine the graphics. Can you see trends? Can you see why certain types of graphics are preferred in various fields or technical specialties?*

- *Examine the textbooks or magazines in your area of interest. Here you should be able to easily identify strong graphic preferences to which you probably are already accustomed. Notice the way in which your graphic perceptions are conditioned by your technical and engineering interests. You adopt the graphics that match your perception of your technical world.*

- *Think ahead to other projects you will be writing this semester or quarter, or think ahead to projects for work. You can do more than try to anticipate the objectives and organization of upcoming projects. Start to visualize the project! Look at the activity in terms of models and think about how you intend to "show" the presentation to the reader.*

Share a Project

The instructor will ask for those students who are familiar with diagraming software applications for electronic schematics and other fields. These students will act as guides to take small groups of four to various campus labs to illustrate how the drawings are constructed.

The instructor will survey the class to see which students are familiar with software table features. Those students will act as guides to take small groups of four to the computer labs to walk through the construction of a table. Each student will construct a table as an exercise by identifying four varieties of a product and the criteria for evaluating the various brands. (The data in the table can be fictional.)

If appropriate, the instructor will assign similar groups to students who are familiar with equation editors or Matlab *or* Excel.

Share the Preparation of a Document

Collaborate with three other members of the class to build a technical document with a full text. This project will develop over a period of several weeks, and you will have opportunities to meet to discuss design, research, production, and progress.

- *Select a familiar technical topic that involves either a set of instructions or a description of procedures.*
- *Develop a text of 1500 words that will include a cover page, table of contents, list of illustrations, glossary, and appendix. You may also want to include warnings, parts lists, tool lists, or a bibliography.*
- *Delegate parts of the task—including graphics—to group members.*
- *Discuss the target audience, design, research, production needs, and progress at each meeting.*
- *Submit the final project to your instructor.*

Work in Progress

10. The Second Draft

With my rough draft in hand I contacted Tom, the primary engineer for production and asked him to meet me and review the draft on the screen. We discussed the changes and a few writing issues that involved positioning the graphics where they would work best. For example, the test setup schematic figures needed to be linked to the flowcharts for readers in both Inspection and Assembly. I had also overlooked linking the charts to Assembly. Tom thought the changes looked fine and gave the go ahead to submit the document after I updated the draft for a final copy.

If flowcharts are not added at this time, the approval changes will be held up when upper management reviews the final project. Any time a new document is created, it must be linked to a flowchart of the production process. Then when an action (needing instructions) is done, as stated in the flowchart block, there should be a support document for that block (containing the instructions) to describe what is to be done.

A couple of the photo graphics also needed clarification. Arrows were used to identify what action was occurring in the photo. To do this I needed assistance from the photo shop department to apply the arrows to the photos. I do not mind asking for help and I recommend that you ask for help when you need it.

Although color looks good on a computer screen, I decided against using a lot of color in the documents. For the most part color is not used in our company manuals, usually due to the cost of color LaserJet printers. Nor is it cost effective for everyone to have a color printer at their station just so they can print a color document. Any color I use for effect is a light shade. I have found that when you use a dark background color and use a black-and-white printer to print the text, that area is not readable.

After correcting the draft, I assembled a second draft document packet. I printed this out and this time I wrote "Second Draft" on it for identification. With my second draft in hand, I paid a visit to the departments to do a trial run of the changed process. I found I needed to clarify the flowcharts again to realign the stockroom assembly flow. These were minor corrections and did not require much additional time. For the most part all went well and I was fairly sure the changes would be functional and would work.

I updated the flowcharts and prepared to assemble my final project document packet. I printed out the changes I had done and wrote "To" (meaning "change to") on these to identify them as the changed documents. I assembled these into a final project packet.

S. B.

Targeting Your Illustrations

You have no doubt noticed the frequent use of the word *tools* to refer to graphics of any kind. Graphics are often underutilized, and since technical documents make frequent use of graphic elements—from photographs to schematics, from drawings to tables—it is necessary to understand the role of illustration. It is important for writers to deploy graphics strategically and with a specific sense of purpose. In a marketing portfolio, "pretty pictures" do serve a purpose, but engineers and technicians do not use illustrations just because tech writing is supposed to be vivid. Graphics are *tools;* they are used to do a job. They have work to do, or else they are not necessary. Even a cover illustration on a front page should have a practical function.

In this chapter I will explore methods that are available to you for putting graphics to *work*. Text layout can, for example, make the graphics work effectively. In addition, the graphics themselves can also be designed to ensure that readers see what you want them to see. I will examine both approaches.

The Target

The first problem you will encounter will have to do less with locating graphics than with finding the *appropriate* graphics. Since *you* are the author of the text of your document, you can carefully control your written communication and make every effort to target the readers. Graphics pose a different dilemma because you will *not* usually have the support of a graphics team, and you may not have the skills for artwork. This book addresses college students who will be looking at limited resources: rough sketch work, templates, CAD or software productions, and borrowed graphics.

The challenge you face in developing the graphics is somewhat different from the one you face in writing the text. The graphics can be amateur or professional or somewhere in between. For example, schematics can be done freehand, but they look more professional if you have the green templates that have been traditionally used in a number of disciplines. The templates are available for many diagram and symbol systems that may be of interest to technicians and engineers.* If you have the skill to generate CAD drawings and diagrams, or if you have software for schematics, by all means use these support systems for your work (see pp. 275-276 in the *Writer's Handbook*.)

For all your illustrations, you should be sensitive to the reader's ability to grasp the meaning of the visuals you introduce in your text. Drawings can operate at many levels of complexity. A floor plan, for example, is the least complicated of all the drawings used by a construction engineer or an architectural firm. In fact, apart from representational illustrations (that configure the finished appearance of a space), the floor plan is usually the only architectural graphic a layman *can* understand. The engineering drawings are much too specialized to communicate information to the uninitiated. This is a particularly tricky problem for architects and construction-related engineering firms, since the "amateur" is often the investor. The people who finance the project are often the *least* likely to understand the fine print on the drawings.

Similarly, this "reader" problem is apparent anywhere that diagramming utilizes symbols and establishes an engineering language of one kind or another. One popular, highly simplified graphic is the block diagram, such as the following one, which is commonly used in any discussion that contains schematics in electronics. The block diagram is composed of connected blocks that demonstrate a pattern, usually a movement (see also Sample 10A).

** Templates are available for math symbols, architectural symbols, landscape symbols, electronic symbols, power and light symbols, electrical symbols, piping systems, flowsheet diagrams for process controls, fluidpower, computer diagramming, and so on.*

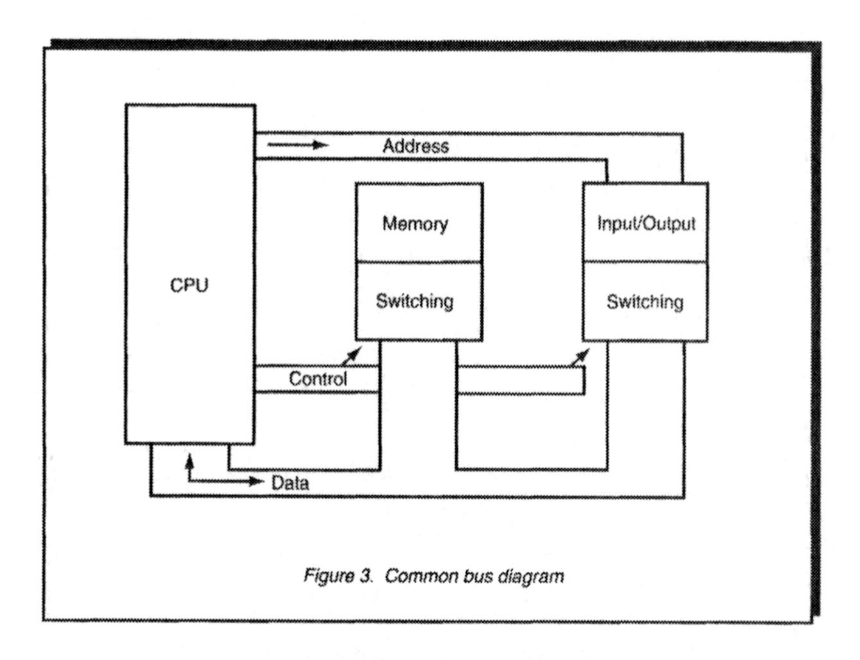

Figure 3. Common bus diagram

What you see in the common bus diagram is an architectural floor plan, but the architecture is that of electronics. Electrical engineers and electronics technicians often use block diagrams as a handy point of reference, and entry-level students prefer them because of the simplicity. At the simple level, graphics of any sort must be free of symbols that are not familiar to the reader. At the opposite end of the scale, the schematics that are up to industry standard read like a language unto themselves. The following schematic can speak only if you know the language of electronics.

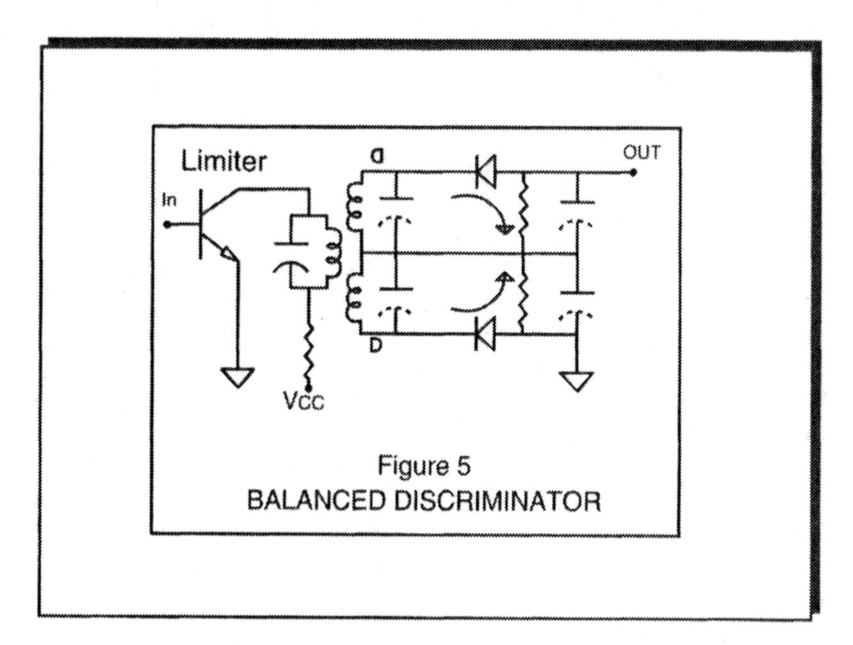

Figure 5
BALANCED DISCRIMINATOR

One other targeting issue to be aware of in the use of graphics is the matter of *omniscience*. Recall that *you* always know what you mean and what you plan to do next. This is omniscience. This knowledge is a burden to you as a writer because this vision of your work may not be clear to a reader. This holds true of graphics, as well. From time to time, I see an illustration that is a real puzzler. Whatever was apparent to the author certainly got lost in the shuffle. Symbolic systems, such as schematics, pose the most apparent problem. More traditional illustrations—photographs, sketches, assembly drawings—are also problematic. Because these illustrations appear to be obvious, you may overlook the fact that a reader may still miss the point. Watch for this problem. Notice, for example, the bar graph below. There are very peculiar optical effects used that the author did not notice and the graph is difficult to interpret.

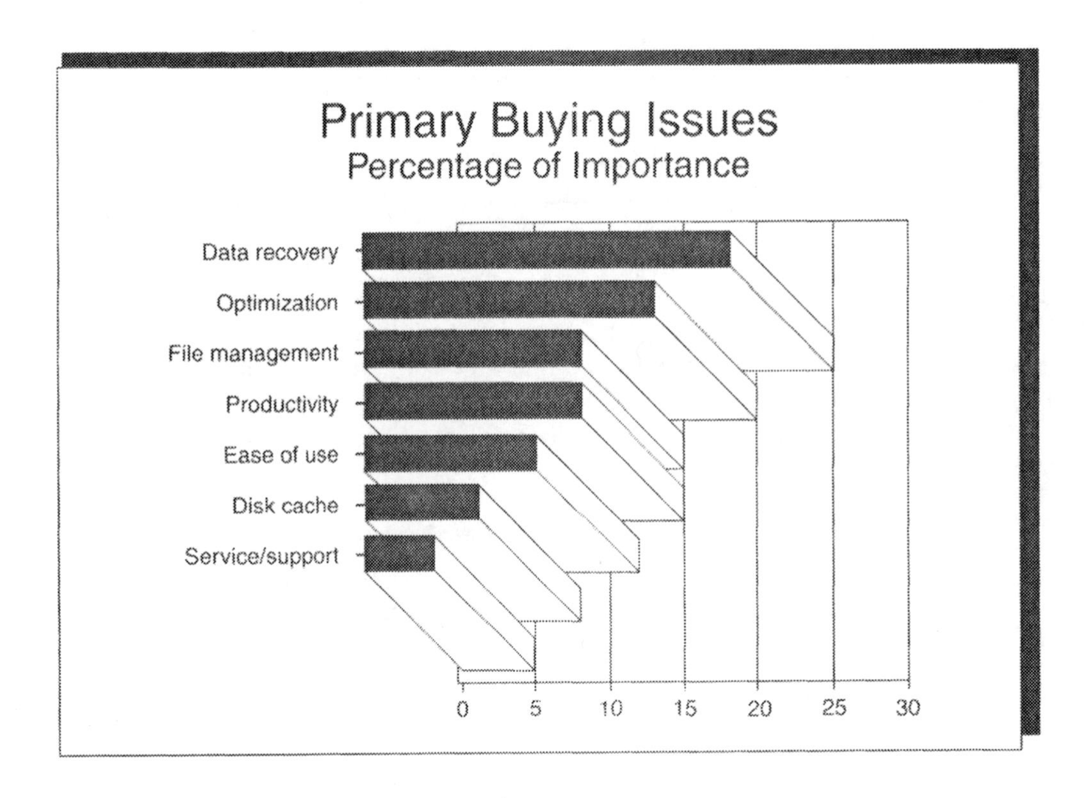

I read a paper on plant genetics that used a series of unexplained photographs. One illustration showed a researcher holding two fruits. They were probably two sizes of the *same* fruit. I could not be sure, but it was obvious to me that the author's intention was to use two of the same with one much larger than the other. Genetic research, being as it is, I could guess that I was supposed to marvel at the larger fruit. But large size may not be a desirable trait, and I could not be sure what was what. By the way, the fruit *appeared* to be tomatoes, but it was not clear. You can see the problem: the *author* knew the obvious, but that did not particularly help the reader! Another time, I put together an assembly kit for a bicycle component. I constructed it upside-down because the assembly drawing was upside-down—at least from my perspective. The list goes on. Here is another puzzle:

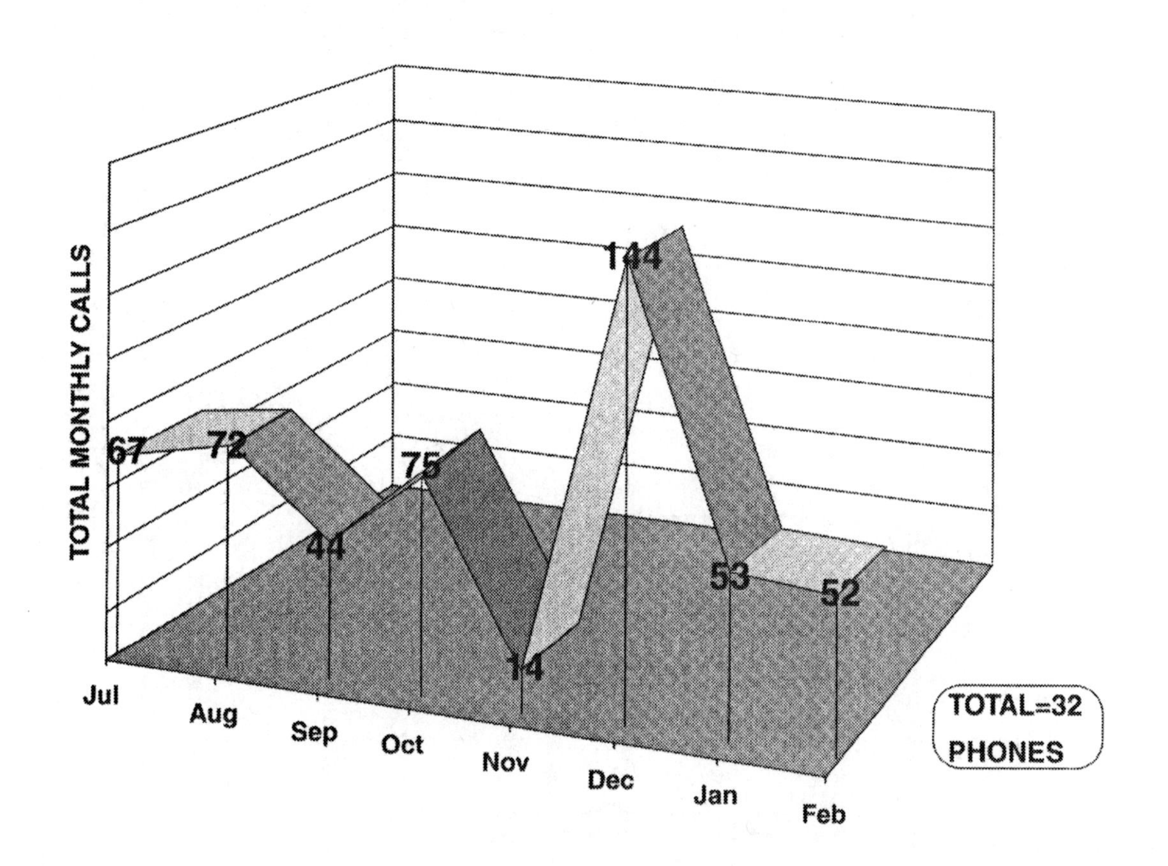

Sample 10.A

-6-

BASIC MECHANICAL REFRIGERATION

The mechanical refrigeration cycle uses a closed, air-tight system containing a refrigerant. The refrigerant acts as a transportation medium to move heat. If the system were not closed, it would use up refrigerant by dissipating it into the air. Because the system is closed, the same refrigerant is used over and over again each time it passes through the cycle, removing some heat by evaporating and discharging it elsewhere by condensing.

The basic mechanical refrigeration cycle, which has four stages, is shown in Figure 11 below.

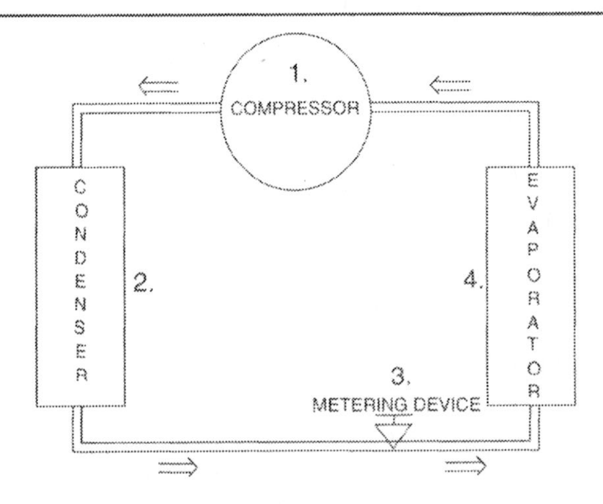

The refrigeration cycle is composed of four stages:

1. *Compression:* Gaseous refrigerant is compressed to a high pressure so that the subsequent condensation does not require impracticably low temperatures. (It also facilitates evaporation on the suction side.)
2. *Condensation:* High pressure gaseous refrigerant is condensed to a liquid (still at high pressure). This condensation can be accomplished by passing the refrigerant through tubes across which air is blowing.
3. *Expansion:* High-pressure liquid refrigerant is allowed to pass through a small orifice, which creates a spray or mist, which is projected into a larger vessel. The activity in the larger vessel is the fourth stage of refrigeration.
4. *Evaporation:* As low-pressure liquid refrigerant is sprayed into this fourth stage the spray evaporates and absorbs heat from the surrounding air.

Figure 11. Four Stages of Basic Mechanical Refrigeration

Textual Integration

If you develop an integrated text—one that positions graphics in appropriate locations in the text—you create the work the way you want it to be seen and understood. If the graphics are in the back in an appendix, you lose that discretion, and the two components of your text certainly will not hold together. If there is a particular reason for separating the graphics, the separation can be done in a matter of minutes. But if you did *not* integrate the text and then choose to do so, you are back to the drawing board.

Graphics are tools, and they often function as languages, like math, that speak if they are rendered in symbols. The math is not demoted to the rear of a text. No part should be removed, except as a practical convenience to publishers, who have distinct concerns about the illustrations. Graphic elements are less effectively used if the reader has to go find them.

Consider that the text explains the graphics, and, vice versa, the graphics help explain the text. An author uses them to visualize, to add reality to, what the text describes. They are used to bridge the distinction between theory and application by modeling. The text is the scenario; the calculations test the dynamics; the illustrations visualize and model the engineering concept. Graphics are most effective if they are *joined* with the concepts that evolve in a technical document. You can reference them if they are in the back, but you know from experience that this is clumsy, at best. A reader does not want to do the work that should have been done by the author. If the graphics will fit in the text at *exactly* the point at which they are under discussion, you have the best option that will properly serve the reader. The reader will expect text and visuals to be side by side.

In addition, if there is any "animating" to be done, the graphics should be on the page in question. If you have a set of gears to explain, then you need to *discuss* the movement shown in the illustration. In this situation, the movement of the gears, for example, can be developed in sequential illustrations of some sort, perhaps a narrow column of three or four illustrations, each of which positions the gears differently to represent the movement. The reader's eye refers back and forth from text to graphic. The writing in the text helps animate by creating the movement in the mind's eye.

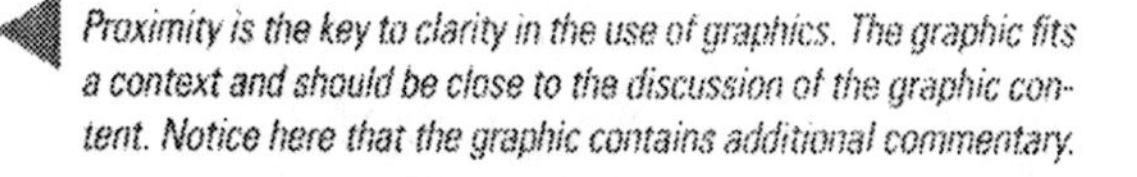

Proximity is the key to clarity in the use of graphics. The graphic fits a context and should be close to the discussion of the graphic content. Notice here that the graphic contains additional commentary.

Sample 10.B

This is because the video is AM and the audio is FM. Another advantage of FM reception in commercial broadcast is that the receivers have broader bandpass and therefore better fidelity. Both AM and FM stations broadcast good fidelity, but the AM receivers are designed to use the narrow bandpass to prevent the detection of two stations simultaneously.

The FM Wave

Figure 1 below shows a frequency-modulated wave. First we will discuss deviation. *Deviation* (see Figure 2 below) is the frequency distance from the center frequency (carrier frequency) to the upper or lower extremes of frequency shift. The deviation does not determine the frequency of the intelligence but the amplitude of the intelligence. The frequency of the intelligence is encoded on the carrier by the speed at which the carrier is shifted above and below its center frequency.

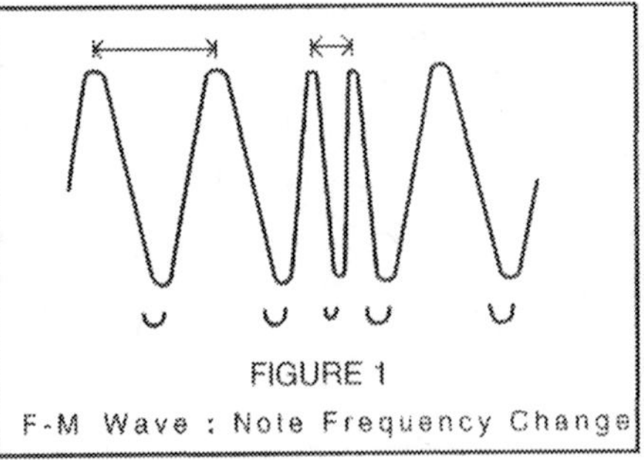

FIGURE 1
F-M Wave : Note Frequency Change

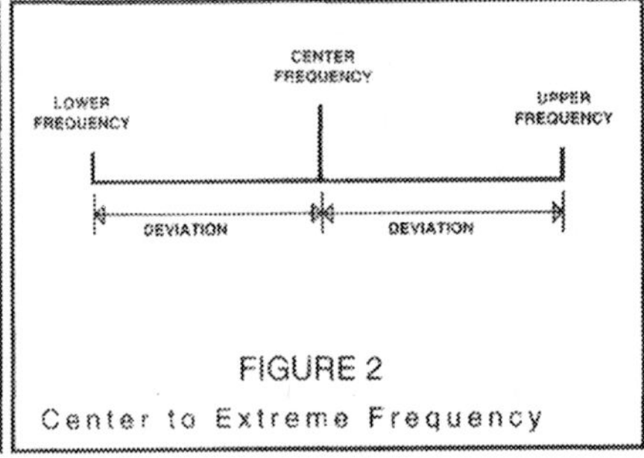

FIGURE 2
Center to Extreme Frequency

The preceding two-dimensional drawing does not move to the tick of time, but note that this text *does*. The text is read word by word in a progression that takes up time—just enough time to animate the graphic in your imagination if I describe it. In sum, having the graphics in a text is valuable for explaining any sequential logic or movement.

Because the text *uses* the graphic, there is only one ground rule: proximity. Place the graphic near the appropriate discussion if at all possible. The text may explain the figure before a reader sees the graphic, or the text may deal with the figure after the presentation is made. The related text may also be developed above and below the illustration.

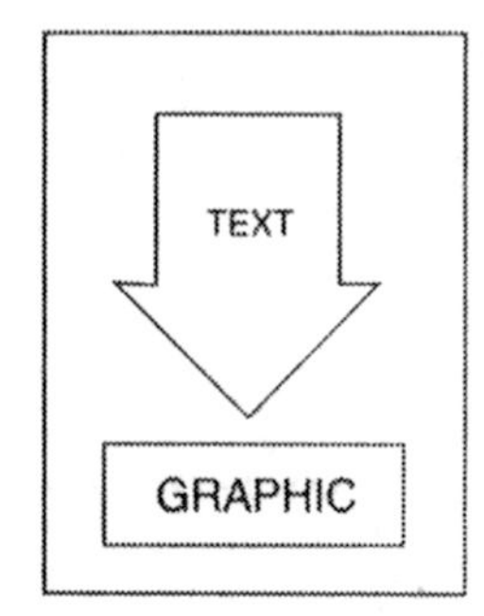

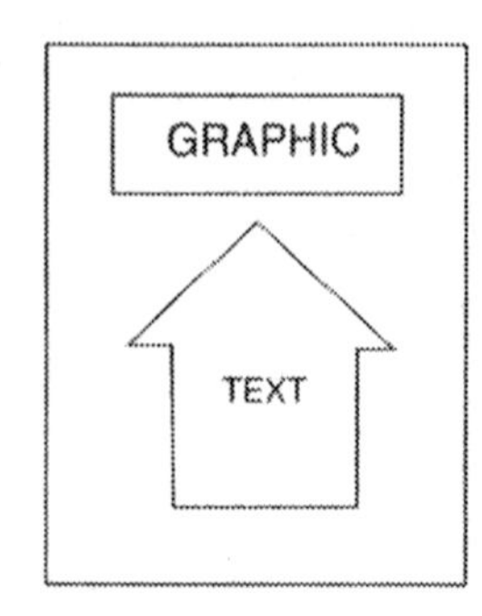

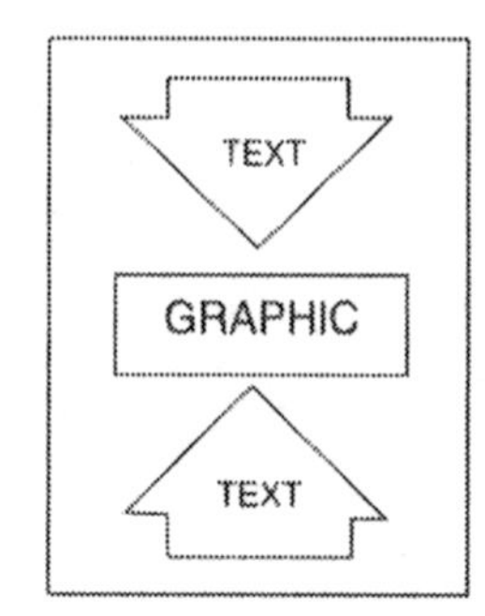

All three strategies are common. What matters most is that the text should *always* explain the logical intention of any graphic element you introduce. A conscientious author or magazine editor will even explain the graphic on a cover page. The explanation usually appears on the reverse side or near the table of contents.

The text and the illustrations are equal helpmates in expressing technical information. It is worth thinking about how the technicians in a video department would create your documents. They would consider the *text* secondary to the visualization, you can be sure. You might not even get a talking head. A "voiceover" is often their preference, and the text

The graphic features can be placed above, below, or in the middle of the text content that is related to it. At times the positioning has to do with the size of the graphic or necessities of the page layout.

becomes the accompanying element to the *visual* presentation. There are times when the text that accompanies your graphics should function as a voiceover that explains and animates the illustration.

Notice that this book is nearly 50% illustration (if you include all the models, samples, examples, and graphics). Most of the samples and model are illustrations of principles and practices that are under discussion. The longer student project samples you have seen are essentially three, four, or five parts of one illustration. This is somewhat unusual as an illustration concept, but in order to show you a significant part of a project, some type of multipage illustration became important.

Assembling Your Graphics

There are features you can add to illustrations to enhance or clarify their meanings. For example, as a rule you should add a title to an illustration, which can greatly assist a reader who is reading your project. You can also assign labels to parts of the illustration, whether you name the components of an assembly drawing or provide the identification of genetic structures in a DNA molecule. A number of these features are extremely useful.

Captions

In order to be sure that a reader understands the intent of an illustration, the illustration is usually titled. The title—commonly called a *caption*—identifies either the content of the visual or explains the intent of it. You may never have noticed, but you are quite dependent on captions. You also take them for granted and may forget to use them, particularly in the case of schematics and certain other diagram conventions.

The length of a caption will vary. It may be far less than a sentence, or it may involve an entire paragraph of fine print that explains the illustration. I call the long version a "talking caption." In between is a vast middle ground in which various publications use different levels of captioning. *National Geographic* seems to have a policy of using one to three complete sentences for most illustrations that accompany their articles. Because of the magazine's commitment to succinct captioning, most readers can enjoy an article or two in any dentist's office in a matter of minutes: they read the captions and not the article.

Samples 10.C

Also, we can use a lens. To avoid the difference of refraction between air and the material of the lens, a lens is formed out of the optical fiber material itself at both ends, as is shown in Figure 3.

The dielectric is stressed by the invisible force of the electric field and can be ruptured by a sufficiently intense potential difference and flux field. This rating is influenced by spacing between the plates, the dielectric material used, and the type of capacitor. Obviously, the closer the plates are to each other, the lower voltage that can be applied (Tontsch 372). Refer to Figure 2 above for a list of dielectric strengths of various materials used in manufacturing capacitors.

1. Disconnect the cut cylinder hose from the cutter cylinder, unscrew the cutter cylinder, and clean the blade with alcohol. See Figure 1.

Figure 1.
Cutter Cylinder and Blade

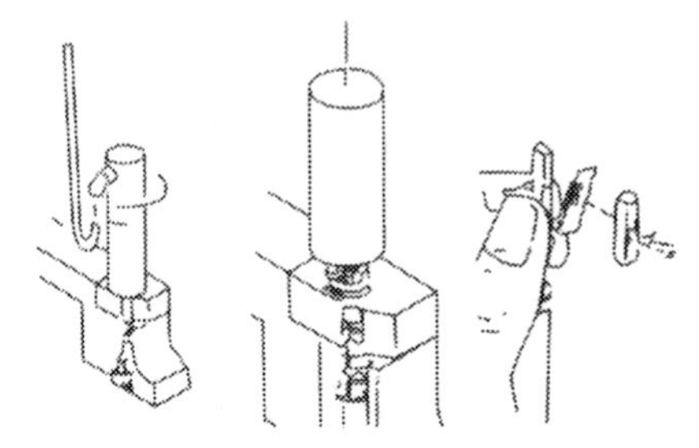

There is a response by most ratings to *temperature, frequency,* and applied *voltage* levels. To account for this, many manufacturers will provide a capacitor rating as a dependent variable and one of these performances characteristics as the independent variable on a graph or table. (See Appendix B.) Responses to the frequency level and the temperature level are the most critical.

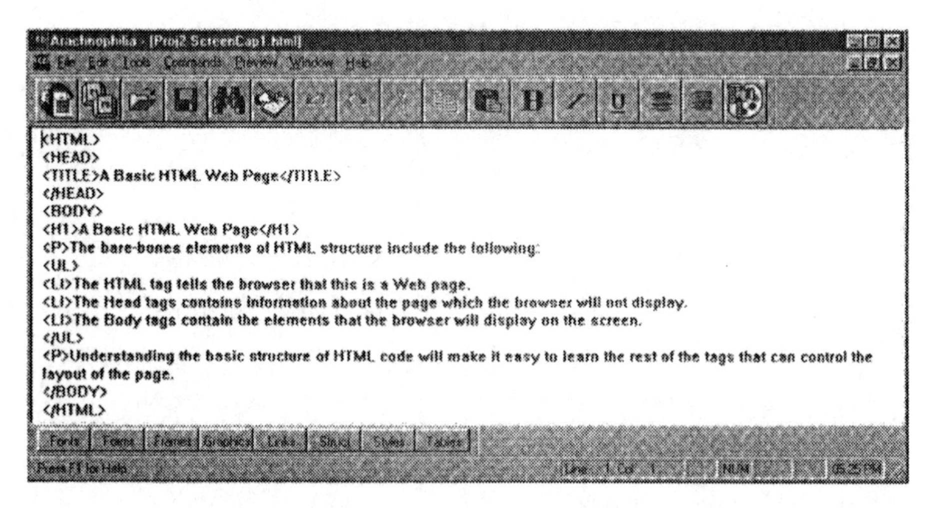

Figure 6 A Basic HTML Page

A minimal HTML page marks up the page itself, the head of the page, which often includes the title, and the body of the page. Inside the body, such things as headers can be specified with the <H1> and </H1> tags, paragraphs can be marked up with the <P> tag (which requires no </P> closure), and unordered lists can be created with the <UL> and </UL> tags. Unordered lists are bracketed by and include line items listed by the <LI> tag. (An unordered list is a bulleted list, whereas an ordered list is a numbered list.)

(Screen image courtesy of Arachnophilia, <http://www.arachnoid.com/>.)

Place the caption below the illustration unless it is a table. You might box a caption if you box the illustration. In this case, box the caption as a narrow banner across the base of the graphic box. If you are using a word processor, you might use italics, or you might change to 10-point type or shade the box lightly or utilize a similar method to identify the caption. Be sure to single-space the caption so that a reader does not mistakenly read it as part of the text, which is a common problem with manuscripts. If you have *Word for Windows* 7.0 or subsequent versions, there is a caption feature.

To Use the Caption Feature on Your Computer

1. Go to **Insert** on the top toolbar.
2. Click on **Caption.**
3. To change the type of caption, click on the **Label** box.
4. To change the caption wording, click on **New Label.**

Figure Numbers

Although popular magazines often omit figure numbers, most technical articles in scientific and engineering magazines and texts make extensive use of figure numbers. This feature is very simple to use, and you will notice that most of the sample and the model documents in this text are designed to incorporate figure numbers. (The line-art drawings

As the sample suggests, figure references can be handled with variety, although the repeated use of the "see Figure X" convention is widely used in technical documentation.

Sample 10.D

Bracketing Tags

The most common type of tag structure is one that brackets an element or section of a page by marking its beginning and end with partnered tags. The beginning or opening tag is bracketed, as usual, by a lesser than (<) symbol and then a greater than (>) symbol. The closing tag is identical with the opening tag, with the exception that a forward slash (/) is inserted between the lesser than sign and the command. For instance, to horizontally center text on a page, the HTML tag </CENTER> must appear just before, and its closing version </CENTER> must be placed immediately after the element or section to be centered, as in Figure 3.

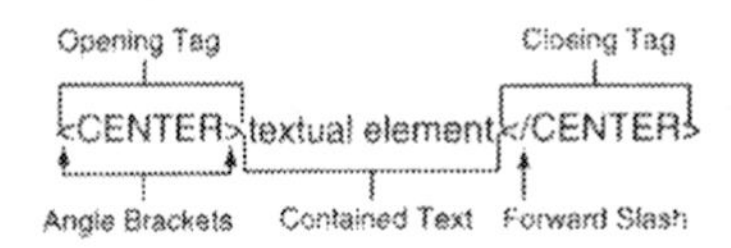

Figure 3. HTML Markup Structure

Nesting Tags

If an element or section of a page needs to be marked up with multiple tags, the tags must be nested. *Nesting tags* refers to the bracketing of internal tag constructions by outer tag constructions; that is, the closing tag of the *first* opening tag must be placed after the closing tag of any opening tag that brackets the same element or section. It is easier to see (as in Figure 4) how to nest tags than to explain how to do it.

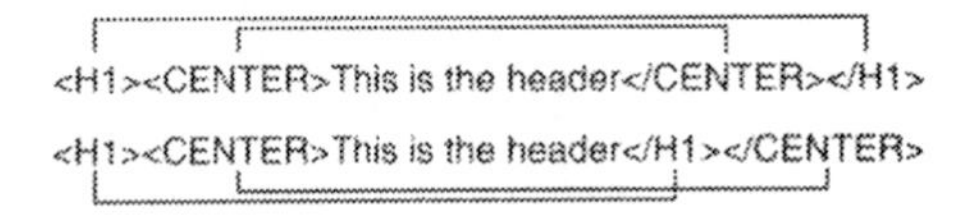

Figure 4. Correct (top) and Incorrect (bottom) Syntax

and smaller samples that appear in the text body are not numbered because that would have introduced a clumsy third set of figure numbers.)

There are a number of advantages to figure numbers, all of which simply serve to link the reader to the correct visual at the proper moment. As I noted earlier, the text should animate the illustration. Place the figure number directly above or directly to the left of the caption. Do not use the symbol #. Avoid decimals unless the document has large chapters. Write out the word *Figure.* During the discussion in the text, the attention of the reader is drawn to the graphic with a simple cue. Here are popular variations you will frequently see.

> The circuit was closed (see Figure 1).
>
> The circuit was open. (See Figure 2.)
>
> The circuit was faulty. See Figure 3.
>
> The circuit was restored (Figure 5).
>
> The circuit shorted, as we see in Figure 4 below.

There are variations on the theme, but these are the most common forms. For the first, and most popular, style—the parenthetical reference—the "see" is *not* capitalized, the "Figure" *is* capitalized, and the period goes *after* the closing parentheses. The second example is a separate but parenthetical sentence, so it is capitalized and contains a period *inside* the parentheses. Notice that you can also use a simple sentence that stands by itself, as you see in the third sample. The fourth example is a simplified reference without a command. You really do not need to tell readers to "see," but using the command is a popular practice.

The last device points the reader forward or backward, up or down. Writers are a little uncertain about this device because the word *below* or the word *above* may really mean "on the next page" or "pages behind." The precise option would be to say, "See Figure 9 on page 15." You will seldom see this particular reference in manuscripts because it is inconvenient for the author. One printing of the document might have used page 15 for Figure 9, but any subsequent printing (especially on other computer equipment) might shift the location. If you use page numbers, you must then edit the text to properly identify all the pages for *all* such references! The terms *above* and *below* are practical solutions to this problem. You might also say, "See the following figure" or "See the preceding illustration." To see the graphic references in a written context, look at Samples 10.C.

If the graphics have been removed from the text discussion, then the figure numbers are particularly helpful. For example, if you have visual contributions for a section of the text that are not critical parts of the discussion, you can move several of them to an appendix.

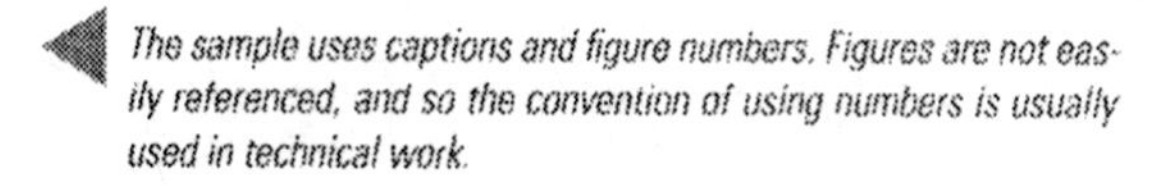
The sample uses captions and figure numbers. Figures are not easily referenced, and so the convention of using numbers is usually used in technical work.

Sample 10.E

♦ Installing The Drives

We will be installing three drives. The hard drive, the CD-ROM drive, and the 3½″ floppy drive. First is the floppy drive. The typical midtower case has two 3½″ drive bays and three 5¼″ drive bays. The floppy drive will use a 3½″ bay. Choose the bay in which you want to install the drive, and remove the plastic face plate. It snaps out. Behind the faceplate you will find a metal plate that is partially punched out. Pry this piece out and dispose of it. Save the plastic faceplate. You may want to use it again later.

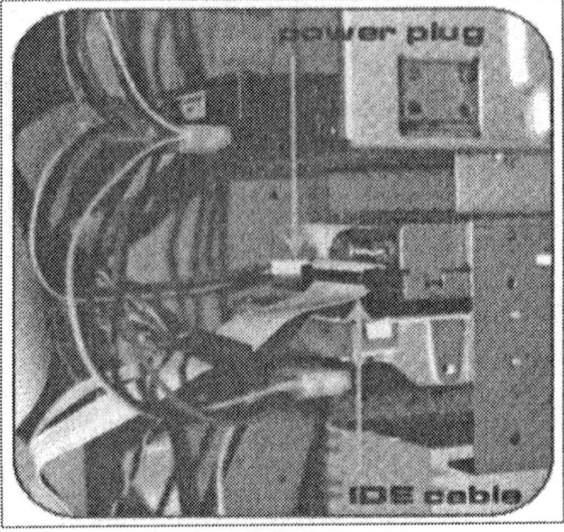

Figure 9. Floppy Drive (Courtesy of Microtell, <http://www.microtell.co.uk/>)

Fit the drive into the bay and fasten it with four screws. There are two types of four-wire power connectors for the drives. The smaller of the two is used for the floppy drive. It is keyed to go in only one way. Next, connect the ribbon cable. Connect the end with the twist to the drive unit. This end of the cable will be marked with an A, designating the A drive. Connect the end without the twist to the floppy controller on the motherboard. Figure 9 shows the floppy drive in place with its connectors.

Tip: When installing the hard drive, pay attention to where it is placed Hard drives generate heat, especially the newer 7200 rpm drives. With this in mind, place them as far away as possible from other components. Also, some cases have a spot next to the power supply for the hard drives. Do not put it there. Keep it to the front of the case.

Frequent use of illustrations, without text to hold them together, becomes confusing. If three diagrams are, therefore, moved to Appendix C, you refer the reader to the appendix (see Appendix C). If the appendix is not dedicated to only these illustrations, you have to be more specific: (see Figures 9, 10, and 11 in Appendix C). As you can imagine, if the text is separated from the graphics, the figure numbers are *very* important to anyone who reads the article. In general, in your college work you can simply number figures sequentially from first to last. You can include graphics from the appendices in the numbering. The standard practice is to provide a separate number sequence for tables.

The MilSpec protocol has a more complicated numbering system that you may have to use for work. The graphics are separated from the tables—always—so you have two lists in front of a document, one for illustrations and one for tables. The number of each standardized subsection is included as part of the figure and table numbers, which are hyphenated. For example, let's say that Section 4 has a figure that is the third in the section; hence, it is Figure 4-3. If Section 2 has a table that is second in the section, it is Table 2-2.

For short documents (a chapter or less) or self-contained documents or for any use other than MilSpec, the *simplest* system is probably the best system: 1, 2, 3, 4 . . . . Remember the goal: simplify complexity; do not complicate simplicity.

Courtesy Line

Since you may have freely borrowed the illustrations in your text, you must identify the sources. The easiest acknowledgment is a "courtesy line" or credit. Below the caption, in parentheses, indicate the author and the article or book from which you borrowed the graphic (even though the author is seldom the illustrator). You will notice that a great many of these acknowledgments simply say something like (Courtesy of Dave Rigby, *Technical Document Basics*). Underline or italicize a book title; put a magazine article title in quotes, and add the magazine title in italics. For Websites, identify the site and the URL (see Sample 10E).* Courtesy lines are usually placed below the illustration, as you will note in Sample 10.E. See the *Writer's Handbook* for discussion of full bibliographic acknowledgement if that is desired.

Be sure that you are consistent in your use of the figure number, the caption, and the courtesy line. Consistency will help a reader understand your text. Inconsistency creates the obvious opposite effect. Use the same size and font for all captions. Also use the same location for the captions if at all possible. Do not move the caption box around to the sides or above the graphic if you can avoid it. Whatever the location, use it as consistently as possible.

** Study the simple acknowledgments you see in the models and samples. For student projects, you need not use the precise acknowledgments that you see me use from time to time in the text. These longer acknowledgments are formal recognition of permission to reprint material.*

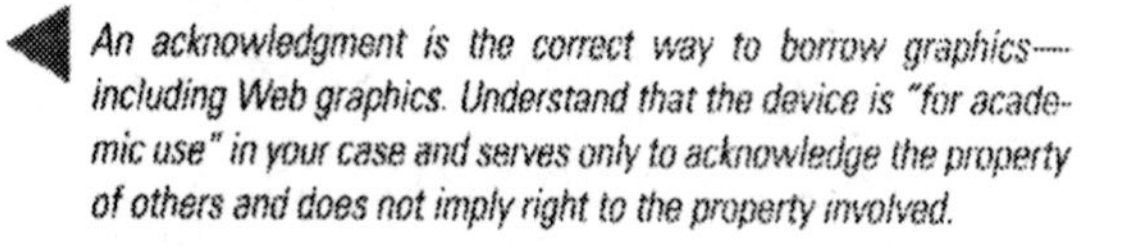

An acknowledgment is the correct way to borrow graphics—including Web graphics. Understand that the device is "for academic use" in your case and serves only to acknowledge the property of others and does not imply right to the property involved.

Sample 10.F

2. Using two tuning forks ($341\frac{1}{3}$ and $426\frac{2}{3}$ Hz), locate two heights in the air column that produce primary resonance for each tuning fork.
3. Using one tuning fork (500 Hz), determine three heights that produce primary resonance.
4. Using two tuning forks (l000 and 2000 Hz), determine five heights that produce primary resonance for each tuning fork.

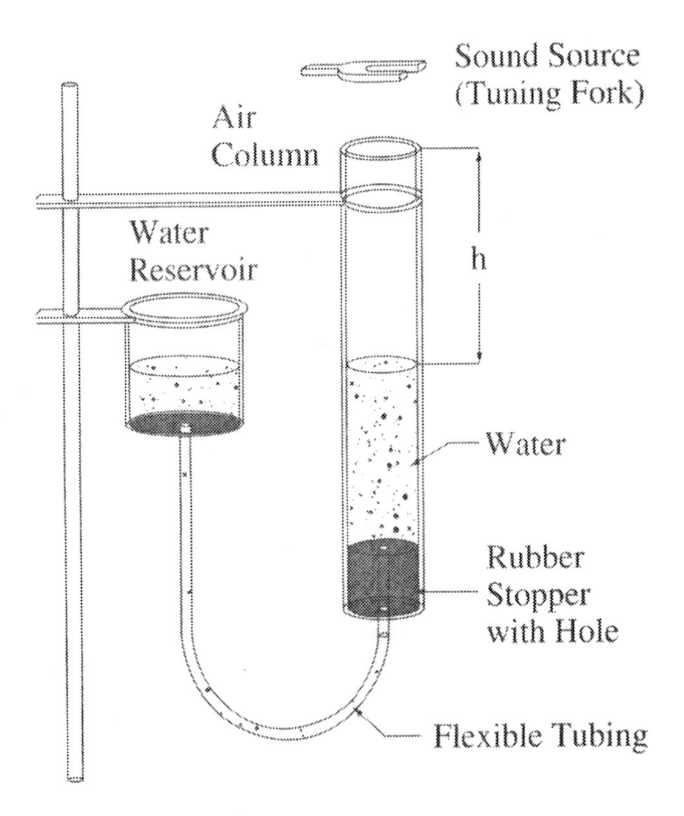

Figure 1 Apparatus for Measuring the Wavelength (λ) and Velocity (v) of Sound in Air

Callouts

Suppose that you have a document developed to the mock-up stage. The graphics are integrated. You have captions, and the text refers to the graphics by figure numbers. Under many circumstances these measures will allow the reader to comfortably move back and forth between whatever visual media you have used and the written text. If there is some remaining confusion, you may need to use additional devices to help the reader. These are usually internal; that is, they are *in* or *on* the environment of your illustration. Callouts are particularly useful. Any identification that names the elements or parts of your graphic is a *callout*. A callout is usually accompanied by an arrow (called a *leader line*) that points at the reference part in question, as you will notice in Sample 10.F and in the following illustration.

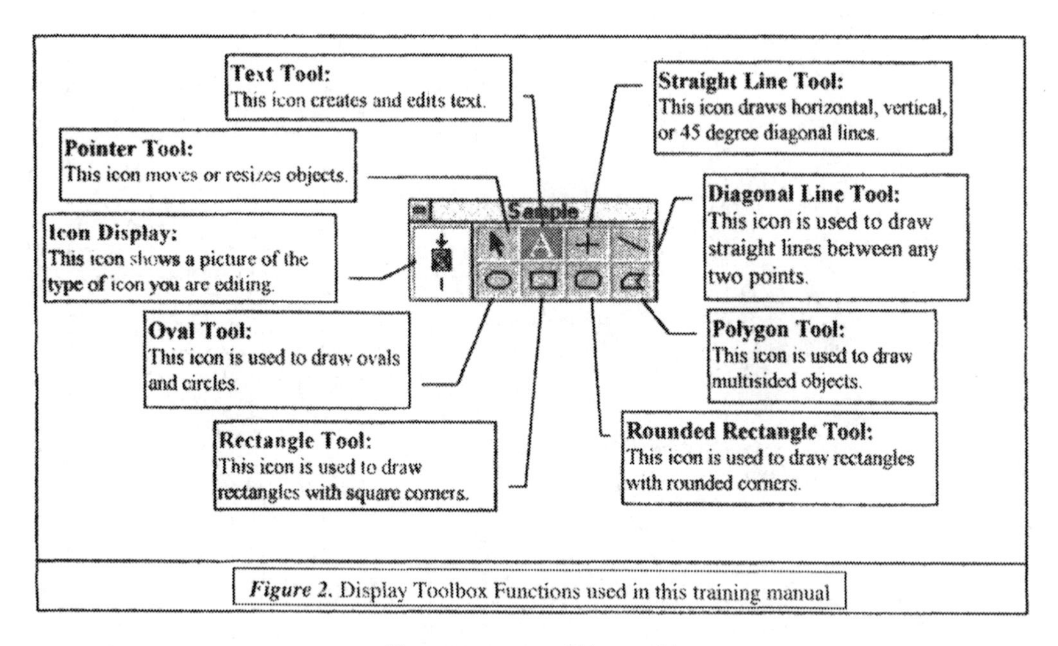

Figure 2. Display Toolbox Functions used in this training manual

(*Screen scan courtesy of Macromedia*)

Computers provide the tools for neat and orderly use of callouts. Note the author's decision to use arrows for the smaller components.

Sample 10.G

3

(Quantity) and Equipment

(1) Büchner funnel with a rubber gasket; (1) filter flask; (1) trap bottle with a three-holed rubber stopper, inlet, outlet and relief tubes (include rubber hose and clamp on relief tube); (1) 125-ml round-bottom or Erlenmeyer flask including a rubber stopper with a thermometer inserted in it; (1) hot plate; (1) steam bath;(1) vacuum aspirator assembly; (1) stand; (1) flask clamp. See Figure 2 below, and Figure 3 on page 6.

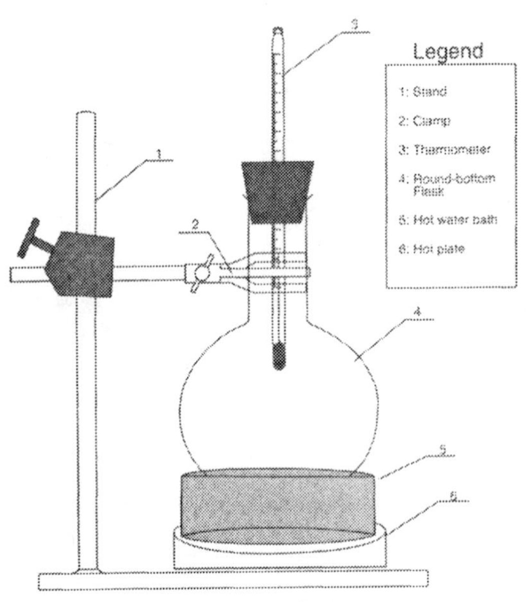

Figure 2. Synthesis Flask

Reactants and Reagents

The reaction is an acid-catalyzed dehydration of an alcohol (salicylic acid, 4 g), with acetic anhydride (8 ml), in an S-2 reaction using H_2SO_4 (5 or 6 drops) as the catalyst. Acetylsalicylic acid is the product, but it is not pure. It is reacted with 50 ml of saturated sodium bicarbonate ($NaHCO_3$) solution to complete the reaction in order

Notice that callout lines are used to connect the callout to the part or point of interest that is being identified. Unless you are trained in drafting you might not realize that there are a dozen or more types of "lines" that are used in drawings: stitch lines, phantom lines, section lines, chain lines, and so on. You are familiar with at least two of them: the extension line and the dimension line.

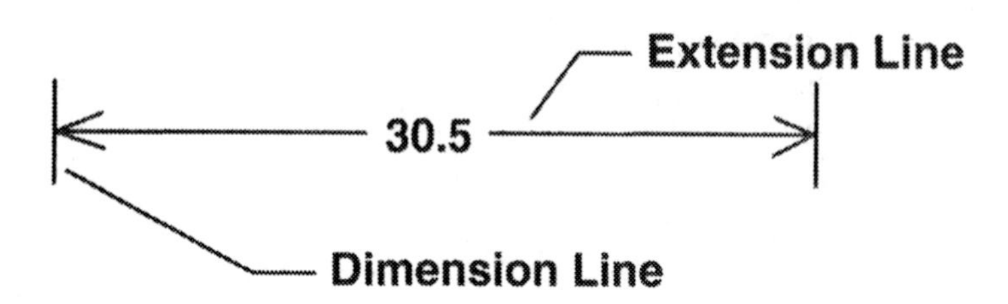

The reason that callout lines are usually diagonal or doglegged (as in the preceding diagram) is so that the lines are not confused with other types of lines that may be used in the graphic, particularly in drawings.

The point is simply to identify the parts under discussion by tagging the elements that match the vocabulary or discussion of the text. As I observed in the analysis of vocabulary in Chapter 8, illustration is a definition technique. By adding callouts to a graphic you provide visual identification for the names used in the text, which is a method of defining the components.

To Add Callouts to Your Graphics:

1. Make sure your drawing toolbar is displayed. If it is not, go to **View** on the top toolbar, then **Toolbars**, and click on **Drawing**.
2. On the drawing toolbar, click on **Autoshapes**.
3. Click on **Callouts**, and select the type of callout you need.
4. Draw the callout by clicking on the point where you would like the callout to be, and then drag away from that point until you get the shape and size callout you want.
5. To move the callout, click on the callout border and move it.

A legend can replace callouts. If a legend is attached to a graphic, then the locators serve to clarify both the graphic and the text. Notice the extension lines.

Sample 10.H

3

The heart is a hollow organ weighing less than a pound, and it is a little larger than a person's fist. Its tough, muscular wall (myocardium, 1) is surrounded by a fiberlike bag (pericardium, 2) and is lined by a thin, strong membrane (endocardium, 3).

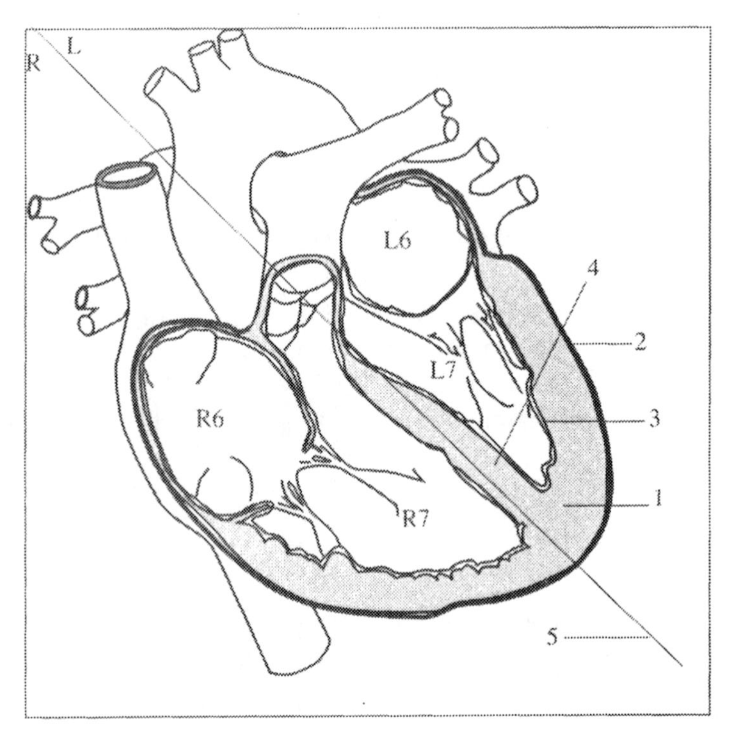

Figure 2. Vertical Slice Through Heart

A wall (septum, 4) divides the heart down the middle of the long axis (5). Each side, right (R) and left (L) is divided again into an upper chamber (atrium, 6) and a lower chamber (vertricle,7)

At its restful state, this four-chambered muscular pump is capable of outputting approximately five liters of blood per minute and of circulating a total volume of five to six liters reconditioned with oxygen. The cardiac (heart) muscle fibers differ from the other

Callouts also can be added to or deleted from borrowed photographs or drawings, but use this procedure only if you are unable to construct the graphic yourself. If you borrow an illustration of a telescope lens configuration, for example, you might want only 5 of the 10 callouts that are on the graphic. This situation suggests that your text is relatively simple and the graphic is relatively complex. To aim directly at your target audience, use correction fluid or tape or scissors or Photoshop to doctor the graphic to suit your needs. You can hand-letter or type in callouts, also. Transfer lettering will cover photographs with callouts in a very professional way. You can also scan and edit the illustration electronically.

The altered illustration should be acknowledged. Instead of saying "Courtesy of . . .," use the phrase "Adapted from . . ." (see the illustration and credit on pp. 30-31). Ellipses and brackets are the conventions for indicating that a quotation has been changed. You should maintain the same integrity in the use of graphics. College writing projects depend heavily on quoted sources, but your needs are likely to concern graphics, particularly illustrations that are important features of your project. You are allowed to borrow the materials because it is a privilege extended to you solely for educational purposes. You have to respect copyright law and you must acknowledge a borrowing of any kind.

Locators

A variation on the callout is the use of simple indicators— usually the alphabet or numbers. In electronics these are referred to as *component identifiers* or *reference designators*. They are placed at the appropriate locations in the graphic, and in the explanatory text, they are placed after the words that otherwise would be used for a callout. Because schematics do not usually need words, locators are often used, as for the following diagram.

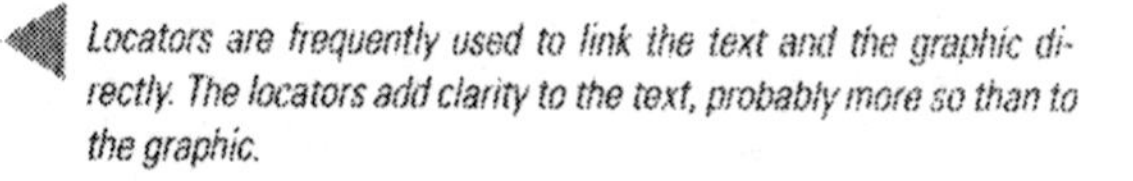

Locators are frequently used to link the text and the graphic directly. The locators add clarity to the text, probably more so than to the graphic.

Sample 10.I

5

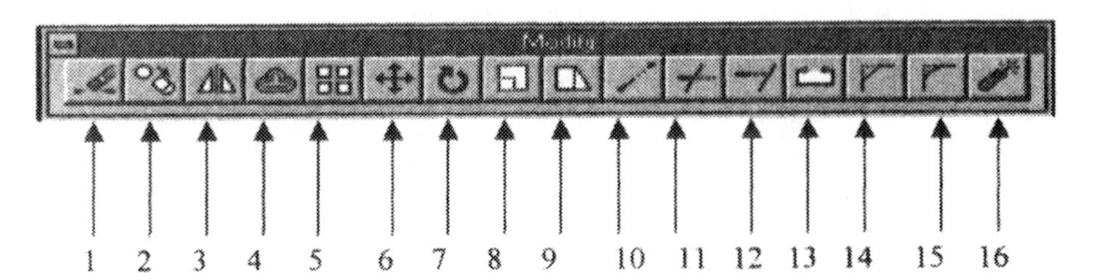

Figure 4. Modify Toolbar (Screen scan detail courtesy of Autodesk, Inc.)

Modify Toolbar

Mode		*Description*
1	Erase	Removes objects from a drawing
2.	Copy object	Duplicates objects
3.	Mirror	Creates a mirror-image copy of objects
4.	Offset	Creates concentric circles, parallel lines, and parallel curves
5.	Array	Creates multiple copies of objects in a pattern
6.	Move	Displaces objects at a distance in a specified direction
7.	Rotate	Moves objects about a base point
8.	Scale	Enlarges or reduces selected objects equally in X, Y, and Z directions
9.	Stretch	Moves or stretches objects
10.	Lengthen	Lengthens an object
11.	Trim	Trims an object at a cutting edge defined by other objects
12.	Extend	Extends an object to meet another object
13.	Break	Erases parts of objects or splits an object in two
14.	Chamfer	Trims the intersecting lines to the endpoints of the chamfer line
15.	Fillet	Rounds and fillets the edges of objects
16.	Explode	Breaks a compound object into its component objects

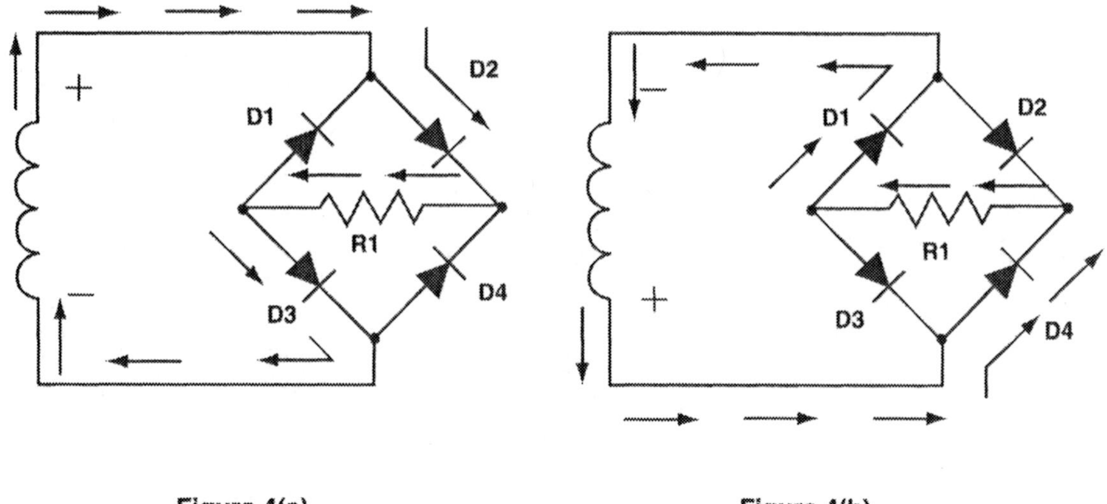

Figure 4(a)
Positive half-cycle of conventional current flow (not electron flow)

Figure 4(b)
Negative half-cycle of conventional current flow (not electron flow)

The text will utilize the locator cues to explain the diagram.

> The direction of current flow on the negative half-cycle of secondary voltage is shown in Figure 4(b). Here, current is again forced to flow right to left through the load resistor R1 because diode D2 is reverse biased. The current then continues up through the forward-biased D1 and back to the source.

Another example of this method of reference appears in Sample 10.H.

Legends

Another popular convention for handling callouts is to list them in the graphic environment (inside the graphic box if you used one) but separate from the illustration. Sometimes the legend is more orderly and easier to understand than the radiating callout arrows going here and there. Certain graphics lend themselves nicely to legends, exploded views, for example. Others, such as anatomical drawings, are not easily altered to create neat, orderly graphics. A legend often omits the arrows, also, and this can greatly improve the appearance of an illustration. A graphic that looks like the crisscrossed spokes of a bicycle wheel because of callouts is not going to serve your purpose.

This very geometric construction neatly handles sixteen icons by location, identification, and description. This was constructed as a text page but could have been single-spaced and boxed as a self-contained graphic.

Sample 10.J

7

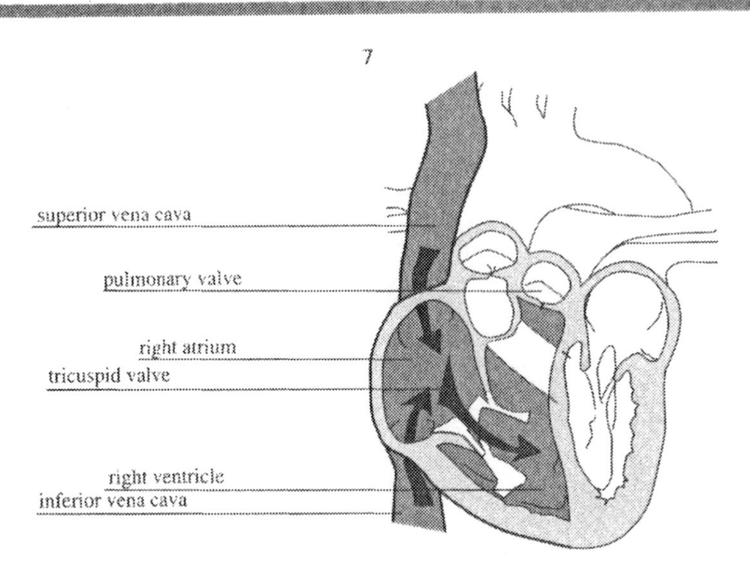

Figure 4. Diastole Phase

As the SA node action potential is relayed to the AV node, the pressure and volume increases within the right atrium and ventricle. The wave of depolarization causes contraction of the atrial walls, and the atrial systolic phase occurs.

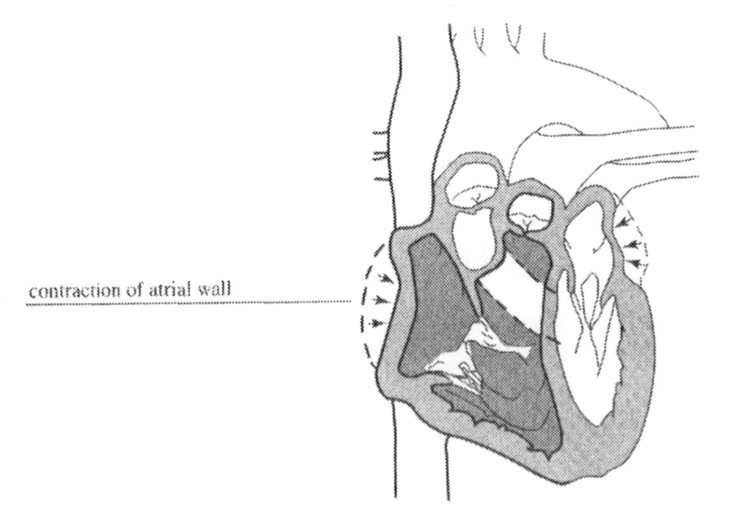

Figure 5. Atrial Systolic Phase

The legend lists a group of items alphabetically or numerically, together with a location code (see Samples 10.G and 10.I). The organization of the list is by order of appearance in an exploded view, but may be alphabetical if you simply want to construct a locator that will point readers to the position of a part. Each item is assigned a letter or number that is positioned as a tag in the illustration.

The organization of one legend, then, may be by order of appearance in a graphic. The organization of another legend will be alphabetical if you want to point readers to the position of a part or item. Readers look down the list for the alphabetical position of a word and locate the part by number or letter.

The legend has one other use that is a convenience in complicated text descriptions: the location code can be mentioned as the parts are being mentioned in the text. The legend numbers then do double duty by functioning as locators in the running text. This is an unobtrusive device even if you use it dozens of times. It also allows a wider audience to follow the text. Those in the know will consult only a few of the locator references, whereas those in the dark busily get to work piecing the text and the graphic together by using the legend and the locator references.

A legend is likely to alter an original illustration. If you alter the illustrations you borrow, indicate that the readers are *not* looking at the original. Replace the courtesy line with the expression "Adapted from," as explained earlier: (Adapted from David Rigby's *Basic Composition Skills*.) Acknowledge all sources for borrowed graphics.

Animation

It was mentioned early in the chapter that the text can help animate the graphic. You often need to walk through the activity of what the object in a graphic is supposed to be *doing*. The illustrations are not animated, of course, so you need devices that will allow the reader to imagine and visualize the behavior of the object. The text is in a position to *explain* the animation, and this option is quite practical. Language is in motion when a reader reads it. The illustration, on the other hand, hangs in space. If the systems are matched, they collaborate, and readers can imagine the motions in the graphics as they are described in the text. You can also create either video packages for genuine animation or software simulations. In between are devices you will find practical and accessible. There is a grab bag of tricks for constructing the fourth dimension for your two-dimensional paperwork. Just as you can create the illusion of depth to add a third-dimension, you can create the illusion of time (movement) to add a fourth.

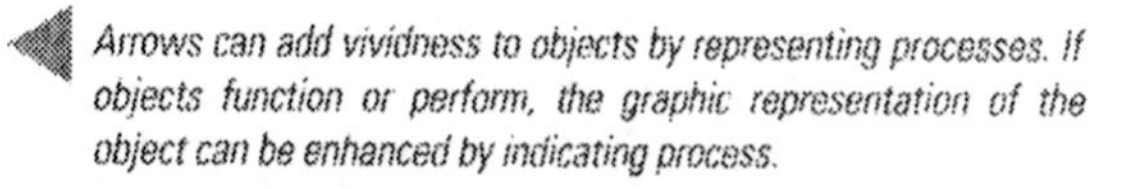

Arrows can add vividness to objects by representing processes. If objects function or perform, the graphic representation of the object can be enhanced by indicating process.

Sample 10.K

The Principle of Cancellation:

There are two ways to prevent electromagnetic interference (EMI) and radio frequency interference (RFI): either by shielding or cancellation. UTP employ cancellation. Cancellation is a rather elegant approach to preventing interference compared with the brute-strength approach of shielding by wrapping it in foil. Each wire in a cable acts like an antenna to RFI and EMI. The longer the cable the greater the interference. As current flows through a wire it creates a small electromagnetic field around the wire.

The direction of the electromagnetic lines in this field is determined by the direction of the current flow (see Figure 2). When two wires are in the same circuit (paired), electrons flow to the destination from the negative voltage source in one wire and from the destination to the positive source in the other wire.

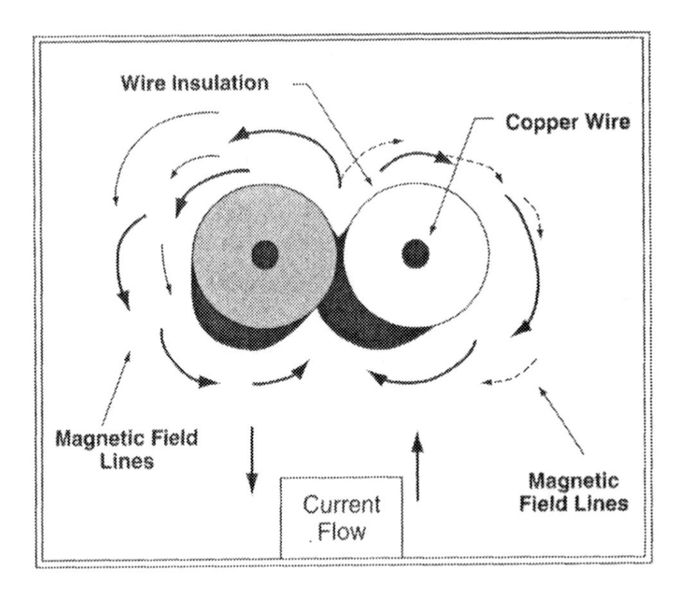

Figure 2. Magnetic Field of UTP

When the paired wires are close to each other, these opposing magnetic fields cancel

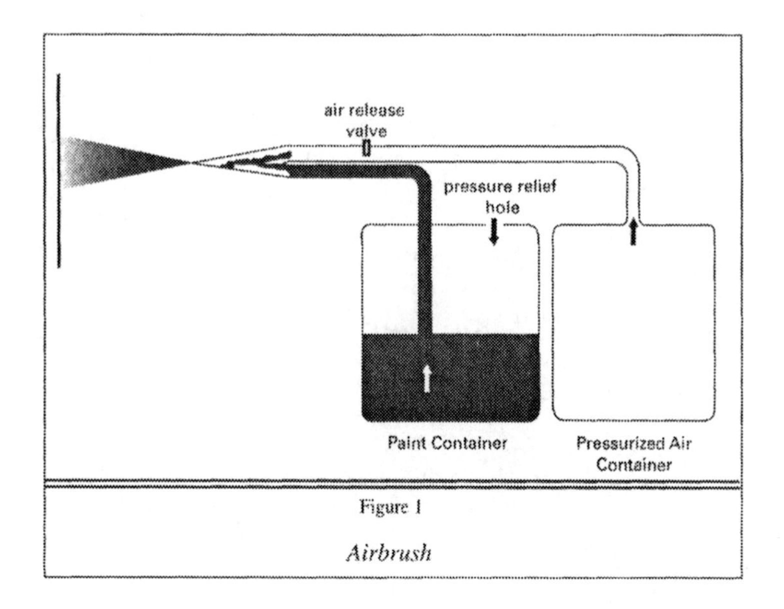

Figure 1

Airbrush

Action Arrows

Certainly as old as writing itself, the universality of arrows and their meaning dates to cave paintings. They remain a handy tool and still signal direction and movement, although not that of a primitive projectile. Do not hesitate to use them to indicate the "movements" of graphic elements. Of course, the movement is not always visible (see Sample 10.K). Apparent movement or theoretical movement enters the picture in many engineering considerations, but the direction of static force, as in the case of a loaded beam, for example, *can* be visualized or conceptualized. Similarly, you can show the "flow" of the dynamics of a circuit, or you can easily render actual movements (see the preceding and following illustrations).

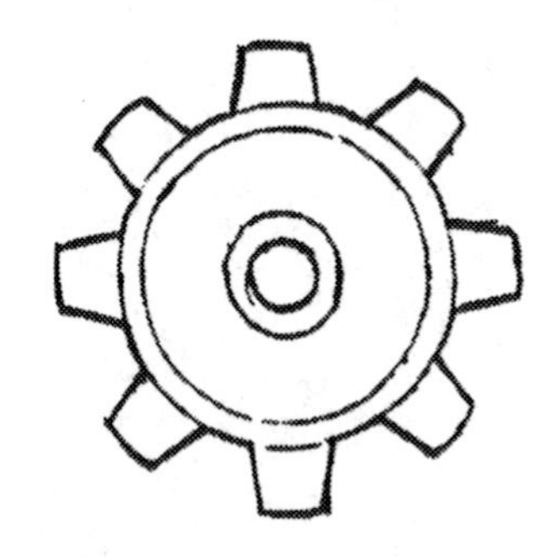

TWO DIMENSIONS

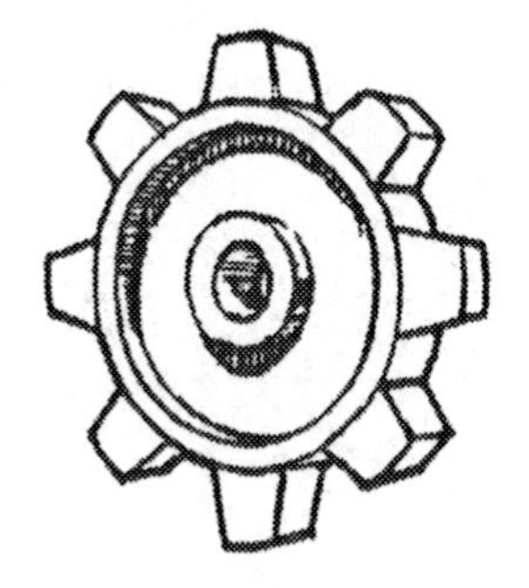

THREE DIMENSIONS

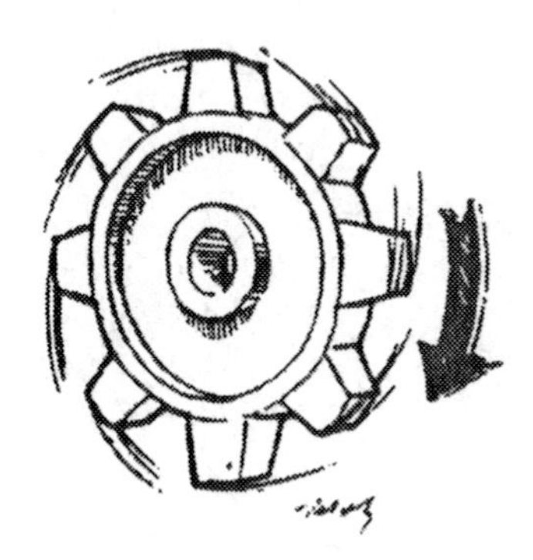

FOUR DIMENSIONS

Whether organic or mechanical, many movements can clearly be indicated with arrows. If the text discusses the movement, the arrows complement the text.

Overlays

I mentioned transparent plastic pages in Chapter 9 in the context of developing cutaways. They can also be used to create the suggestion of movement. Transparencies are always a bit of a luxury due to the expense, but they have the advantage of being able to impose changes on the *original* position or original movement or moment of your object. For example, an arson investigator for a large suburban fire department near Seattle found that the transparency idea was the best method he could use in court testimony.

Suppose a large fire has questionable origins, and a legal contest begins between the insurer and the owners of a business. Attorneys defer to the arson investigation team for evidence. One important issue is always the point of origin of a fire. For this part of the hearing, the arson investigator uses overlays. For evidence, he constructs a transparency series to visualize the location and size of the fire, one transparency for perhaps every hour of the five hours that it burned. The spread of a fire could otherwise be hard for an uninitiated group of jurors to understand. The investigator overlays the transparencies and tells the story backward. The concentric circles of the fire diminish one by one until the point of origin is visible. The movement is a movement in time, and the investigator presents it so that people can understand what happened. The overlays are quite vivid.

Overlays can be used for sequencing time, but they are common for illustrating layers for a variety of applications that visualize buildups—of electronic schematics, let's say, or floors of a building. They can also be used to overlay lines on line graphs and color-coded surfaces on a surface chart.

Cartoons

Cartoons are sometimes used to suggest animation in technical documentation, but cartooning is also an accomplished skill few authors are likely to have. Mark is a professional illustrator with an interest in technical writing. He did the sketches on pp. 413 and 433, so I asked him to draw a speeding car. Notice the obloid tires and other animation tricks.

Cartooning conventions can be adapted to our technical work. Cartoons often add a friendly note to software manuals and other documents where there is a concern for simple and clear representations of tasks. The popular magazine *PS* is a preventive maintenance magazine published by the U.S. Army. It uses cartoon illustrations, comic-book graphics, and dialog balloons, but the subject matter is technical.

Airbrush Drawing

The exhaust in the speeding-car illustration could have been handled with an airbrush. An airbrush is a unique drawing tool that is often used to animate apparatus by adding vapor or steam or exhaust or, perhaps, an explosion in a piston or in a rifle chamber, for example. The airbrush is a miniature spray gun that is used to create extremely realistic renderings of cutaway views of machines that *appear* to be working but could not possibly be once they are cut open. If a cutaway piston is functioning in an illustration of photographic quality, it is probably a photograph touched up with an airbrush, or else the entire illustration is airbrush work.

Some of these animation tools are conveniences you may use sooner or later. The world is certainly not static. A writer often develops documents as though the focus of the analysis is frozen in time, but this is purely a fiction an author designs as a convenience to study the subject. At times, the subject is best understood in an animated environment. Readers *learn* from and *use* the animation if an author has provided the "movement." Simple indications of the movements can be added to an illustration and discussed in the text.

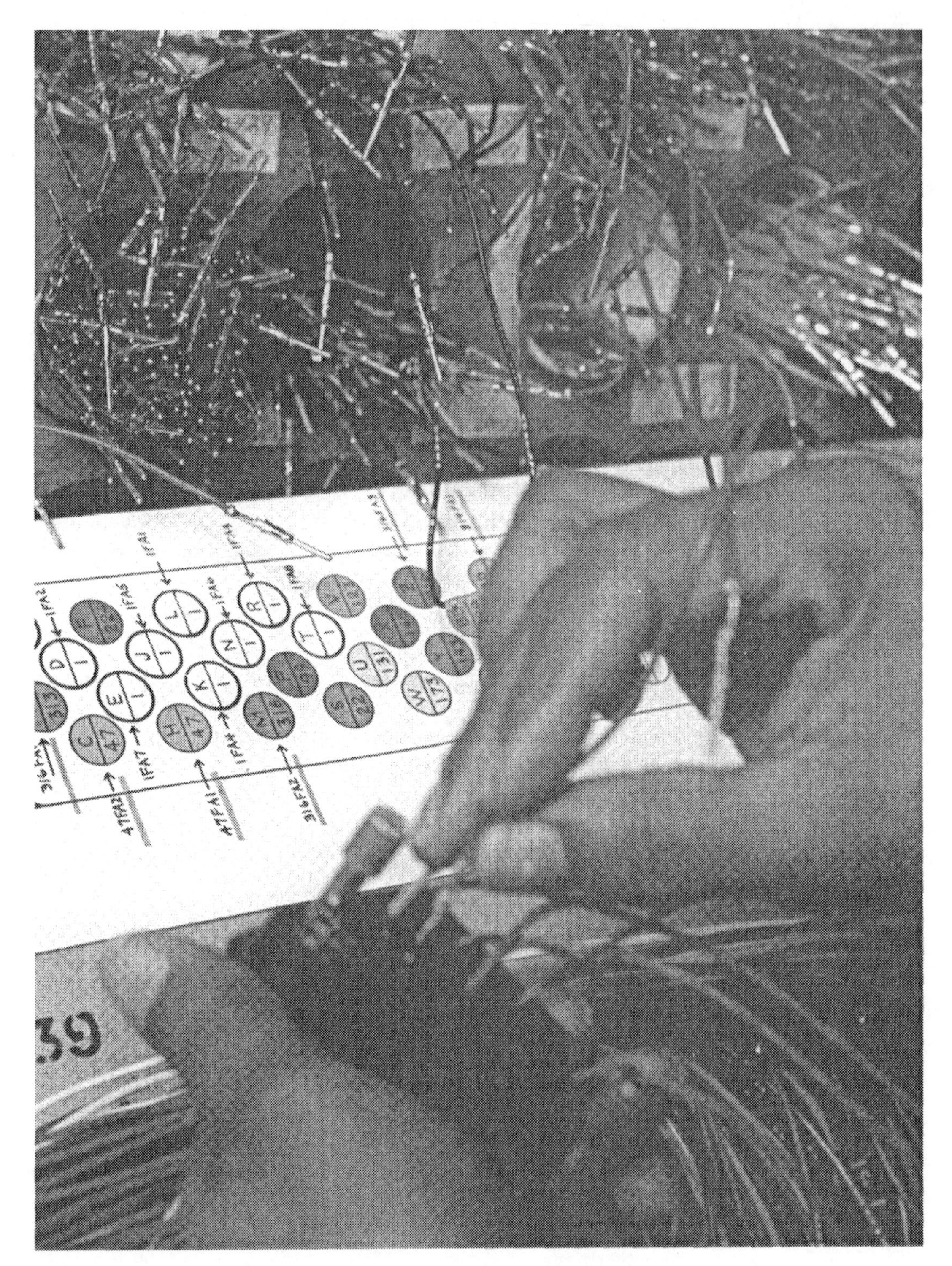

(Illustration courtesy of the Xerox Corporation)

The Use of Color

Color is conspicuously absent in this text.* I did not *need* to use color. The usefulness of color depends on the subject matter. The original photograph of the illustration on the left was in color. The assembly in the photo is constructed entirely by a color-coded installation process that is common in electronics industries. The message is quite clear: black-and-white graphics have a limited utility.

In the twentieth century, color became a tool that was used in many industries, particularly for production-line assembly. Although color photography became popular in magazines due to the four-color processing that gained popularity in the 1930s and 1940s, the use of color was considered to be something of a luxury in industry, at least in technical documentation. The assemblies were often brightly colored, but the schematics were black and white.

Manuals and the various written products that accompany manufactured goods are considered utilitarian and generally receive low production standards. Four-color processing—the inking procedure for printing illustrations in color—is usually reserved for catalogs, sales literature, and an assortment of promotional printed materials. Any reasonably large company will have a publications policy, and the rule of thumb seems to be that color printing is reserved for either high-profile products that are designed to attract attention or, when necessary, technical graphics that call for precision. It is this second use that is of primary interest here.

Determine your need for color by examining the typical practices in your field of study. Is instrumentation heavily influenced by color in some way? Are the devices assembled by color? Do the devices render images in color? Is color the point of an analysis of some sort? Also look at your specific needs. Did you design a graphic to use color as a point of clarification? For example the last illustration in this text (see p.486) is designed to show a digital schematic, but it is unique in that it is actually sectioned by color in the original version. The discrete parts are not visibly distinct in black and white, which is precisely the point of the use of color. The following illustration shows the well-known color-band identification code for resistors. This illustration is a detail from a catalog.

** At my request, the publisher was willing to omit all color processing so that I could bring the Wordworks series to students at an affordable price.*

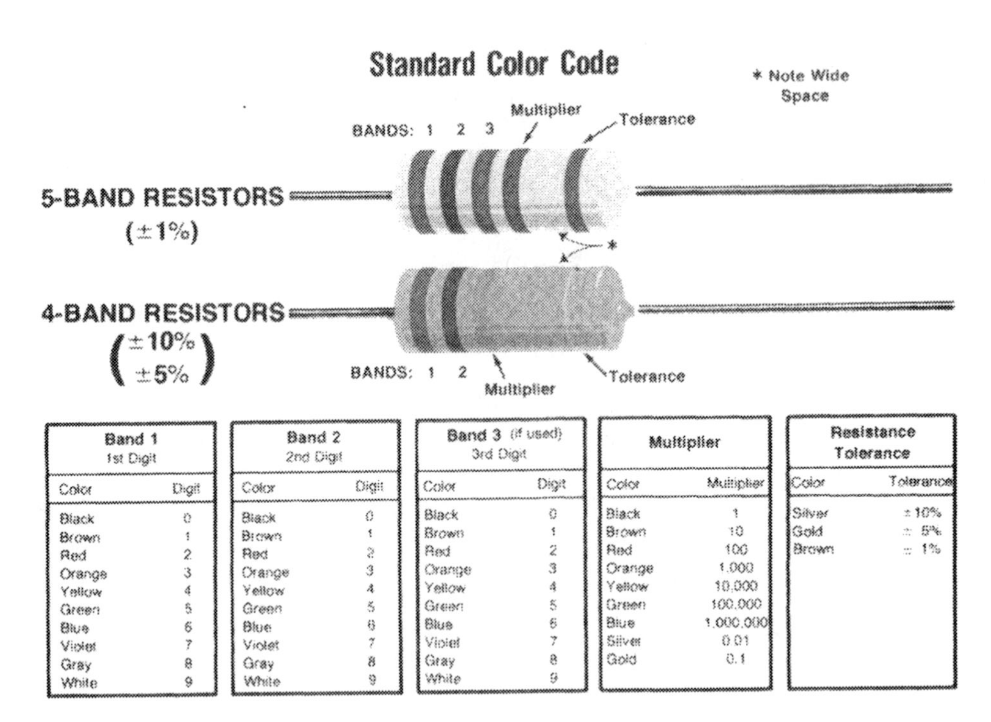

Band 1 1st Digit

Color	Digit
Black	0
Brown	1
Red	2
Orange	3
Yellow	4
Green	5
Blue	6
Violet	7
Gray	8
White	9

Band 2 2nd Digit

Color	Digit
Black	0
Brown	1
Red	2
Orange	3
Yellow	4
Green	5
Blue	6
Violet	7
Gray	8
White	9

Band 3 (if used) 3rd Digit

Color	Digit
Black	0
Brown	1
Red	2
Orange	3
Yellow	4
Green	5
Blue	6
Violet	7
Gray	8
White	9

Multiplier

Color	Multiplier
Black	1
Brown	10
Red	100
Orange	1,000
Yellow	10,000
Green	100,000
Blue	1,000,000
Silver	0.01
Gold	0.1

Resistance Tolerance

Color	Tolerance
Silver	±10%
Gold	± 5%
Brown	± 1%

(Illustration courtesy of the Tandy Corp.)

The assembly that is being constructed in the illustration on p. 436 involves various colored wires. The black-and-white version of the photograph certainly speaks to the significance of color. Throughout the many industries of the engineering fields and the sciences, the use of color is an important complement to our understanding of concepts and realities. In astronomy, a discussion of the Doppler effect or a table of star types will be enhanced by color. A book concerning precious and semiprecious gems would be very limited without color.

In addition, there are a number of color standards that are used as material codes. Among them is the very important OSHA Hazard Diamond developed by the department of Occupational Safety, Health, and Welfare. The color code of this well-known symbol is used by emergency crews when responding to accidents involving various chemicals.

Scanning represents the high end of color reproduction for student productions. In my classroom experience, I am increasingly seeing student projects that involve electronically transferred color graphics. With the color scanners in your campus computer lab you can import photographs, or transfer digital photographs from a digital camera, directly into the project. Photographs and other color graphics can also be taken from the Web. The quality of these reproductions is directly related to the quality of the equipment. Web graphics are generally 72 dpi and do not copy very well when printed. The typical affordable scanner now operates at better resolution and better color recognition than the older models. Color printers are also useful. I also see projects in which such features as boldface have been replaced with color in the running text.

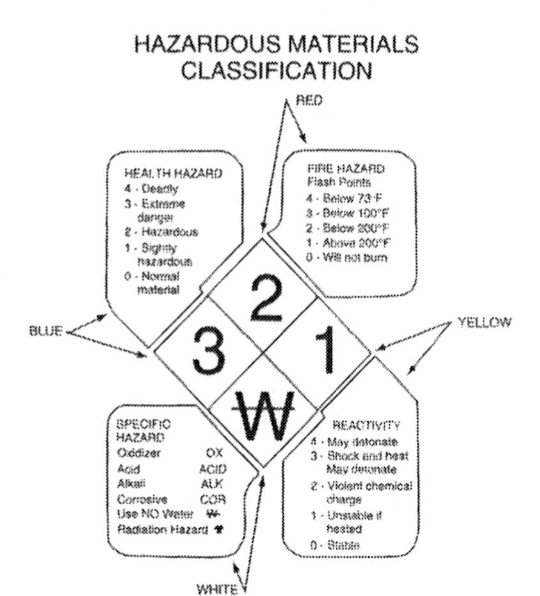

Consider color as a practical tool in your work. Because color xerography is readily available, it is now convenient to use color in your technical documentation. At a production run of 500 pages, copy shops remain a pricey resource for color printing, but for a college student who may be reproducing three or four illustrations, the cost is not prohibitive—usually less than a dollar a page. Xerography is a handy approach to color, particularly if you use pasteups that include a page of text and an original color photo or a borrowed illustration in color. Two details are of possible concern. If you cut out a borrowed illustration, develop a pasteup, and make a color copy, the result will be reasonably faithful to the original. If, however, the original cannot be cut out, you must make a color copy and use it for the pasteup. The copy you then make of the pasteup is a second-generation reproduction and the color quality is likely to deteriorate. A second detail is a minor disappointment I often hear: a page of color copy does not quite match the look of the rest of the text because the text of print, although black, is actually being printed in color, and the paper stock is also slightly different in appearance.

As I have said before, use whatever works. Low tech will often do the job. Colored pens can make an attractive graphic if you can get by with line art, meaning simple drawings. Avoid colored pencils unless you want a very soft appearance in the coloring. Inks saturate the line on the paper and have considerably more intensity than pencil lead. Again, consider the pasteup as your layout method. If you leave a space for a graphic that you plan to develop in ink, risks are quite high. One smudge and the page of text is ruined. Of course, if you print extra copies of the page you can draw the illustration in position until you get a neat version. However, you might find that a more professional looking graphic will result from a larger original that is then color copied to scale to fit the page. Create a pasteup with the scaled version and recopy. In this case the second-generation effect is not significant because the colored lines are quite easy for a good copy machine to reproduce.

Use color in any form that serves your needs. If color will in some way increase the success of what an illustration is designed to communicate, then it should be used. If you use color for a pie chart or a bar graph, the significance is probably minor since the same effect can be achieved with the various fields of black-and-white patterns on your computer. However, if there is a color code involved (electronics) or if color is the point of interest (spectrometry), or even a point of value (gem stones), color is indispensable. If you have a sick plant at home and you take a leaf to a nursery, the specialists there will diagnose the illness largely on the basis of exactly one feature: the color distribution of the leaf structure. They will consult a reference book and match the leaf to a color photograph of a diseased leaf and make a diagnosis.

Graphic Components

Comments

The samples in this chapter illustrate the flexibility of layouts that can be used for both simple and complex presentations.The elements in the graphic environment—captions, callouts, and so on—can help clarify simple graphics and are indepensable in complex graphics.

Graphics must be *used,* which means that the text must explain and explore the visualized material. The refrigeration and waveform projects (Samples 10.A and 10.B) illustrate the way in which authors integrate the discussion and the graphic.

The brief quotes in Samples 10.C from various texts serve as models of the ways in which the text can refer a reader to a graphic.

The use of callouts and animation arrows and the idea of sequencing a group of graphics are illustrated in Samples 10.F, 10.J, and 10.K.

Sample 10.E shows an example of a credit for a source for borrowed graphics. The source also should be identified in a bibliography if you develop one.

The legend is a very orderly device that is commonplace in technical graphics (Samples 10.G and 10.I).

Summary

- Graphics speak their own languages.
- If you have any choice, select graphic tools by readership needs. Use simple, somewhat complex, or complex visuals that fit the audience.
- Put the graphic elements to work: *write* what you want the reader to see.
- Label the graphic with a figure number and a caption.
- Use arrows to animate parts that are supposed to be moving.
- Refer the reader to the graphic by using figure references: (see Figure X).
- Add desirable callouts.
- Build legends if they will help clarify the text or the graphics.
- Acknowledge borrowings with a courtesy line to credit the source.

Activities Graphics Assembly

Locate one of your old laboratory reports that included tables, graphs, waveforms, schematics, or other types of illustrations. Is it up to your new standards for graphics? Make a copy of the document and add the following conventions where they are needed. (Use a colored pen.)

Figure numbers	**Callouts**
Captions	**Locators**
Figure references	**Legends**
Animators	**Courtesy lines**

Use the Chapter Summary as your guide.

Select an illustrated technical article of interest to you. Copy the article and identify all the uses of the conventions listed in the first exercise. Use a highlighter. You should find many of the devices in the first three or four pages of the text.

Share a project

The instructor will survey the class to see which students are familiar with various popularly available graphics applications. These students will act as guides to take small groups of four to various campus labs to walk through a drawing and demonstrate the software. For example, if you have a network technology lab, it will have Visio, which has become an important tool for network engineers, network designers, and others.

Students who have had experience with computer-generated callouts might be asked to form groups of four for another excursion to the labs. The group leaders will demonstrate the procedure for constructing callouts to identify important features in a graphic.

Share the preparation of a document

Collaborate with three other members of the class to build a second technical document with a full text. This project is developed over a period of several weeks and you will have opportunities to meet to discuss design, research, production, and progress.

- *Select a familiar technical topic that involves either a product or mechanism, or a group of such products or mechanisms (for comparison).*
- *Develop a text of 1500 words that includes a cover page, table of contents, list of illustrations, glossary, and appendix. You might also want to include warnings, component lists, or a bibliography.*
- *Delegate parts of the task to group members— including graphics, and add a table if you develop a comparison.*
- *Discuss the target audience, design, research, production needs, and progress at each meeting.*
- *Decide what graphic components will be critical and how they will be created (or borrowed).*
- *Discuss the graphic features (captions, callouts, and so on) that should be used.*
- *Submit the final project to your instructor.*

Work in Progress

11. Final Project Review

I assembled my final project document packet and prepared for the final project review. This consisted of the copies of the existing process documents, which I had identified as "From" (meaning formerly), copies of the new process changes, which I had identified as "To," and the Quality Manual "change record forms" that were to be signed off during the approval process.

Because the changes affect more than one department, I met with Mitch, my supervisor, and Tom, the primary engineer for production, to do a review of the changes. We discussed adding foldouts, a table of contents, glossary, tools list, and training records. The readability was accepted and all went well during this review. They signed off their sections as approved. This review went smooth partly because I had involved the primary engineer throughout the process, so Tom knew the history of this change proposal long before this final review.

Usually I walk the changes around for sign-off on my projects. The managers who have knowledge of the changes will review them and sign it as "approved" while I am there. For those managers who are not familiar with a project, I go over the document with them, answer any questions, and leave the original with them for their review. When they are done they contact me; I pick up the original packet and forward it to the next department, and eventually to Document Control for entry in the on-line Quality Manual.

A few of the managers had questions, which I welcomed and answered as well as I could (this is where it pays to have an understanding of the subject). The packet was approved and forwarded on time to Document Control. Usually the time frame for submitting process changes is 10 workings days. I had 15 working days as my limit due to the number of departments involved. The total time for this project was 13 working days, so obviously I was happy to be under the 15-day limit.

Thank you for your time in following along with me on my technical writing journey, and I wish you many writing journeys of your own. Engineering businesses constantly change, and someone has to be there to record the changes. There is a lot of writing to be done.

S. B.

Positioning Your Illustrations

Now, let's turn to the size and location of the graphic material you might utilize in a document. You have many layout possibilities that will depend on the text, the amount of detail provided in each of the graphics, and other practical considerations. If you cannot locate a scanner to edit the scale (size) of your graphics, I suggest that you produce a number of sizes of each graphic on a copier. With a copy machine, you can reduce or enlarge the scale as you wish. Copy the graphic at the scale you think you want, and then copy three or four more, perhaps 10% and 20% larger and 10% and 20% smaller. With these optional sizes, you have a lot more flexibility when you are designing page layouts. Toss anything that scales down to the point at which it is not easy to read. The use of 8-point print is cutting it thin. Use nothing smaller. The intended reader should be able to *use* the graphic, so he or she must be able to read it.

Sample 11.A

-6-

8. Remove the lockwire from the detector's high- and low-pressure vent plugs with the lockwire pliers.
9. Remove the high- and low-pressure vent plugs using the 9/16 inch box-end wrench. (See Figure 3.)
10. Install the high-pressure quick-disconnect coupling from the calibration rig to the detector's high-pressure vent threads.
11. Install the low-pressure quick-disconnect coupling from the calibration rig to the detector's low-pressure vent threads.
12. Slowly open the nitrogen supply valve until the valve is fully back-sealed.
13. Slowly turn the nitrogen regulator valve clockwise to obtain a pressure of approximately 2 psi on the regulator-pressure gauge.
14. Crack open and then fully open the right tank pressurization valve while observing that fluid is flowing from the right tank to the detector and returning to left tank via the connecting lines.
15. When air bubbles are no longer visible in the connecting lines, shut the right tank pressurization valve.
16. Shut the calibration rig's tank bypass rig.
17. Check the calibration rig and detector valve lineup against Figure 4.

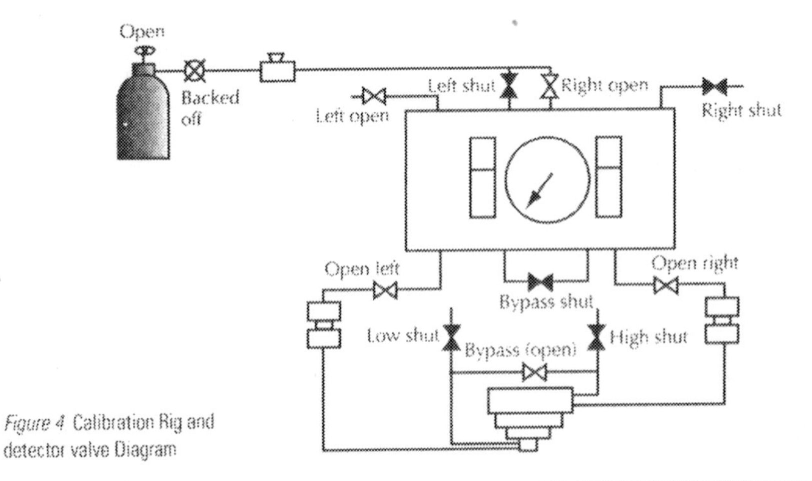

Figure 4 Calibration Rig and detector valve Diagram

Full Width

The handiest size of graphic is one that you can center on a page between sections of text. The illustration does not have to be the full width of the page. It is centered, and the space to the left and right will vary depending on the figure. The text is *not* wrapped on either side.

RADIUS = .125 REFERENCE RADIUS
ADD INVERTER BUBBLE
OR
ROTATE 120 DEGREES
BUFFER
INVERTER BUFFER
INVERTING INPUT
GATED OUTPUT

Figure 6 Modifying the DRIVER symbol

As a practical convenience to the author, the full-width presentation is certainly the easiest to compose. Be sure to leave appropriate space for the graphics. Of course, Murphy's law will catch up with you. There will be times when the space is not big enough because the graphic is too large. (See pp. 255 and 257, for example.)

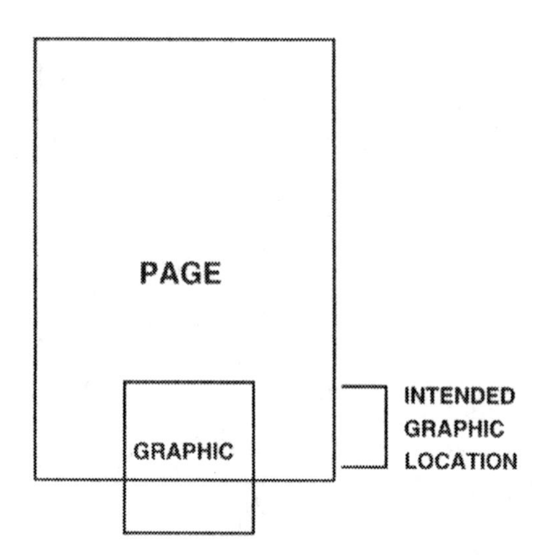

At this point you look at the graphics you copied at different sizes and visualize their location. I am addressing this as a glue stick procedure. If you import graphics on a computer, the space relationship is no different. You can scale with the scanner, but the graphic space still may not be appropriate at times.

Full-width graphics are the most convenient graphic feature for lay outs. They are easily positioned and can be large enough to show considerable details.

Sample 11.B

6

There must be low distortion for a clear, crisp output. The first audio amplifier contains a de-emphasis circuit. At the broadcast station, the upper audio frequencies have been emphasized to overcome noise that is present in the circuitry (usually caused by thermal heating). This overamplification is reversed in the receiver by a de-emphasis circuit in the audio amplifier, which results in an overall flat audio response. The designed bandpass of the audio circuits is 50–15,000 hertz for very good fidelity.

Power Audio Output

The audio stages amplify the audio signal to an adequate voltage level, but the speakers need a good amount of current to drive them. At this stage, the gain is 1, but the current capacity is very high. The amplifier can be either Class A biased or Class B with two active elements in a push-pull configuration (see Figure 6).

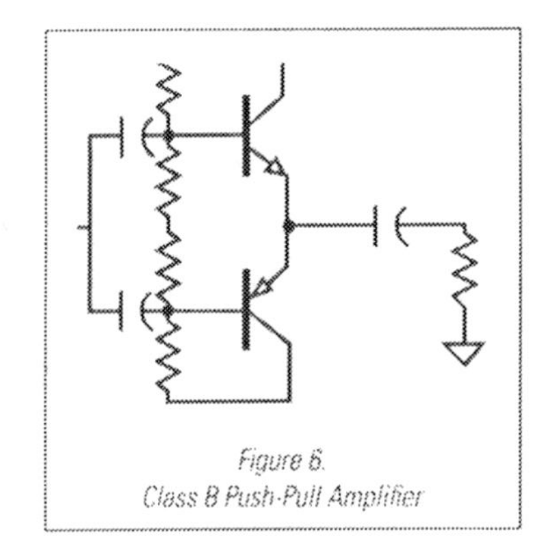

Figure 6.
Class B Push-Pull Amplifier

Conclusion

The FM scheme for carrier modulation is widely accepted as the best method for impressing analog intelligence on a carrier. This discussion showed what FM is and how it is detected. Every stage of this system can be analyzed in greater depth: however, you now have a basic understanding of FM reception.

To Build a Box in Position on a Page

1. Make sure your drawing toolbar is displayed. If it isn't, go to **View** on the top toolbar, then **Toolbars**, and click on **Drawing.**
2. Click on the icon that looks like an empty square.
3. Draw a box by clicking on the point where you would like the box to be and then drag away from that point until you get the shape and size box you want.
4. To move the box, click on its border and drag the box where you would like it to go.

If you use graphics only in the full-width style, the document is easily adapted to a "full paste up" situation in which you paste the parts of the *text* itself. Take a chunk of text and glue it on. Paste in the graphic. Cut more text into position. Move on to the next blank piece of paper, and so on. This is an emergency option, but it is effective if a document is not too long.

Half Width

Graphics that are half the width of the page are designed or used in such a way that the graphic sits beside the text. This is a more complicated layout. The half-width graphic is not the problem; the issues concern the other half: the half width of text. It can be a little difficult to make the text neatly fit the space you assign to it. The layout configurations can be as complex as you wish.

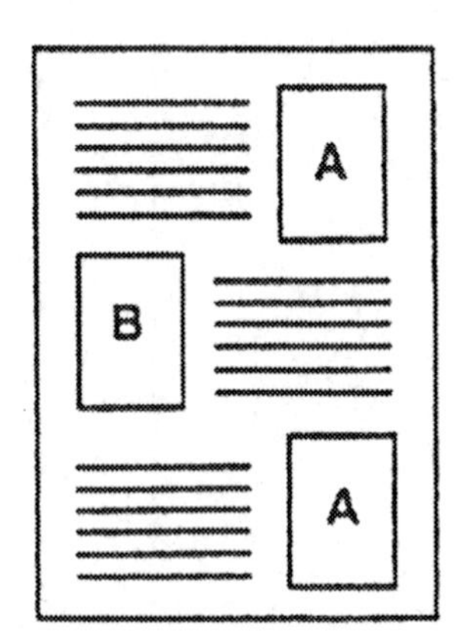

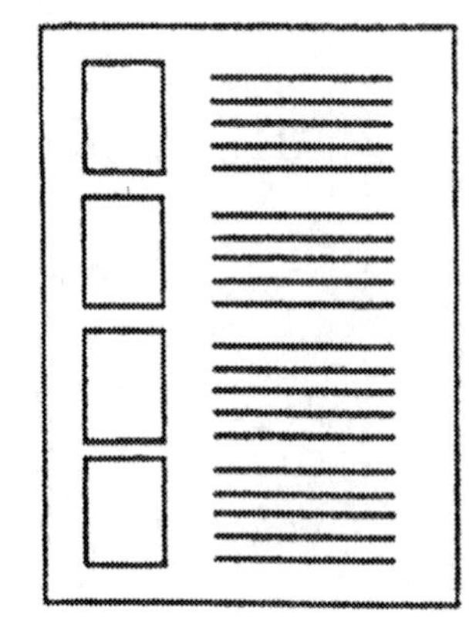

The text can utilize a graphic on the right side of the page (A) or on the left side (B). You can develop a sequence down one side or stagger the graphics. The sequential format that uses a line of illustrations in one column is particularly effective for instructions, explanations of mechanical processes, and discussion that calls for phases.

To Wrap Text around a Graphic

1. Click on the graphic to select it.
2. Click on **Format** on the top toolbar.
3. Click on **Borders** and **Shading** and then click on the **Wrapping** tab.
4. Select the style of text wrap you need.
5. Select the side on which you want the text to wrap.

The half-width graphic is handy because the text discussion of the graphic can be placed beside it. Note that the actual size may be less than or greater than a half-width.

Sample 11.C

7

The Combinations of Logic Gates

When you add* an inverter to an AND gate, you create a new logic gate called a NAND gate. The operation of a NAND gate is the inverse of the AND gate. This means that the NAND gate's output is the complement of the AND gate's output. Figure 4 shows a true† two-input NAND gate and its equivalent NAND gate, which consists of an AND gate and an inverter. The output, C, is binary 0 only when both inputs, A and B, are binary 1 (this is the opposite of the AND gate's operation). Inversely, the output, C, is a binary 1 whenever either input, A or B, or both are binary 0. Table 4 lists the differences in the logical operation of the NAND gate compared with that of the AND. The NOR gate is a more practical logic element for implementing a logic function.

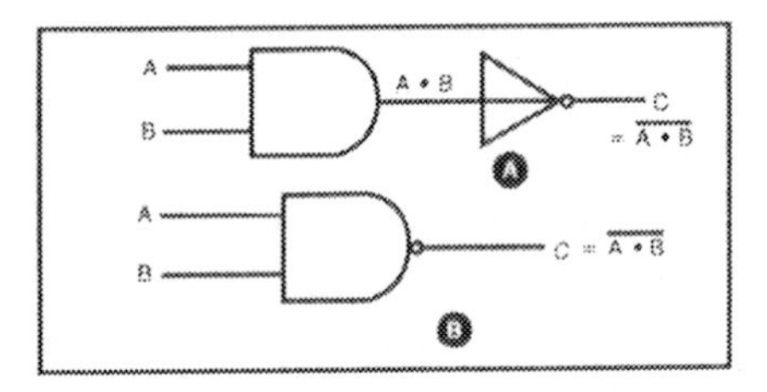

Figure 4. *(A)* Equivalent NAND Gate. *(B)* True NAND Gate

Table 4. NAND Gate Truth Table

Input		*Output*	
A	*B*	*A*B* (AND)	$\overline{A*B}$ (NAND)
0	0	0	1
0	1	0	1
1	0	0	1
1	1	1	0

The word NOR is a contraction of the term NOT OR. This means that the NOR gate can be created by using the same logic as applied to the NAND gate, combining the function of an OR gate and an inverter, as shown in Figure 5. Similiar to the NAND gate's output, the NOR gate's output is merely opposite that of the OR and can be used as

** To add is to connect the AND gate's output terminal to the input terminal of the inverter.*

† A true gate is a logic gate that is already in the package. It does not need any other gates but still operates like its equivalent gate.

It is often desirable to have the graphics and the text side by side. *Technical Document Basics* uses a facing-page format for the samples and the text. On a single page of text, the practice is similar, as you can see in the following sample. This is one page from a series of pages that demonstrate block-lettering practices. It is important to have the illustrations near the guidelines that are developed in the text.

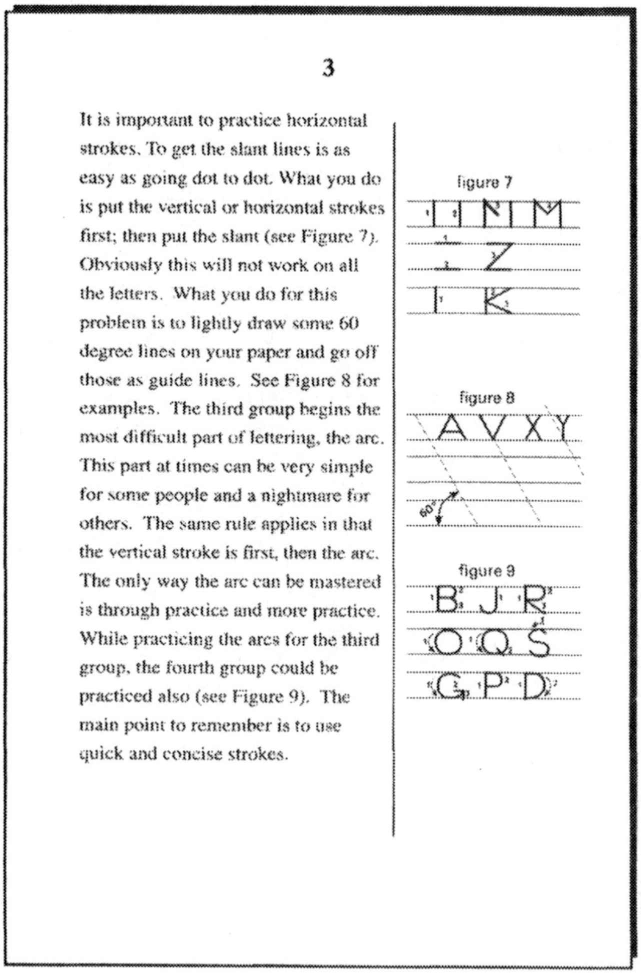

3

It is important to practice horizontal strokes. To get the slant lines is as easy as going dot to dot. What you do is put the vertical or horizontal strokes first; then put the slant (see Figure 7). Obviously this will not work on all the letters. What you do for this problem is to lightly draw some 60 degree lines on your paper and go off those as guide lines. See Figure 8 for examples. The third group begins the most difficult part of lettering, the arc. This part at times can be very simple for some people and a nightmare for others. The same rule applies in that the vertical stroke is first, then the arc. The only way the arc can be mastered is through practice and more practice. While practicing the arcs for the third group, the fourth group could be practiced also (see Figure 9). The main point to remember is to use quick and concise strokes.

figure 7

figure 8

figure 9

In order to develop a two-column format with one column of text and one column for graphics, you construct a one-column format with a side bar. You can use generous margin settings for a narrow column of text that will leave half the page blank, or you can set the margins and place a textbox in the blank space and use it for written or graphic content. The Sample on this page was a pasteup attached to blank space.

Of the disciplines that use side-by-side graphics with regularity, digital computer technology often discusses gate logic in pairs of graphics that combine a gate diagram and a truth table.

Sample 11.D

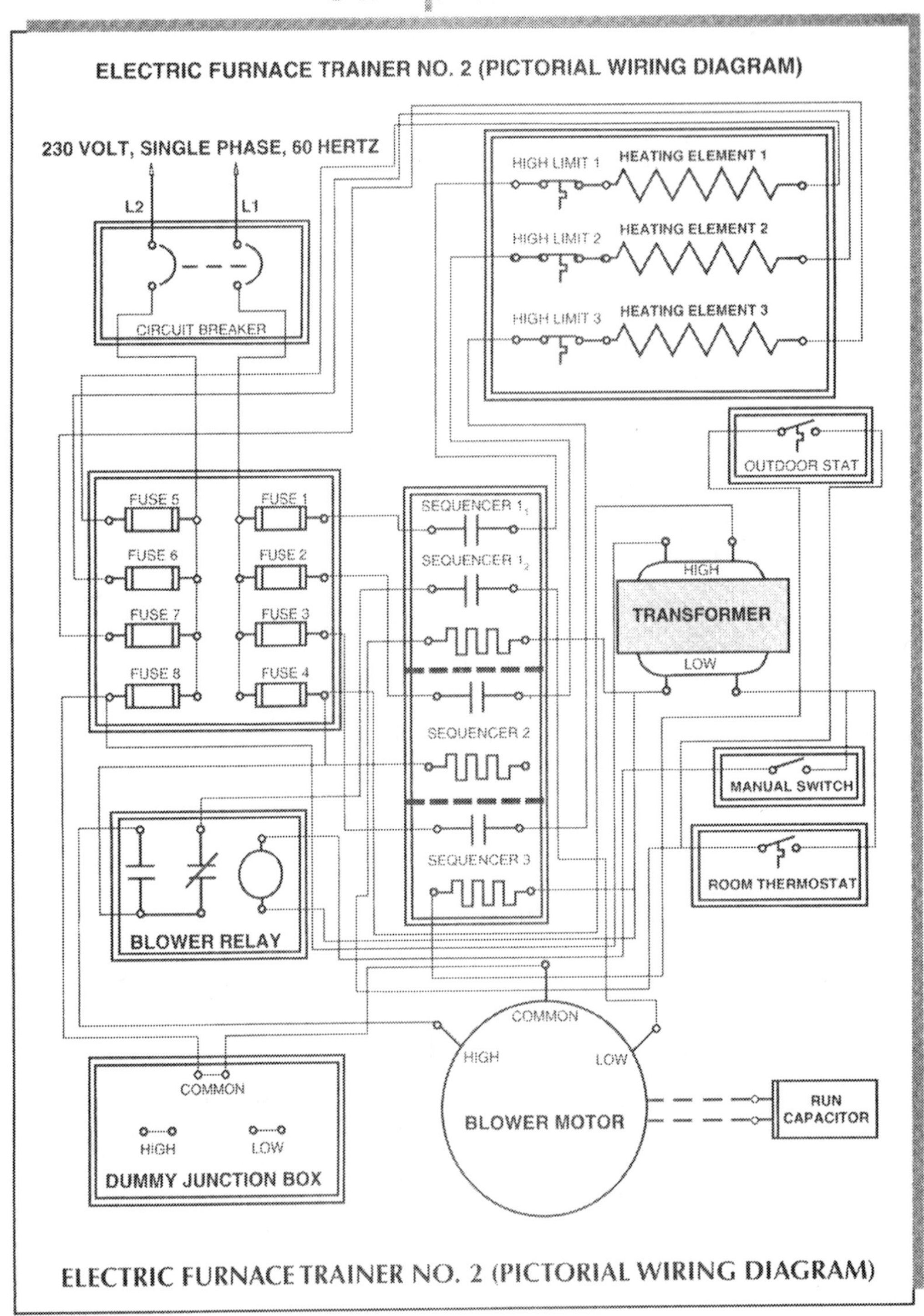

ELECTRIC FURNACE TRAINER NO. 2 (PICTORIAL WIRING DIAGRAM)

To Create Text Box Sidebars

1. Select the text where you will be placing the sidebar by highlighting it.
2. Click on **Format** and then **Paragraph** on the menu bar.
3. Choose the desired indentation. All of the selected text will be indented.
4. Deselect the text by clicking anywhere on the page.
5. To build the text box, click **Insert** and then **Text Box** on the menu bar.
6. The pointer will turn to a cross. Click and drag the mouse to size the text box.
7. Release the mouse to fix the size of the text box.
8. Move the text box into the space created by the indent by clicking on the border of the text box and dragging it into position.
9. Click inside the text box to add text.

Side by Sides

Several engineering disciplines favor specific layouts for graphics. Side-by-sides involve positioning two graphics next to each other. This style of presentation is frequently used in electronics. The left illustration will identify a circuit diagram, and the right illustration will identify the appropriate waveform for the diagram. The side-by-side format is also used in digital electronics. The left graphic box will contain a gate diagram and the right graphic box will contain a truth table. The graphics are often linked as, for example, 2.a and 2.b.

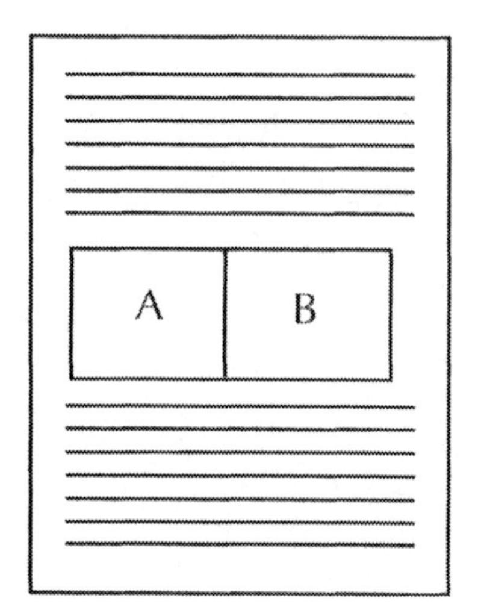

Full-page graphics allow the author to present a greater amount of material. Readability determines the scale. Large illustrations can be handled in full-page and foldout formats.

Sample 11.E

PROPOSED NETWORK INSTALLATIONS: AN OVERVIEW

The purpose of any network installation is to provide connectivity between computers and resources. When the school district decided to implement a new network installation at Jefferson Elementary, it was decided that this installation had to meet certain criteria, including:

- Providing for 3 student computers and 1 instructor computer per classroom
- Providing **3 Mbps**[1] of bandwidth throughput to each networked computer
- Providing 2 computer labs each with 24 computers
- Control of student access to resources on the network
- Allowing **WAN**[2] connectivity to the district offices
- Providing access to the Internet
- Supplying application and file servers with 10 Mbps network access
- Allowing for future growth, including the expansion to 24 student computers per classroom, and the increase of throughput to 10 Mbps to each networked computer.

1. Mbps
Megabytes Per Second— the amount of data (bytes) that can pass through a cable or device within one second.

2. WAN
Wide Area Network—a network that typically can span between local area networks, or can span between entire cities.

Full Page

Full-page illustrations are quite common. This book uses over one hundred of them that I have adapted from the work of students like yourself. They are convenient because they are easy to read at 80% of scale. The full-page insert is your best solution when large graphics cannot be scaled down. You will often need to insert entire pages to maintain a readable scale for graphics. In fact, some can demand more than a page. The one drawback is that full-page inserts may somewhat lose the effectiveness of side-by-side relationships in your document, since you use only one side of a sheet of paper. The graphic will not be next to the text. I do not have that problem because a book has facing pages, and there is always text beside (that is, facing) the graphic page. You will almost always compose one-sided documents, so you lose the vividness of the side-by-side effect when you have to use full-page graphics.

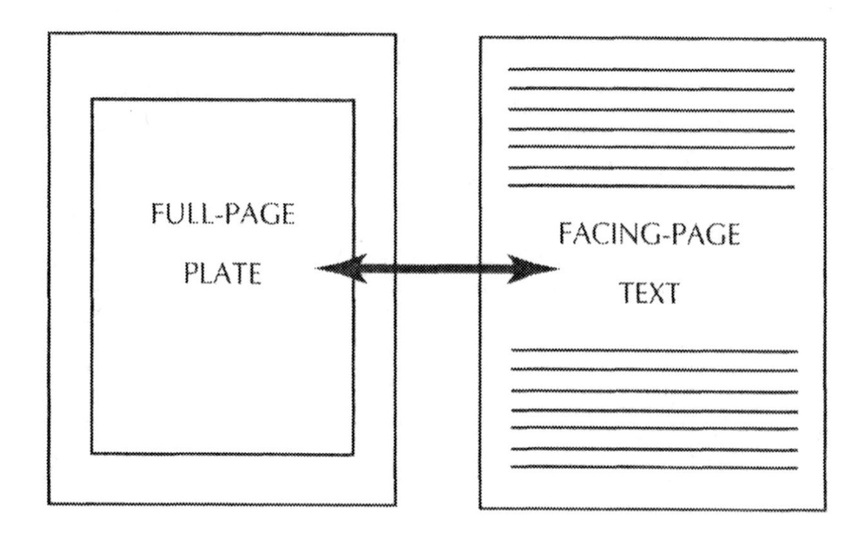

Windows

Another graphic layout that is characteristic of specific technical and engineering disciplines is the "window," a small text box that frequently accompanies computer applications, Internet productions, and hard-copy manuals for systems and networks. Most engineering fields have little or no use for windows, but the graphic device is popular in computer-related industries.

Windows are not, strictly speaking, graphics. They often contain commands, definitions, clarifications, or other points of interest. The real graphics—the screen captures—are usually centered on the page. A network system design engineer who was recently at the college had projects full of windows. Each engineer speaks his or her own graphic language.

Notice that windows create a layout pattern that is not always desirable. Because the window fits **inside** *the outside margin, the text is moved in to make room for window displays. In this instance the narrow text width was not a problem.*

Sample 11.F

Similarly, the **actual mechanical advantage** (AMA) is defined as the ratio of the force of the load to the effort force applied:

(5) $$AMA = F_R/F_E$$

Figure 1 shows that 1 lb of force is required to lift the 1-lb load; therefore, its AMA = 1. But if our black-box machine would require us to apply only 100 lb (F_E) to lift a load of 400 lb (F_R), then

(6) $$AMA = 400\ lb/100\ lb = 4$$

The **machine efficiency** is defined as follows:

(7) $$\text{Efficiency} = \frac{\text{actual mechanical advantage}}{\text{ideal mechanical advantage}}$$

In this example, where the AMA = 4 and the IMA = 4,

(8) $$\text{Efficiency} = AMA/IMA = 4/4 = 100\%$$

However, friction must normally be taken into account in pulley systems; thus, the effort force will normally be higher than the load. For the sake of this example, let's say 200 lb was actually required to lift the 400-lb load instead of the 100 lb in a frictionless machine. Thus,

(9) $$AMA = 400\ lb/200\ lb = 2$$

and if the IMA is 4, then

(10) $$\text{Efficiency} = 2/4 = 50\%$$

For the fixed-pulley system of Figure 1, IMA = 1 and AMA = 1, neglecting friction. The machine is 100% efficient, but there is **no mechanical advantage.** The user must pull up with a force of 1 lb to raise the 1-lb load. Can the effort force be reduced?

To Create Text Box Windows in the Margin

1. Select the text where you will be placing the sidebar by highlighting it.
2. Click on **Format** and then **Paragraph** on the menu bar.
3. Choose the desired indentation. All of the selected text will be indented.
4. Deselect the text by clicking anywhere on the page.
5. To build the text box, click **Insert** and then **Text Box** on the menu bar.
6. The pointer will turn to a cross. Click and drag the mouse to size the text box.
7. Release the mouse to fix the size of the text box.
8. Move the text box into the space created by the indent by clicking on the border of the text box and dragging it into position.
9. Click inside the text box to add text.

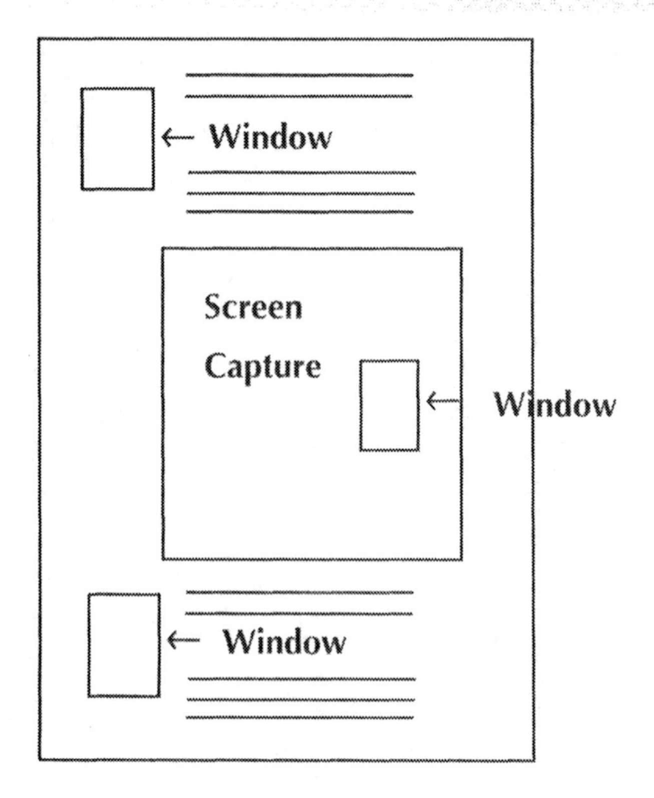

Math Displays

One of the most common yet distinguishing features of technical writing, the math display holds a unique position in layout procedure. The math (or chemistry, physics, and so on) is handled much like a graphic. The math is positioned in the center of the page and is removed from the running text. In one example in this text, the math is boxed, but this is uncommon (see p. 180). There is one practice that distinguishes math displays from

This sample illustrates the convention for math displays. Notice that they are also numbered in sequence, much like graphics.

graphics. Whereas math can interrupt a sentence at one or more points, the general practice regarding graphics calls for a completed sentence before the graphic. Graphics are very awkward if they are placed within the context of a sentence. In this text I created only one such construction (on p. 105). For a discussion of math displays see the *Writer's Handbook,* Chapter 8).

To Use the Equation Editor

1. Go to **Insert** on the menu bar.
2. Click on **Object.**
3. To open the program, scroll to **Microsoft Equation** and *double*-click.
4. For help with the editor, return to the menu bar and click on **Help.**
5. The help options list **Equation Editor: Help Topics.** Click to open.
6. Select from the choices in the dialog box with a *double* click to call up instructions.

Sequencing and Segmenting

Sequences

Photography greatly encouraged interest in animation and helped us understand the sequential nature of moments of movement. Among other things, nineteenth-century painters got galloping horse movements correctly painted at last. You have probably seen some of the famous early photographs of a galloping horse or a running man. By the late nineteenth century, the first "video games" were being marketed in the video parlors of the day. These penny arcades were filled with nickelodeons instead of jukeboxes, and the video games were hand-cranked flip cards of cowboys and daring ladies and corny cops. They were called "mutoscopes." Whenever you see a sequential set of graphics in a technical document, you are seeing this method for taking a look at short, quick takes on the subtleties of change in something. Time-lapse photography is a handy scientific example. You have probably seen the amazing work that has been done on clouds and storms or the work on plant-growth cycles. The daffodil's life cycle comes and goes before your eyes in a matter of seconds.

The most practical technique for rendering sequential graphics in technical documentation is to create a two-column format that allows the graphic elements to unroll like a length of motion picture film. It usually is less than half a page wide, and the layout looks something like this:

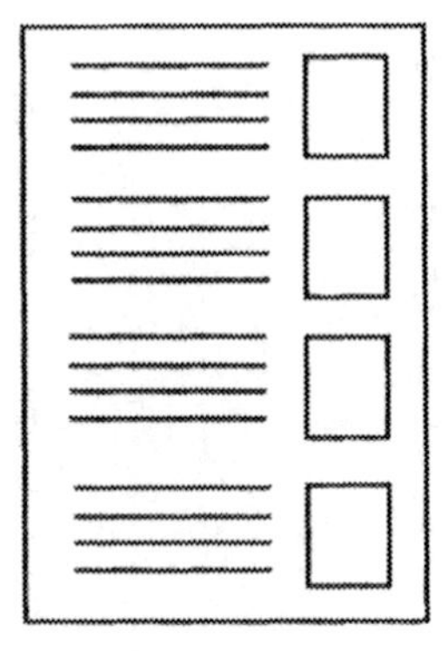

Readers gain a sense of movement from the proximity of one illustration to the next. See Sample 11.G for an illustration of this popular format.

Sample 11.G

MISCELLANEOUS PARTS OF THE REGULATOR

Because of the feedback loop generated by the way R5, Q3, and Q4 are connected, there is a possibility that, at some frequency, the loop gain of points A, B, and C could equal 1 and that the phase shift could equal 180°. See Figure 14.

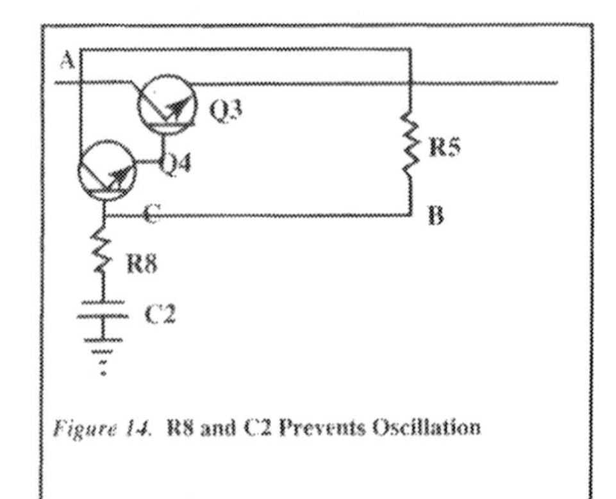

Figure 14. **R8 and C2 Prevents Oscillation**

This situation would result in oscillation. R8 and C2 make sure that the gain is not 1 when the phase shift is 180°. R8 and C2 keep the circuit from oscillating.

Figure 15 shows C3 in parallel with the zener diode. Zeners are noisy devices, and C3 is there to cut down on noise to the rest of the circuit.

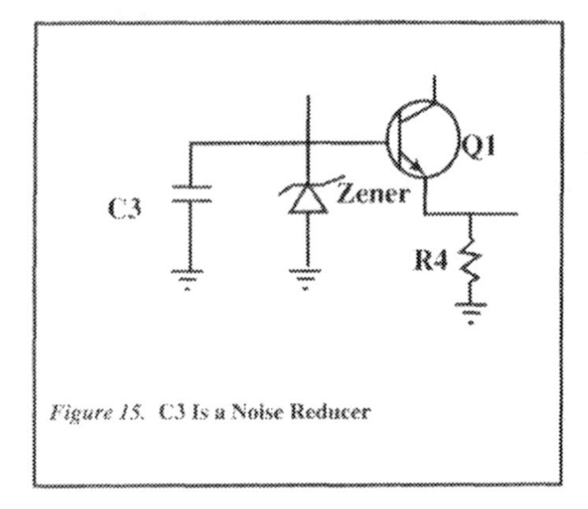

Figure 15. **C3 Is a Noise Reducer**

When the output voltage tries to increase or decrease quickly, the feedback circuit tries to correct the change just as quickly. This can make V_{out} look unstable. To stabilize the output it is necessary

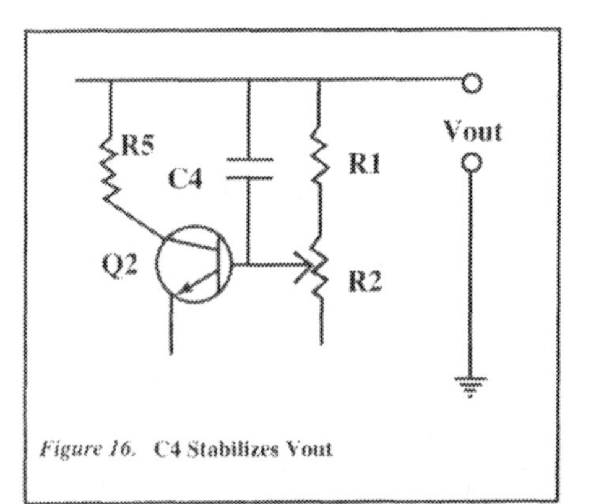

Figure 16. **C4 Stabilizes Vout**

Segments

A common difficulty with graphics is that some are too detailed to be clearly understood. A complicated computer screen image of icons, pull-down menus, and text is one example. A federal tax return form is another. To handle complicated illustrations such as these, present the material in pieces. The first appearance of the image should be the full object so that the reader perceives the appearance—and the problem. The image can be segmented with a dotted line, colored lines, or colored sections. The image can also be segmented with scissors or software (Photoshop). The tax return might look something like the figure on the left. The screen capture might look something like the figure on the right.

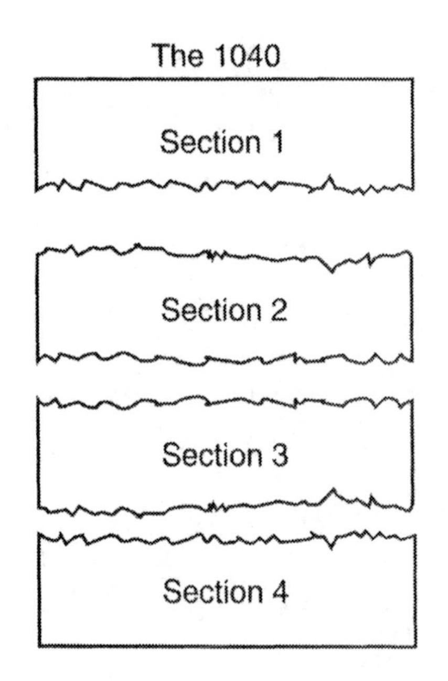

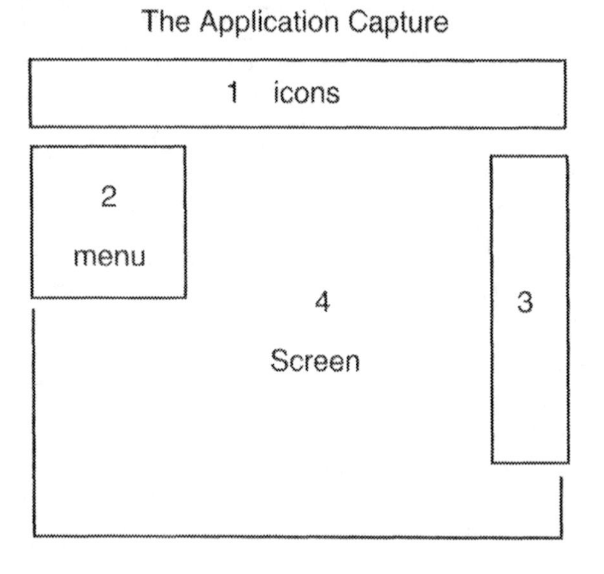

These illustrations will usually be full-page plates, but be sure to maintain your regular margins. In other words, the illustration of the 8½″ × 11″ tax return is scanned at 80% of scale or copied at 80% of scale and cut up and pasted on a page to create the appearance you see in the figures. The sections are then keyed to the subsequent text. The following pages deal with the subsections. The item numbers are used to organize the IRS form. The pull-down menu is handled with arrows. These are variations of legends and callouts.

A sequence of visuals is useful at times, particularly if there is, in fact, a sequence logic involved: a, then b, then c, then d. The text, to the extent possible, is positioned beside the appropriate figures.

Sample 11.H

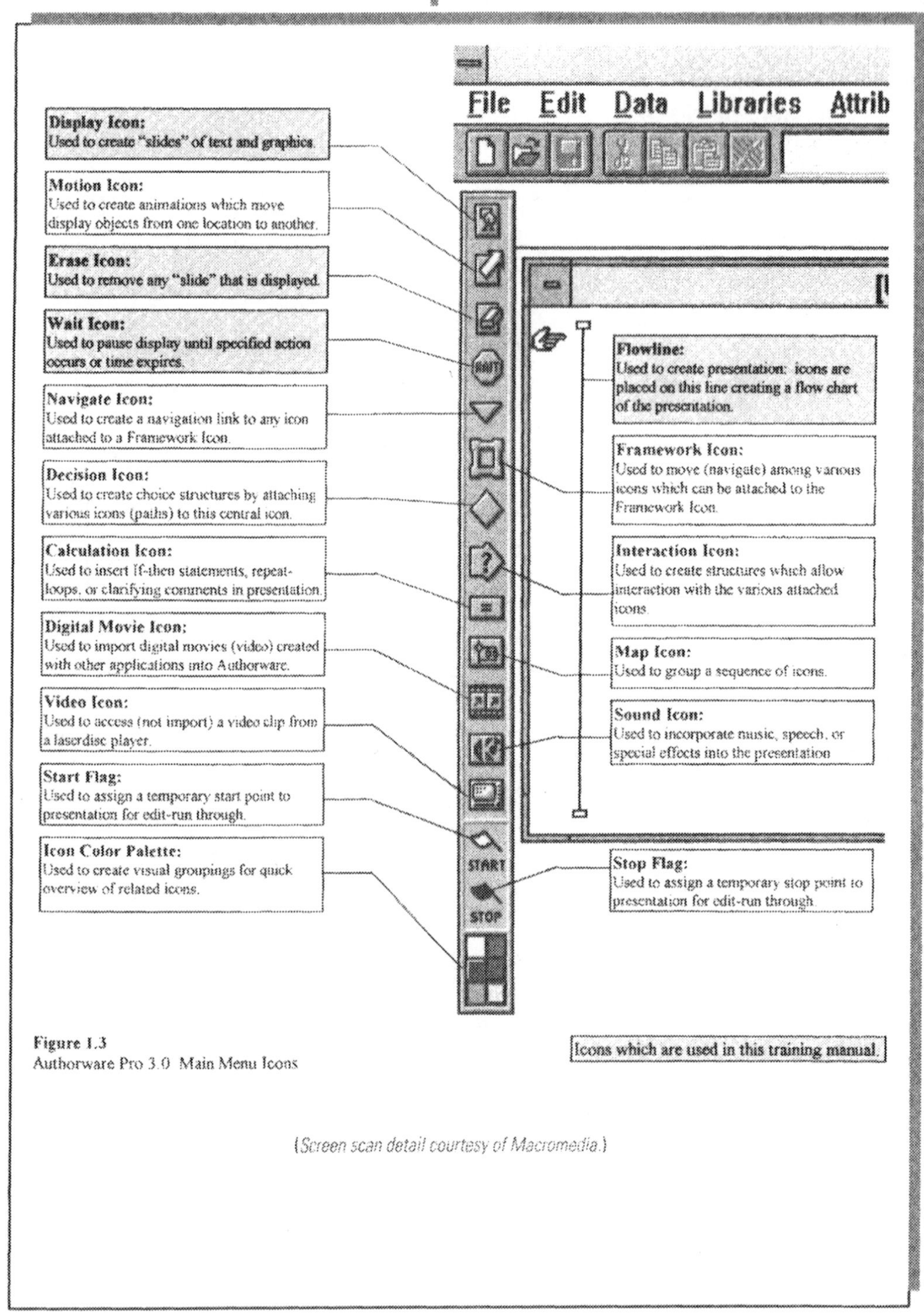

File Edit Data Libraries Attrib
Display Icon:
Used to create "slides" of text and graphics.
Motion Icon:
Used to create animations which move display objects from one location to another.
Erase Icon:
Used to remove any "slide" that is displayed.
Wait Icon:
Used to pause display until specified action occurs or time expires.
Navigate Icon:
Used to create a navigation link to any icon attached to a Framework Icon.
Decision Icon:
Used to create choice structures by attaching various icons (paths) to this central icon.
Calculation Icon:
Used to insert If-then statements, repeat-loops, or clarifying comments in presentation.
Digital Movie Icon:
Used to import digital movies (video) created with other applications into Authorware.
Video Icon:
Used to access (not import) a video clip from a laserdisc player.
Start Flag:
Used to assign a temporary start point to presentation for edit-run through.
Icon Color Palette:
Used to create visual groupings for quick overview of related icons.
Flowline:
Used to create presentation: icons are placed on this line creating a flow chart of the presentation.
Framework Icon:
Used to move (navigate) among various icons which can be attached to the Framework Icon.
Interaction Icon:
Used to create structures which allow interaction with the various attached icons.
Map Icon:
Used to group a sequence of icons.
Sound Icon:
Used to incorporate music, speech, or special effects into the presentation
START
STOP
Stop Flag:
Used to assign a temporary stop point to presentation for edit-run through.
Icons which are used in this training manual.
Figure 1.3
Authorware Pro 3.0 Main Menu Icons

(Screen scan detail courtesy of Macromedia.)

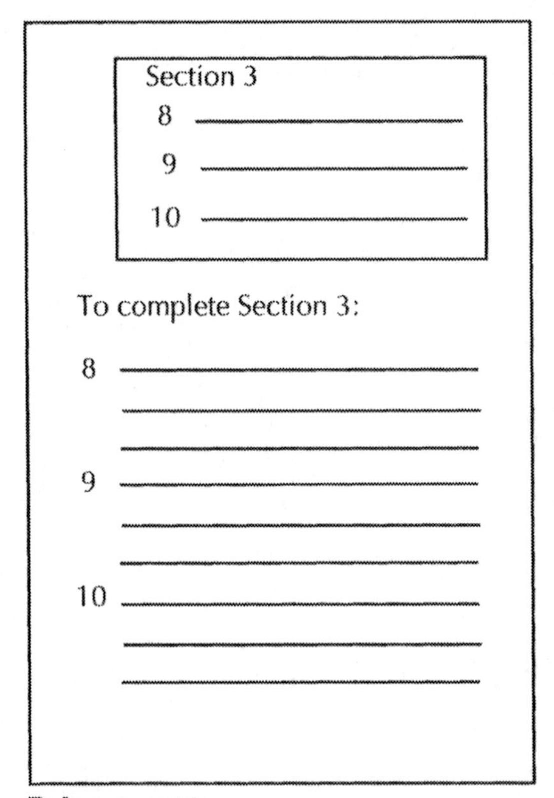

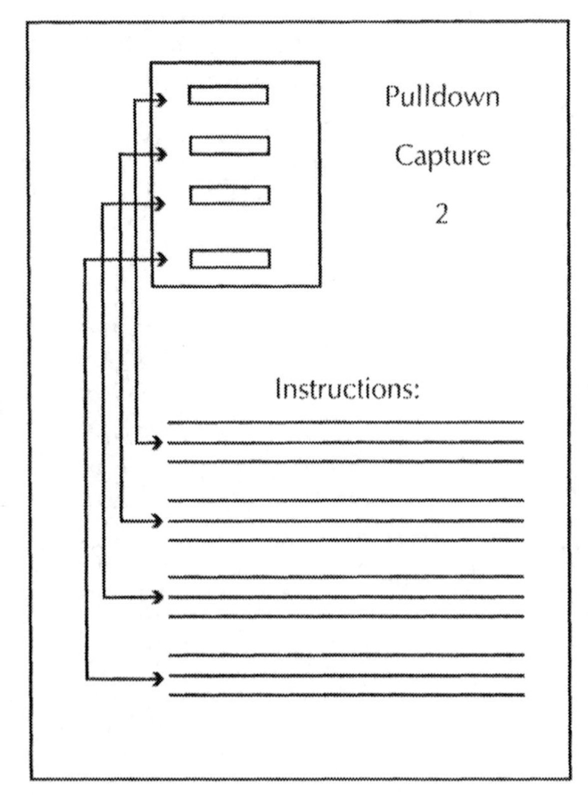

Enlargements

Sample 11.H illustrates the left side of a screen display. The icon keys are isolated so that all the explanations can be developed for all the icons. In the following example, the author uses a screen capture that illustrates two selection bars, each three keys wide. His concerns are limited to very few of the functions. Once the overall screen display is illustrated, the author displays only enlarged segments, such as this one.

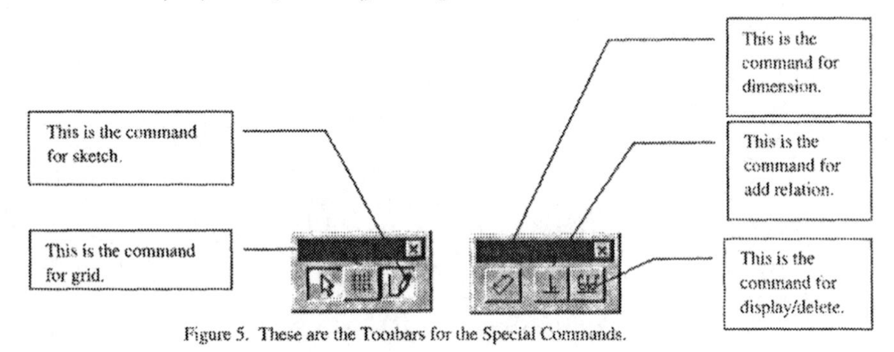

Figure 5. These are the Toolbars for the Special Commands.

(Screen scan details courtesy of Autodesk Inc.)

This is a segment shown enlarged. Blowups allow an author to come to terms with miniaturization but also put the illustration on a scale that will match the needed callouts.

Sample 11.I

2

The Start Menu

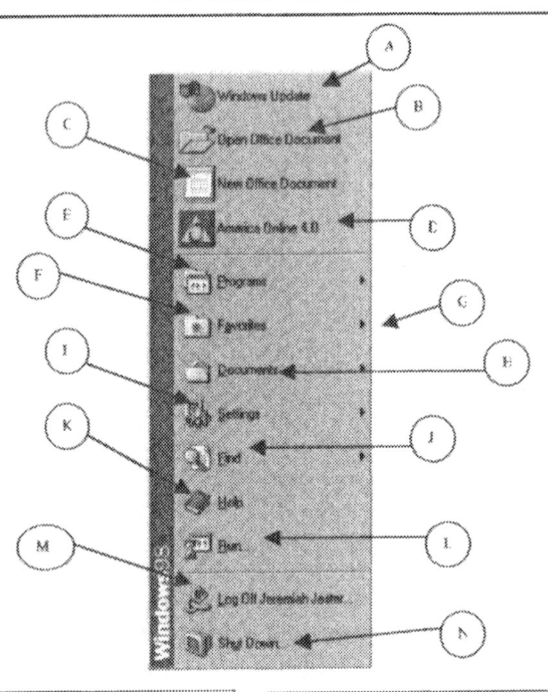

Ⓐ **Windows Update** - Program that, when connected to the Internet, updates your OS.

Ⓑ **Open Office Document** - Icon that prompts the Microsoft Office program into opening an existing document

Ⓒ **New Office Document** - Icon that prompts the Microsoft Office program into creating a new office document.

Ⓓ **America Online** - This icon launches the AOL (America On Line) ISP (Internet Service Provider).

Ⓔ **Programs** - This subfolder lists all registered Windows programs.

Ⓕ **Favorites** - Combined Internet Explorer function that allows user to reference previously viewed sites.

Ⓖ **Command Arrow** - Prompts a subfolder

Ⓗ **Documents** - Category lists all previously viewed Documents and/or files.

Ⓘ **Settings** - Function allows users to change hardware and Operating System settings.

Ⓙ **Find** - Function users use to locate missing files.

Ⓚ **Help** - Function allows users to search the Microsoft Help database using keywords to identify a problem.

Ⓛ **Run** - Function allows users to activate and open programs via a command prompt.

Ⓜ **Log Off** - Function allows users to log off the computer so another user may log in.

Ⓝ **Shut Down** - Function allows user to make the PC go on "Standby," to "Shutdown," "Restart," or to "Restart in MS-DOS mode."

(Screen shot reprinted by permission form Microsoft Corporation)

Segmenting a document or screen image is a helpful practice. Each section can then be explained as needed. Enlargements, or blowups, of the desired material are usually placed next to the desired text.

Sample 11.H and the preceding figure use caption boxes to explain the enlarged icons. Another option is illustrated in Sample 11.I, where the legend is the focus of attention. In other words, Sample 11.H was structured by the order of appearance of the icons on the icon bar, whereas in Sample 11.I the reader consults the legend and not the screen scan.

To Capture an Image of the Screen

1. Pull up the program or dialog box that has the image you want to capture.
2. Use the **Printscrn** button to capture an image of the entire screen.
3. To capture a smaller window or dialog box, click on the smaller window or dialog box to select it, then hold down the **Alt** key as you press the **Printscrn** button.
4. Place your cursor where you want the image to be, then click on **Edit** on the top toolbar, and then **Paste.**
5. Screen captures can be resized by clicking on a corner of the image and moving toward the center of the image to reduce the size, or moving away from the center of the image to enlarge the size.

Text Links

Software applications may involve any number of additional difficulties for which segmenting is helpful. Sample 11.J is from a project that provides instructions for completing an electronic bookkeeping process. The text divides the instructions into segments of what the software application calls "Tabs," each of which is a smaller unit of the larger accounting form. These variations of the segmenting concept suggest some of the clarity that can be added to a text through the use of this visual concept. The screen captures are from the software application, but the page you see was designed and built by the author of the document. The author added the command windows, and they are linked to the captures in a conventional method; they could simply be considered callouts. The list of tasks is then pegged to the callouts, and the text is quite easy to follow as a result. This page is part of a manual that was designed to clarify the confusion that seemed to be resulting from use of the actual software application. It is used to assist business and accounting majors who need to understand the program.

Another variation on the use of the enlarged segment. In this case the author chose to use locators and a legend because there is text in the legend.

Sample 11.J

18

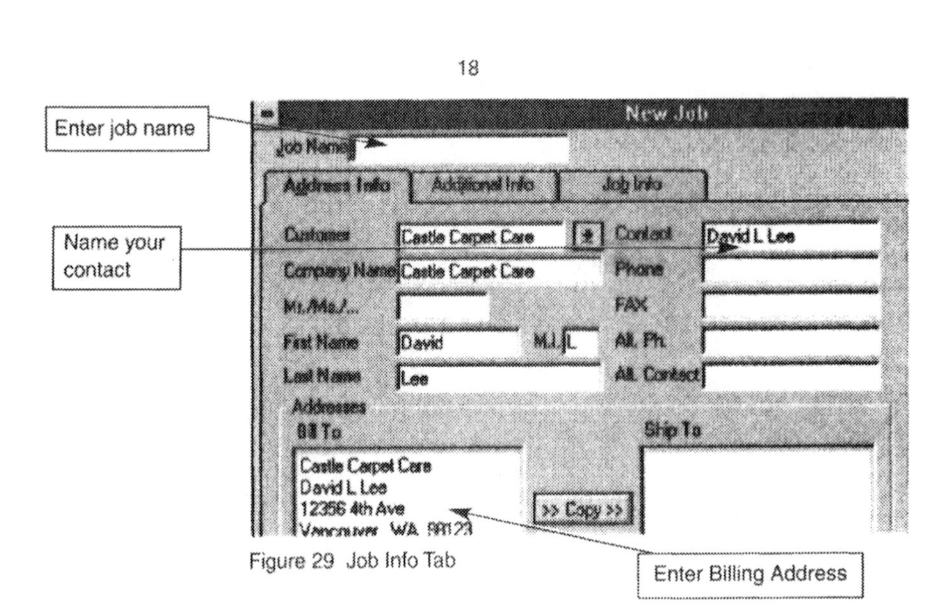

Figure 29 Job Info Tab

4. Give the job a name.
5. Identify the customer.
6. Name your contact, and fill in other relevant information (optional).
7. Give the job's billing address.
8. Give the job's shipping address.
9. Categorize the job (optional) (see Figure 29). Use the customer-type drop-down box list box to give the job type.

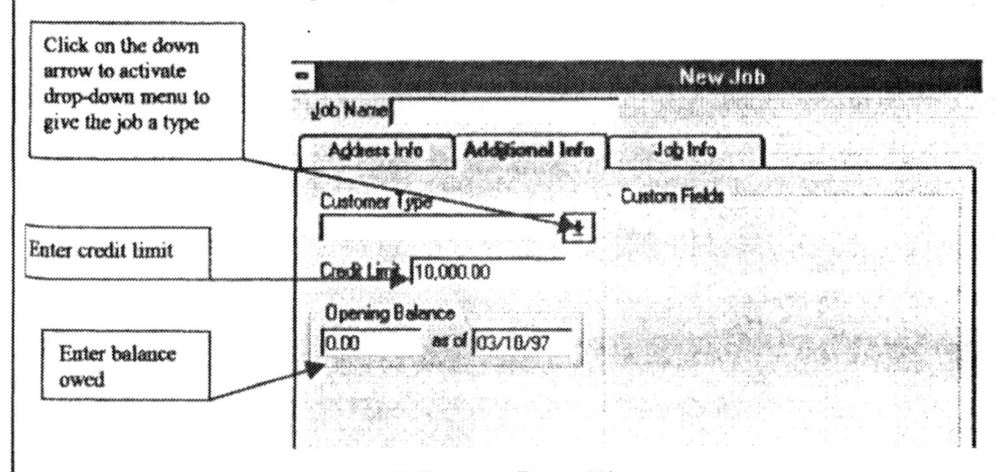

10. Set the customer's credit limit (see Figure 30).
11. Enter the amount that this customer currently owes (see Figure 30).

(Screen scans courtesy of Intuit Inc.)

To Crop a Screen Image in Order to Use Only a Small Segment of It

1. Click on the screen capture that has been pasted into a document.
2. If the **Picture toolbar** does not appear, click on **View** on the top toolbar, then **Toolbars**, and click on **Picture**.
3. Select the cropping tool that looks like two tilted connected **Xs**. If you are not sure which icon is the cropping tool, move the mouse over an icon to get a help balloon that describes the icon.
4. Click on one of the resizing points that look like white squares located on the border of the graphic.
5. Once you have cropped your screen capture, click outside the image to deselect the cropping tool.
6. The screen capture can now be resized if desired.

The text links that can be used to control the graphic and text relationships are variations on callouts and locators. Unlike the traditional caption, which is used as an identifier, the text links are often commands. The following example is handled with locators. Notice the commands on the right. They look like instructions in a running text, but they are linked to the graphic and function as a legend.

Save the data file to your hard drive and change the file type

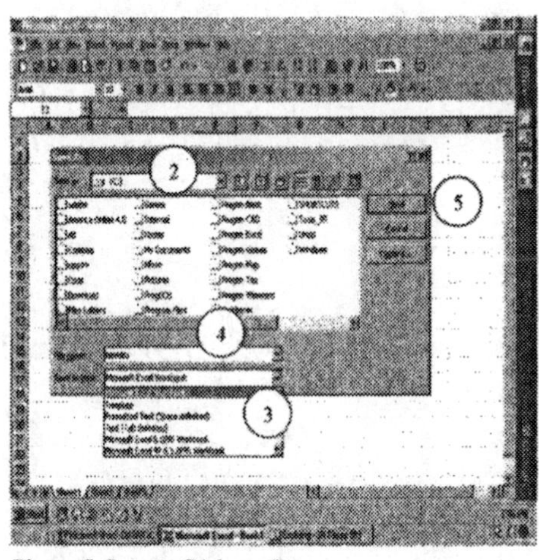

Figure 5 Save As Dialogue Box

1. *Click the **File** menu, then **Save As.** The Save As dialog box will appear (see Figure3)*
2. ***Save** in the **C: drive***
3. *Change Files of type to **Microsoft Excel Workbook***
4. *Filename is **newstu.xls***
5. *Click **Save***

(Screen shot reprinted by permission form Microsoft Corporation)

This page is taken from a thirty-page manual that uses exactly the same format repeatedly. The relationship between text and graphic can be altered for variety or repeated for consistency

Sample 11.K

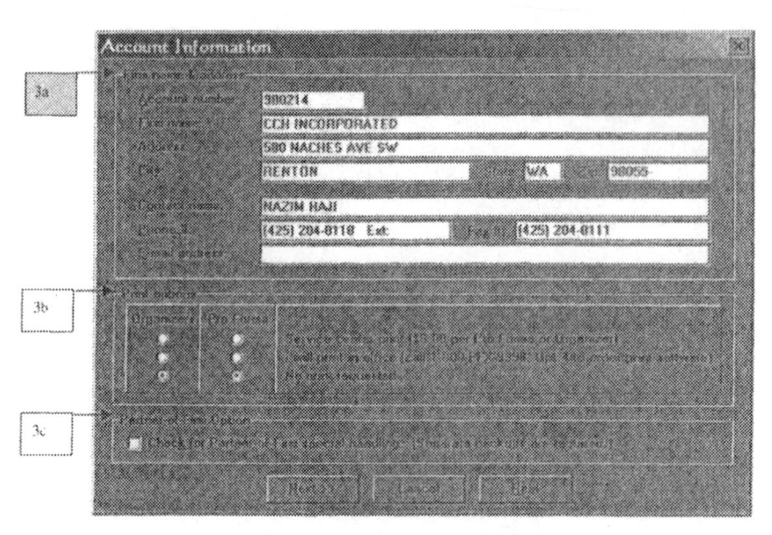

Figure 6. Account Information and Print Options.

3a. Firm Name and address

Your account number, firm name and address, contact name, phone and fax number, and an e-mail address (optional) should automatically appear on this screen.

> ***Note: All fields are required except the e-mail address.***

The firm address will be used to ship the converted returns back to you, as well as any printed Organizers or Pro Forma that you order for the converted returns. This address will be used only for shipping conversion materials

3b. Print Options

You have three options on how you wish Organizers and Pro Forma to be printed.

- For a per return charge, you can request that the ProSystem service center do your printing. Print services for Organizers are available only for individual tax data. Pro Forma is available for all products.
- You may order print software and fill all your print requirements in house. With the print software, you can select returns for which you want to print Organizers and Pro Forma. This is the most commonly selected option. Print software is ordered independently. Please call Customer Service at **1-800-PFX-9998 option 4** to order your print software.
- If you do not want Organizers or Pro Forma printed, you can select "No Print Requested."

3c. Partner of Firm

If your firm has partners who would like to have their returns processed separately, then you would check this option. These files must be backed up and sent on a separate disk with a separate work order form.

(Screen shot reprinted by permission form Microsoft Corporation)

The author created the preceding example for a manual at the campus where she works. She experimented with various options for linking the text and the graphics. In the following example she replaced the locators with arrows, with equal clarity.

Retrieve the DataExpress downloaded file

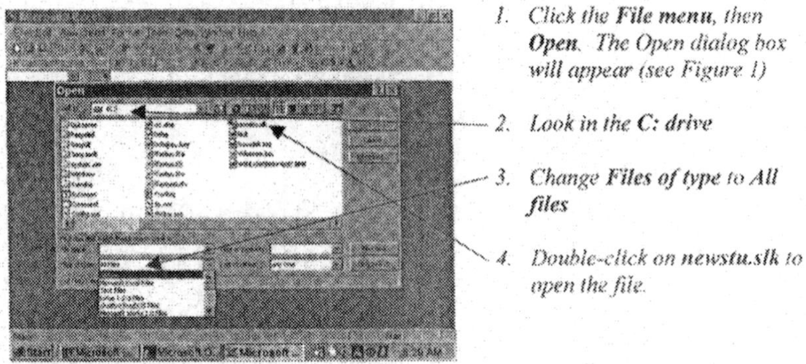

1. Click the File menu, then Open. The Open dialog box will appear (see Figure 1)
2. Look in the C: drive
3. Change Files of type to All files
4. Double-click on newstu.slk to open the file.

Figure 3. Open Dialogue Box

(Screen shot reprinted by permission form Microsoft Corporation)

She used callout boxes also, but note that they are not identifiers. Each one is a command that is related to the text below the scan.

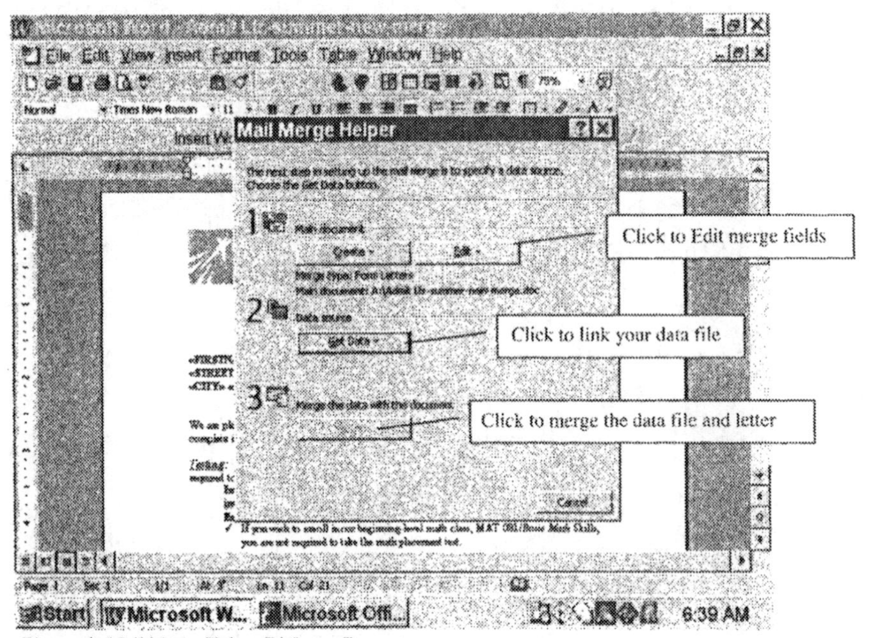

Figure 6. Mail Merge Helper Dialogue Box

(Screen shot reprinted by permission form Microsoft Corporation)

Among the most popular formats for text links is this layout with a graphic placed above numbered instructions. The locators are identified in steps 3 a, 3.b, and 3 c)

Foldouts and Special Needs

There are a number of foldout concepts that can be used. The usual approach is to use 11″×17″ paper that has one or two creases to be unfolded. Take care to make sure that your folds are not stapled closed on the bound edge. The question is whether it is better to use a visual that fills the *entire* width or one that is on the outer or extended page. Consider service manuals, for example. The foldouts in the manuals are either fully integrated or grouped at the back of the manual. In either case, readers flip back and forth with foldouts flapping this way and that. Perhaps there is a design feature here that could be modified in both instances.

First, if the tactic is helpful, a writer might put *all* oversized material in the back to cut down on the confusion.

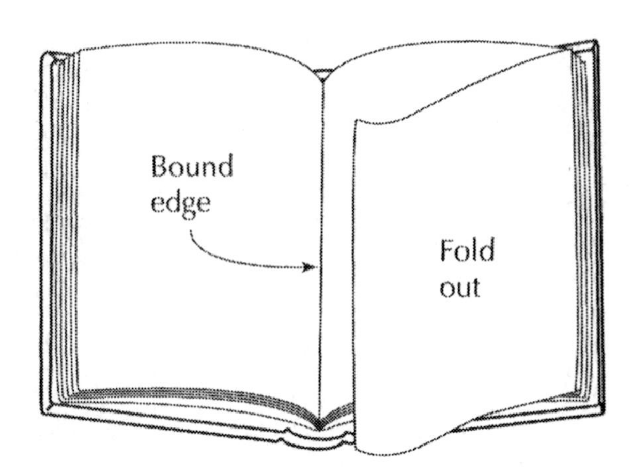

However, an author could also try to position each *entire* graphic to the right of the text so that no part of it is hidden by the text. This is ideal but not always practical. In this situation, the left side of a foldout is a text, a legend, or even blank so that the entire visual—most often a schematic or diagram—is available to the reader at all times while the text is being consulted.

Sample 11.L

Table 3 Priming Problems

Condition	*Possible Cause*	*Suggested Remedy*
Pump fails to prime or loses prime	Air leaks	Clean and tighten all suction connections. Make sure suction hoses and gaskets are in good condition. Use the following procedure to locate air leaks: 1. Connect suction hose to pump and attach suction cap to intake end of hose. 2. Close all pump openings. 3. Open priming valve and operate primer until vacuum gauge indicates 20 to 22 in. Hg. (If primer fails to draw specified vacuum, it may be defective, or leaks may be too large for primer to handle.) 4. Close priming valve and shut off primer. If vacuum drops to less than 12 in. Hg in 10 minutes, serious air leaks are indicated. With engine stopped, air leaks are frequently audible. 5. If leaks cannot be heard, apply engine oil to suspected points, and watch for break in film or oil being drawn into pump. 6. Connect suction hose to auxiliary pump. Open one discharge valve and run in water until pump is completely filled and all air expelled. Close discharge valves, apply pressure to system, and watch for leaks. A pressure of 100 psi is sufficient. DO NOT EXCEED RECOMMENDED PRESSURES. 7. If pump has not been operated for several weeks, packing may be dried out. Close discharge and drain valves and cap suction openings. Operate primer to build up a strong vacuum in pump, run pump slowly, and apply oil to impeller shaft near packing gland. Also, make sure packing is adjusted properly.

10

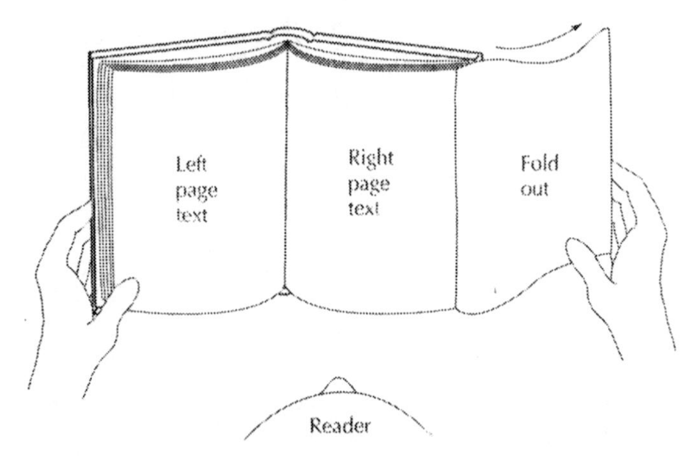

There are two specific reasons for using foldouts. Engineers rely on the foldout as a solution to problems of *scale* and *shape*. If a graphic cannot be reduced in scale because it will be unreadable or unclear, then the foldout provides twice as much room. Rather commonly, the shape is the real consideration because large families of graphics are often rectangular (schematics, some architectural drawings, tables).

To Build a Table

1. Go to **Table** on the menu bar.
2. Click on **Insert Table.**
3. In the dialog box, enter the number of rows and columns desired.
4. Click on **OK** for the basic format.
5. *Or* select **Autoformats** for other styles.
6. There are many options in the autoformats. Select one that suits your needs.
7. See also, the new feature, **Draw Table,** in the pull-down **Table** menu.

Standard 8½″×11″ paper is, of course, used upright. All the primary documentation is in an upright rectangle. Rectangular graphics are clumsy in this environment and are as awkward as an old cinemascope movie on a square television screen. The foldout not only provides twice as much space for the desired graphic but also configures a large horizontal rectangle, which is the desired shape. In addition, a reader does not have to turn the document sideways to read it.

One alternative to the foldout is to turn a rectangular graphic sideways on a page of text so that it is in the "landscape" position. This is a convenient strategy when the graphic is not difficult to read in the 8″×11″ size range (see Sample 11.L). For single-sided student projects, the top or the heading of a sideways page is *always* on the left of the text.

Tables often contain considerable text and may involve graphics as well. These can be handled vertically, in landscape format, or in foldouts.

Engineering drawings are often presented by the square yard, of course, and will not package in terms that can be adapted to writing. Proposals are often submitted with a tube of drawings or a large portfolio folder.

Regardless of the type of layout you select, and most of them are a composite of different graphics in different page formats, there are a few taboos. First, be very alert to the potential confusion of too much graphic material and too little text. Although there is great emphasis on graphics in engineering areas, and you must speak visually, *there is a text*. The text, the graphics, and the math systems you use have inherent strengths and weaknesses. The English language is not very visual, for example, but it is intensely logical, and it is the tool to use for organization. Use the text to explain, analyze, and organize *all* the material that is brought together—including the graphics. A sheaf of drawings does not, and cannot, hang together as anything other than a sheaf of drawings.

Also be wary of inserting three or four graphics back to back or grouped together on a page. The same risk of disorganization applies here. Assume, and it is a safe assumption, that you will serve your reader best if you *write* your way from figure to figure. Use language as the transition. Explain where you are going next, and explain the significance of the next graphic. Visual material holds together rather poorly. Language can bridge all the pieces and make the total much less of a puzzle. This is a problem of continuity, and it is one of the most common problems I see in technical writing.

On a similar note, try not to end a document with a graphic. (Material in the appendix is a different issue.) When you come to the last page, try to end with a paragraph or two of text. This is a particularly strong suggestion if the text ends with conclusions. If there are conclusions, then the graphic should support the final comments and not the other way around. The graphics are best used as components of modeling or as evidence that is incorporated into, and organized by, the language of the text.

The various graphic layouts identified in this chapter are all quite common. There may be company standards or guidelines for page layouts to which all the documentation of a company is expected to conform. At colleges and universities, however, you are more likely to be encouraged to provide the best documents you can for your instructors. There will be no company guidelines. Certainly, at the entrepreneurial level, the writer can be fearless in the design and development of a document. At all times, though, I would encourage the author to take charge—and that, of necessity, involves a lot of graphic management. With practice, your control over this dimension of document engineering will be a key element of your success with technical documentation. The craft of illustration is indispensable because engineering documents need *models*.

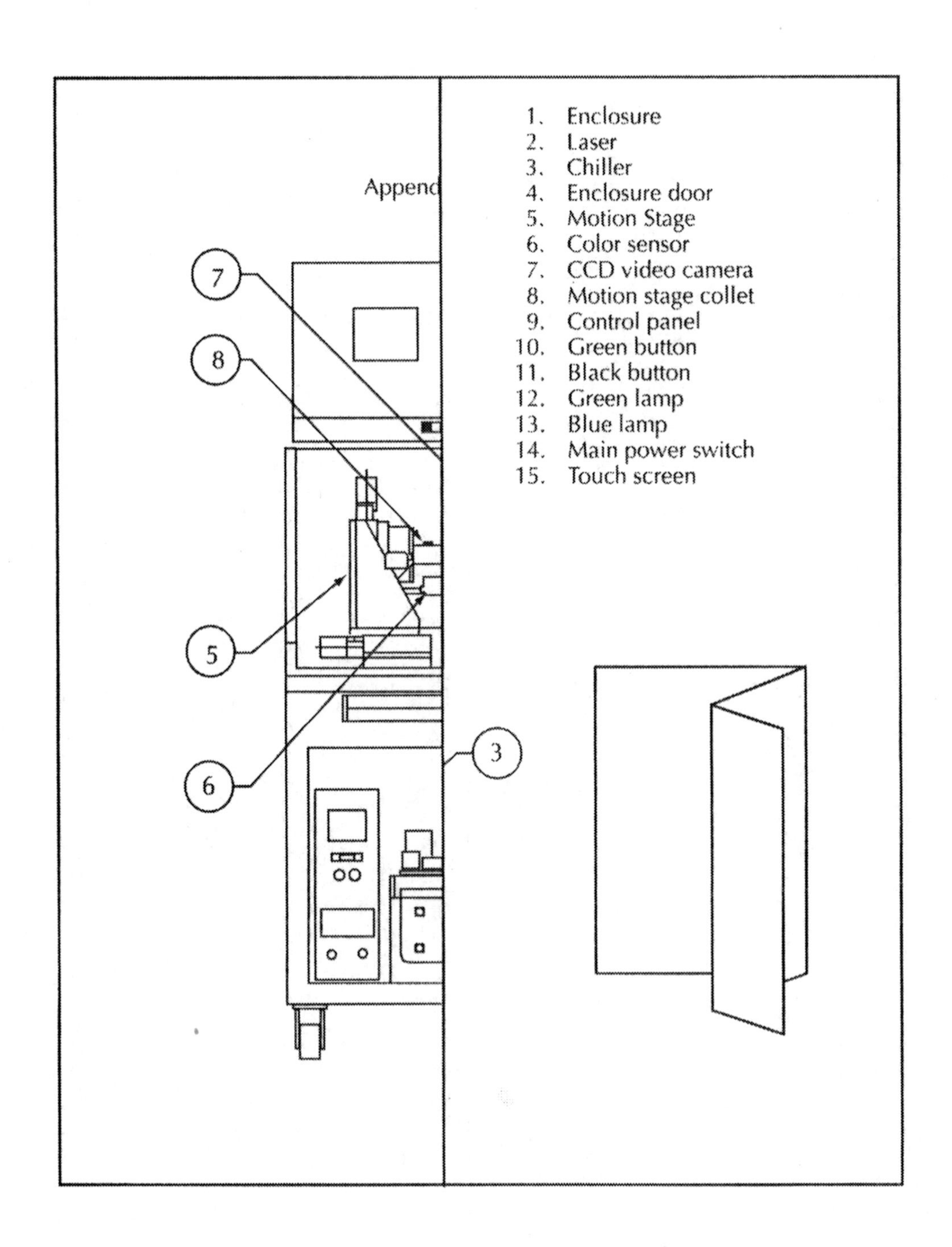

Append
1. Enclosure
2. Laser
3. Chiller
4. Enclosure door
5. Motion Stage
6. Color sensor
7. CCD video camera
8. Motion stage collet
9. Control panel
10. Green button
11. Black button
12. Green lamp
13. Blue lamp
14. Main power switch
15. Touch screen
7
8
5
6
3

Sample 11.M

5

Creating a Shortcut

1. First go to the icon for which you want to make a Shortcut. I have picked the 3.5 Floppy Drive for this example. Locate the icon. In this case the icon is in the My Computer directory.

2. When you have located the icon for which you want to make a shortcut, hover the mouse arrow over the icon until it illuminates.

3½ Floppy (A:)

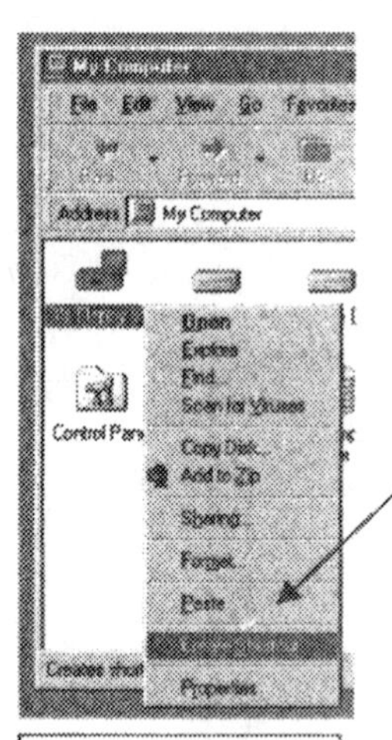

Figure 15-Create Shortcut Location

3. Click on the right mouse button and a drop down menu should appear. Hold the mouse arrow above Create Shortcut until it illuminates. Release the mouse button (Figure 15).

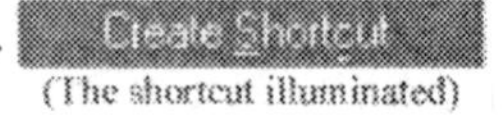
Create Shortcut

(The shortcut illuminated)

4. Since we are making a shortcut from the My Computer directory and shortcuts are not allowed in this location, a box will appear asking you if you want to place the shortcut on the desktop instead. Select the "Yes" button (Figure 16).

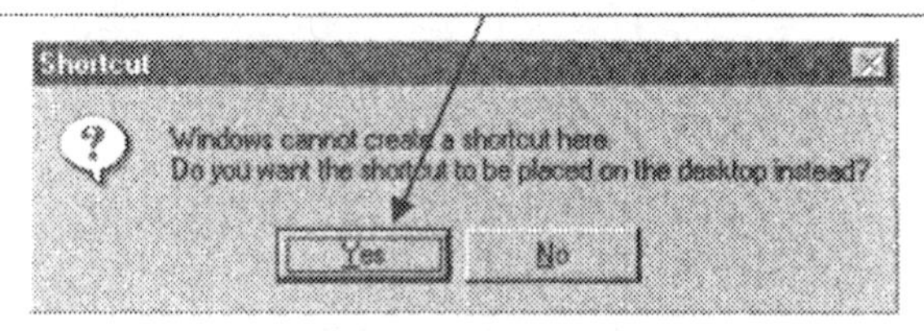
Shortcut
Windows cannot create a shortcut here.
Do you want the shortcut to be placed on the desktop instead?
Yes
No

Figure 16. The Query.

5. Once you have clicked the "Yes" button, the Shortcut will appear on your desktop as illustrated below and is ready for use. As you will notice, the icon is accompanied by an upward turning arrow in the bottom left corner of the icon. This symbol indicates that the icon is a shortcut and is actually referencing the real icon in another location.

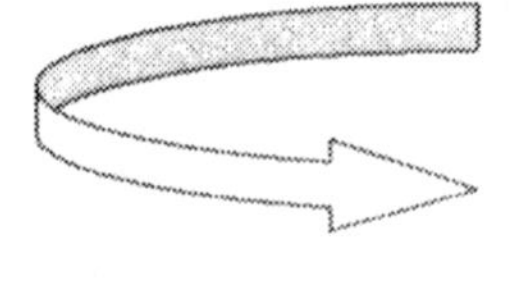

Shortcut to
3½ Floppy (A)

Note: The "My Computer" icon acts as a central point of access to all drives and folders.

(Screen shots reprinted by permission from Microsoft Corporation)

Design Management

Some of the most innovative design and layout work in the publications world appears in the products of engineering firms and corporations in which publications departments are struggling to communicate concepts about products and services. The support documentation that links a client to services or products is critical to the success of industries large and small. Whether a civil engineering firm is preparing a multimedia presentation for a city council, or a cellular phone company is designing a manual for one of its vehicle installations, the creation of the document calls for imagination. Writing is one avenue of communication. Graphics is another. Design work is yet another. Perhaps the engineering firm will present a maquette—a model built on scale—of a sewage treatment plant. Perhaps the cellular phone brochure is in the shape of a phone and consists entirely of full-scale photographs of the phone panel with small command windows to explain the functions.

Technical Document Basics, however much it encourages you to be innovative and however much it might hope to be innovative itself, is a very traditional document. The text you are reading may not follow all the rules, but it abides by most of them. On the other hand, you must respond to whatever demands you encounter in a writing project. Novelty, of course, is not a significant player in documentation, except in merchandising. You have no doubt seen a few extravagances, such as pop-ups or holographs or cutout windows (that allow you to look through one page to another). You are not likely to be in a position in which opportunity will often arise for extravagant document design work, but you do need to be alert to the flexibility that you have in technical writing applications. You are dealing with a mixed-media environment that is often inclined to be unconventional—of necessity. This is why the most traditional of all presentation methods—the essay—is often the least likely tool you will find attached to products and services. Note Sample 11.M. This author has excellent control of his computer's capability, and the result is a unique page design.

It is appropriate to end with a document model that demonstrates the innovative quality available to authors of technical writing. The model contains neither bells nor whistles—but it contains no text either! When you first look at Sample 11.N, the product seems very ordinary, and it is. However, few written media can break away, as this model does, from the conventions of paragraphs and sentences. The model illustrates a point you might overlook. It is a troubleshooting manual without a text; properly speaking, it is a troubleshooting manual with a text of tables. It is a vivid and effective design idea. The design layout is the strategy behind the document. It is not exactly writing. It is the kind of multimedia engineering you have observed from the outset.

Imaginative layout designs are no longer restricted to publishers. The page you see here is a vivid and appropriate approach to its subject matter. This is a college composition done with a computer.

Sample 11.N

EGR231 OUTLINE PROJECT 3 11/200X

TROUBLESHOOTING CECO PRODUCTS

Audience: Retail customers and Cruising Equipment staff who have some knowledge of electricity.

Objective: To provide a document that will help customers through installations. It will also be used by Cruising staff when helping customers with installations.

Form: Tables of problems and solutions and possibly flowcharts.

I AMP-HOUR METER TROUBLESHOOTING

A. No display or display lasted for only short time
   1. Check 12 volt power supply.
   2. Check fuse and fuse holder.
   3. Unit hooked to more than 12 volts (return meter) (most common problem, stress in installation, manual)
   4. High voltage spike (return meter)

B. Counts amp-hours in wrong direction (reversed shunt leads)

C. Does not count a charging or discharging device (device on battery side of shunt)

D. Accumulation of negative amp-hours (adjust meter)

E. Never gets to zero amp-hours (need to condition/equalize)

II AMP-HOUR METER + AND +2 TROUBLESHOOTING

A. No display
   1. Check fuse and fuse holder.
   2. Check positive 12 volt power supply.
   3. Unit hooked to more than 24 volts (return meter)

B. Incorrect battery voltage readings (battery sense leads, fuses, fuse holder, and ground)

C. Displays amps or in wrong direction (reversed shunt leads)

C. Does not count a charging or discharging device (device on battery side of shunt)

III FORM LETTERS FOR COMMON REPAIRS

A. Nothing found wrong with amp-hour meter and quad-cycle

B. Destroying high-voltage protection circuit of amp-hour meter

The next few pages are examples of troubleshooting tables for servicing the electrical systems on pleasure boats, and the author designed the tables as stand-alone documents that do not have a supporting text. The company wanted no text. This points up the uniqueness of technical writing: at times, documents may contain little or no material of a conventional sort. As a result of this company's requirement, the tables had to be quite thorough and vivid. Scott first presented to me the outline you see in Sample 11.N. This was the only "text," and it was a document for my use that was strictly an organizational tool for the project. Because outlines are effective for instructions, it is interesting to compare the outline style with the table format that he then designed as the finished product, three pages of which follow (Model 11.A).

Model 11.A (1)

TROUBLESHOOTING CRUISING EQUIPMENT

PRODUCTS

Audience: Cruising staff and retail customers

who have some understanding of electricity.

Objective: To ease installation problems and

reduce customer service time.

Department: Ceco manufacturing

Department Head: Richard P.

Project 3

Professor: David R.

November 27, 200X

Cruising Equipment Co. Seaview Ave. NW, Seattle WA 98107

Model 11.A (2)

2

AMP-HOUR METER TROUBLESHOOTING GUIDE

This troubleshooting guide will become page 7 of the

Amp-Hour Installation Manual (see Appendix A, page 9).

AMP-HOUR METER TROUBLESHOOTING GUIDE

PROBLEM	CAUSE	SOLUTION
No display	No + 12V on red wire	Check for 12V between red and black wires of meter. Also check fuse and fuse holder.
	High-voltage circuit blown	Return to be repaired.*
Counts amp-hours in wrong direction	Shunt leads are reversed.	Shunt wires: green battery side and orange ground side.
Meter does not count when a charging device is on (e.g., solar panels, alternator, charger etc.).	Too many loads are on.	Turn off loads.
	Device output on the battery side of the shunt	Ensure that all electrical devices are hooked to the ground side of the shunt and that only the batteries are hooked to the battery side of the shunt. See page 4, numbers 2 and 3, of AMP-HOUR METER INSTALLATION TIPS AND TRICKS and the wiring diagrams on pages 7 and 8.
After each time the battery reaches your charged parameters (see page 1), a larger number of negative amp-hours appears.	Charge efficiency underestimated	To adjust, see "ADJUSTING CHARGE EFFICIENCY" and ADJUSTMENT RULE, page 5. Charge efficiency pot may need to be turned clockwise.
After hitting the charged parameters (see page 1), the meter consistently displays the same number of negative amp-hours.	Batteries are not completely charged.	Condition/equalize batteries. See "ABOUT THE QUAD-CYCLE CHARGING METHOD," page 6.

*INSTRUCTIONS FOR MOST EXPEDIENT RETURN
Please enclose a letter with a detailed description of the problem, a shipping address, and a daytime phone number. If the unit is out of warranty, please enclose a Master Card or Visa number. The minimum charge is $XX.XX (including UPS 2nd day shipping), and $XX.XX is the typical maximum charge. We will notify you if the charge is higher.

Note: All Cruising Equipment Amp-Hour Meters have an 18-month warranty.

Model 11.A (3)

5

BACKOFF TROUBLESHOOTING GUIDE

This troubleshooting guide will become page 26 of the Backoff

Installation on manual (see Appendix C, page 14).

BACKOFF TROUBLESHOOTING GUIDE

PROBLEM	CAUSE	SOLUTION
LEDs do not light up. (Unit does not operate.)	No 12V at unit	Check fuse, fuse holder, ground supply to terminal 4, and +12V supply to terminal 5.
	Backoff fuse blown	Take over off unit and place unit so that LEDs are on right. Check for +12V on both sides of F1 (see installation diagram) If +12V is only on left side, then F1 (3 amp fuse) needs to be replaced. Return for repair.*
A sheridable load does not come on.	Load too large	Turn off nonsheddable loads, or sheddable load is shorted.
	No AC at relays	Check 110 VAC supply to relay.
	No 5V at relays	Check 5V supply at Backoff and relays.
	Relay damaged	If there is 5VDC across the DC side of the relay and 110VAC across the AC side of the relay, then the relay needs to be replaced.

*INSTRUCTIONS FOR MOST EXPEDIENT RETURN
Please enclose a letter with a detailed description of the problem, a shipping address, and a daytime phone number. If the unit is out of warranty, please enclose a Master Card or Visa number. The minimum charge is $XX.XX, and $XX.XX is the typical maximum charge. We will notify you if the charge is higher.
If you have any questions that are not covered this secuon. please call Cruising Equipment at (206) 782- 0000.

Note: The Backoff AC Load management system has an 18-month warranty.

Crusing Equipment Co. Seaview Ave. NW. Seattle WA 98107

Peer Review

Early on in this text, I suggested that you develop an outline as a peer *preview*. Once the project is completed, you are ready for the peer *review*. It is likely that whoever did the preview is now an appropriate choice for a close look at the final merchandise. The peer review is an "engineering edit." You are engaged in an engineering process in your writing, and it is your coworker, an engineer, who must review the work.

When a technical project nears completion, it is important for an author to find a reader to review the material. Editing suggestions are particularly valuable if the reader will pick up a pen and volunteer changes and corrections. Editing is indispensable, since authors often fail to see their errors. Unfortunately, the development of technical documentation can be managed awkwardly. Because editing is seen as a language issue, very important areas of technical information may go largely overlooked. Problem areas in your projects may, indeed, involve writing issues, such as organization or grammar errors, but writers also need to be concerned about the accuracy of engineering features, math, chemistry, physics—and the graphics.

Technical editing is a unique type of editing procedure. There are two types of editing that must be completed once a document is drafted. One edit addresses the issues of language, organization, clarity, and other word-based concerns. A very different edit double-checks the math and the engineering, as well as the schematics and other graphic features that may be of a technical nature. For this activity, you will need a peer review.

To be sure that you receive the proper edit, ask your reader to focus on the calculations and the graphic elements and not only on the spelling errors. There is an obvious tendency for even the accomplished engineer to skip over some of the more technical features, possibly because the author is asking the reader to reprocess the complex work embodied in the math and in such graphics as electronic schematics or ball-and-stick visualizations in chemistry. Admittedly, the author is relying on the reader to do a somewhat thankless task. The hope is that the employee can handle the editing on company time. Return the favor, and establish a routine for an editing procedure. If you structure an in-house editing method of some sort, it will be an invaluable assist that you and the members of your editing pool can use to refine and polish documents.

Sample 11.0

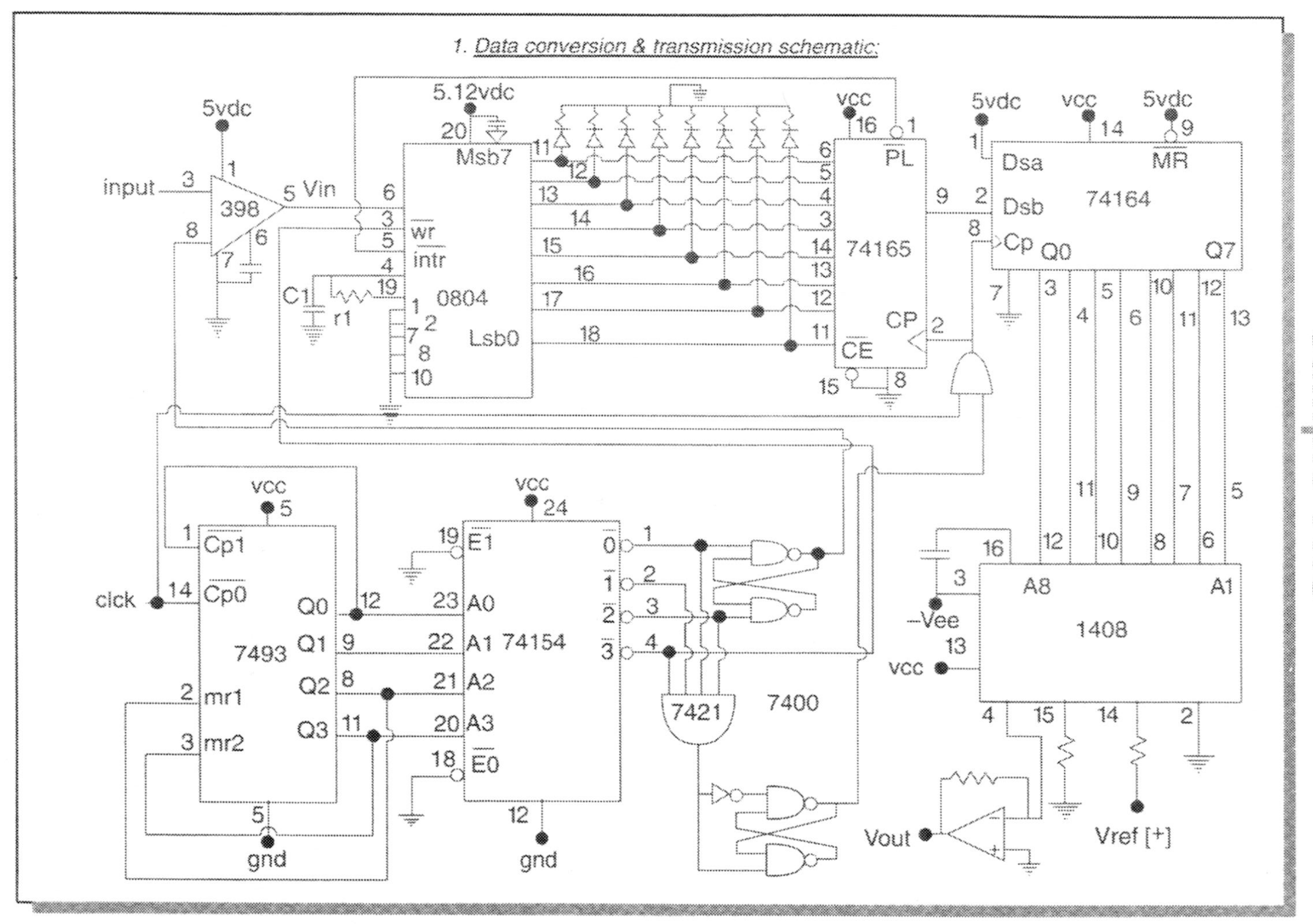
1. Data conversion & transmission schematic:
5vdc
input
398
Vin
5.12vdc
Msb7
wr
intr
0804
Lsb0
C1
r1
vcc
PL
74165
CP
CE
Dsa
Dsb
Cp
Q0
Q7
MR
74164
clck
Cp1
Cp0
7493
Q1
Q2
Q3
mr1
mr2
gnd
E1
A0
A1 74154
A2
A3
E0
7421
7400
A8
A1
1408
–Vee
Vout
Vref [+]

In the case of elaborate details—in architectural drawings, for example— every detail should be reviewed from upper left to bottom right. The same is true of any schematic, such as the one in Sample 11.O. When you look at this elaborate rendering, it is very clear that only a peer review will edit the product correctly. On a more accessible level, you can imagine the reviewing problems related to common symbols, such as the decimal point or the symbol of length for a foot. One little slip and 12′ can become 12″, or 310.131 can become 31.0131. The scrutiny is demanding.

For your own editing purposes be sure to use the grammar checker and spelling checker features on your computer.

1. Go to **Tools** on the top icon bar.
2. Click on **Options** at the bottom of the Tools menu.
3. Click on the **Spelling and Grammar** tab at the top of the dialog box.

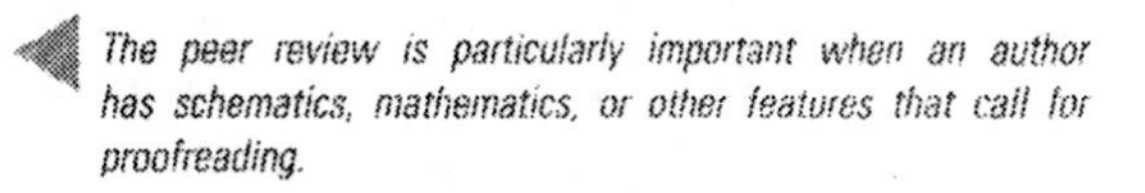

The peer review is particularly important when an author has schematics, mathematics, or other features that call for proofreading.

The Layout Samples

Comments

Each of the full-page samples are sections taken from projects that utilized the typical page layouts we have analyzed. They appear in an order that corresponds to the text: full-width, half-column, side-by-side, windows, and full-page.

The final samples illustrate sequences, foldouts, and innovative design considerations.

Summary

- Full-width graphics are the most convenient choice. The image does not need to fill the width of the page.
- Insert a full-width graphic after a completed sentence in the running text. (Do not divide a sentence with a graphic feature.)
- Half-column graphics allow the text to be placed beside the graphic under discussion, but they involve the inconvenience of text wraps.
- Side-by-sides and windows are unique to certain engineering disciplines but can be used for any purpose desired.
- Sequences can be used to "animate" a progression of graphics.
- Segmenting can be used to subdivide complex images for discussion. Each segment is enlarged and placed on a page with appropriate discussion.
- Sideways illustrations ("landscape") are a convenient way to present large illustrations.
- Foldouts are most conveniently located in appendices.

Activities Graphics Layouts

Your projects give you plenty of opportunity to use each of the page layouts under discussion in this chapter. You now need to see what your classmates are doing in terms of their projects.

Since you also need to be encouraged to use peer review, your instructor can arrange for you to swap papers with another student on the day papers are due.

If possible, trade papers with someone in your field of interest so that you can do a peer review. Take the paper home, read it, add suggestions or corrections (or applause!) anywhere in the paper. Focus on the graphic elements and the technical issues. What topics were selected? Who were the intended readers? How was the paper organized? What page layouts were used for the graphics? Are there technical errors?

Return the paper to the instructor with a brief explanation of your evaluation of the merits of the project you reviewed.

If the class has been asked to develop an appendix that explains how computers were used to construct each project (see pp. 69-71), review these details whenever a graphic or a page layout is of interest to you. Ask the author for details or consider asking for a demonstration of how a certain graphic or page layout was constructed.

Share great moments in technical document oversights. You probably see such blips all the time. Photocopy them to share with others in the class.

Here is a favorite of mine from a set of instructions provided by a federal agency:

> "Mark or label boxes and wrapped packages on end (not top, bottom, or sides) . . . ."

Share a Project

The instructor will survey the class to see which students are familiar with scanners. These students will act as guides to take small groups of four to campus labs to walk through a demonstration and demonstrate the technology. The students are encouraged to try their hand at embedding a graphic in a page layout of their own design.

In teams of three, reconstruct a poorly designed document. Computer manuals or the manuals that accompany imports are likely targets. If you cannot readily find something to reconstruct, have each class member bring in a page or two from popular magazines that could be redesigned. Advertising copy pages might be interesting.

To reconstruct a page, make plenty of copies first. Discuss the changes. Does it need a new layout? New text? Callouts? Figure numbers? Captions: Cut up the page and develop a pasteup. Build the new elements. The finished product can be a photocopy, or the page can be scan-constructed in the computer lab.

Share a Document Preparation

Collaborate with the same three other members who worked with you on the technical document team projects described on p. 402 and p. 444. The final team submission is a project management report *in which you explain how you developed the other team projects.*

- *Account for your time and effort in the projects.*
- *Briefly describe the procedures that were used to tackle each project.*
- *Describe the roles of the individuals and their responsibilities.*
- *Describe the outcomes and explain any difficulties that emerged.*
- *Construct a flowchart to graphically represent the production schedule you followed on one of the projects.*
- *Submit this project as a memo (of 500 to 1000 words) to your instructor.*

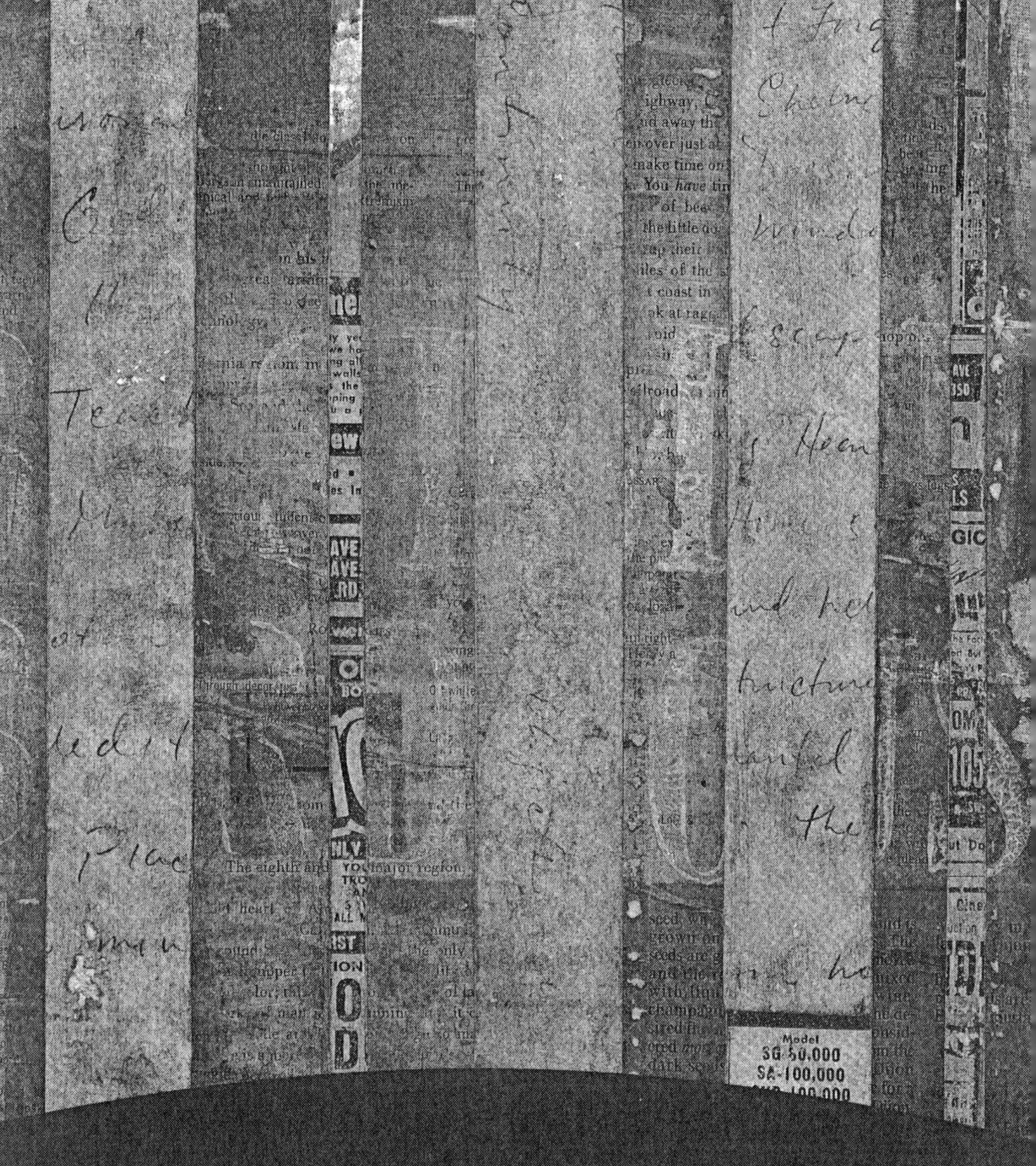

APPENDIX

A

Guidelines For Editing Projects

Peer Reviews

The experience of editing another writer's work is probably new to you, so I would like to establish guidelines for your editing activities. Your particular need to assist or be assisted in a writing project is a work-related matter; it will be a peer review of a document that is in progress. There will be two concerns: the technology and the writing.

The Technical Edit

Concerning the technical content, you are in a better position to assist a student in your field than any reader who is unfamiliar with the subject matter— and this is likely to include your instructor. When you have an opportunity to read student projects for editing purposes, do a *close reading* if the technology is familiar to you. A close reading is one that devotes attention to every detail. Read the text for technical accuracy. Study the mathematics and make certain the calculations contain no errors. Carefully review the graphics for accuracy, and read the schematics or other technical drawings to see if they are correct. Circle problem areas and indicate the issue or add the correction in the text or in the margin. If you think there is questionable material that you cannot resolve, put a question mark in the margin.

If you are reading a text that is not in your subject area, take a different approach and read the project with the intention of seeing how different disciplines approach the writing task. There is no specific method of writing that will serve every engineering field in an ideal way. Similarly, one type of mathematics may serve civil engineering and not electronics. One graphic type may serve structural design and not electromechanical CAD. If you have the opportunity to read the contributions of a variety of students in different specialty areas, the experience will enhance your own skills and perceptions.

The General Edit

The other type of edit that you will need in a work-related situation is the traditional analysis of the writing. The easiest way to develop editing skills is to divide the tasks into neatly perceived areas of interest. For example, I always look at audience targeting as a specific issue, and in my reading I mentally cluster the techniques for targeting. Vocabulary, definitions, sentence complexity, and paragraph length are specific concerns to focus on in this case. If, however, a coworker tells me that he or she is not a strong writer, then grammar basics and spelling become a critical area of attention. Again I would look at the concern as a cluster of related issues: subject-verb agreement, pronoun agreement, sentence fragments, and spelling.

The particular concerns of *Technical Document Basics* represent technical editing clusters that you need to pay attention to as you read peer-group papers. The concerns of the textbook are quite evident if you review the table of contents. Each chapter represents a specific area of interest, and so each chapter presents an identifiable group of editing considerations. For example, examine each document you read for layout practices. There will usually be five or more page layouts to examine: the cover page, table pages for contents and illustrations, text body pages, and possibly a glossary or an appendix or a bibliography. The editing issue is a simple question: are the pages handled correctly? The question can be looked at by the standards presented in your textbook, but they can also be approached from whatever standards are desirable, such as those of your instructor or your supervisor at work. In an entrepreneurial environment where *you* determine the standards, clarity is your primary measure of layout quality.

The specific areas of interest for editing a technical document should include consideration of the following:

Layouts

Formats

Style of introduction

Audience targeting

Graphics

Graphics integration

Type of document

Procedure

As you read through a project keep the two editing concerns in mind: the technical edit and the general edit. Ideally you might read a project several times in order to focus on one type of editing at a time. Regrettably, college is similar to your workplace, and time will be at a premium. Since you will only have one opportunity to read a document, give the project two types of editing at the same time. Edit for *both* technical and general errors. In addition, use two types of editing practices. *First,* amend the text as you read. Circle any problem you see and correct the problem above the circle. *Do not alter the original text in any way because your instructor still has to evaluate the project also.* You can also draw arrows to a problem and explain your perception in the margin. In other words, you can make a bit of a mess in the interest of creating a better project for your coworker, but *add* corrections and thoughts; do not remove anything or make the original impossible to read.

Once you are done with the reading, stop and review your overall impression. Add a paragraph (or more) of your ideas at the end to explain what you think. This summary is your *second* task. Do not confuse the role of editing with "criticism." Tell the reader what was well done! The best way to suggest corrections is to match strengths and weaknesses. As you probably know from experience, strong criticism is seldom an effective learning tool. Peer evaluation means *helping*. The idea is not to decide what is *wrong* but to strive, as a team, to make a document *better*. Sum up with helpful and supportive comments.

I have provided five editing checklists. Use them in order, but use only one for a given set of readings. When new projects are due, several weeks later, move on to the appropriate checklist. Each list suggests technical and general edits. The lists are cumulative and expand as you study more of the text. Assuming the projects are written in the approximate order in which they are presented in the text (Chapters 5, 6, and 7), I add an editing checklist for one or more of the document prototypes to each checklist.

First readings:	**Instructions**
Second readings:	**Product or Mechanism Description**
Third readings:	**Definitions or Classifications**
Fourth readings:	**Causal Analysis or Troubleshooting**
Fifth readings:	**Academic Projects:**

- **Comparisons**
- **Problem-Solution**
- **Risk Analysis**

This list of prototypes may not reflect every student's choice for projects. Engineering technical students are likely to focus on the prototypes identified in the first four readings. Pre-engineering transfer students may elect to spend most of their efforts developing the projects that are the focus of the fifth readings. You may also encounter workplace projects that are unrelated to any of the prototypes identified here.

First Readings
Editing Instructions or Procedures

The Technical Edit

- Are all the facts in order?
- Is all the math properly managed?
- Is the math (or chemistry) smoothly integrated into the language?*
- Are the graphics technically accurate?
- Are the schematics correct in every detail?

The General Edit

- Are there grammar and spelling errors?
- Are the page layouts properly managed?
- Is the introduction effective?

Prototype: Instruction or Procedures

- Has the author chosen the most appropriate format (outline, running text, heading and paragraph mix)?
- Were preliminaries (parts lists, equipment lists, and warnings) properly addressed?
- Is disassembly thoroughly handled?
- Are the tasks in order and clearly explained?
- Are the graphics adequate to the purpose?

* *For questions concerning math displays, please consult the* Writer's Handbook.

Second Readings

Editing Product Descriptions

The Technical Edit

- Are all the facts in order?
- Is all the math properly managed?
- Is the math (or chemistry) smoothly integrated into the language?*
- Are the graphics technically accurate?
- Are the schematics correct in every detail?

The General Edit

- Are there grammar and spelling errors?
- Are the page layouts properly managed?
- Is the introduction effective?

Prototype: Product or Mechanism Description

- Does the document focus on functions, parts or principles?
- Is the focus appropriate to the target audience?
- Is the depth and length of the project adequate for the purposes of defining or explaining the product?
- Do the graphics support the text?

* *For questions concerning math displays, please consult the* Writer's Handbook.

Third Readings

Editing Definitions or Comparisons

The Technical Edit

- Are all the facts in order?
- Is all the math properly managed?
- Is the math (or chemistry) smoothly integrated into the language?*
- Are the graphics technically accurate?
- Are the schematics correct in every detail?

The General Edit

- Are there grammar and spelling errors?
- Are the page layouts properly managed?
- Is the introduction effective?
- Is the target audience clearly identified?
- Does the text language seem appropriate for the reader?
- Are sentence complexity, paragraph length, and document length appropriate?
- Are the graphics appropriate for the readers?
- Are the graphics vivid?
- Are graphic features handled correctly? (Look at the callouts, legends, captions, figure numbers, credits, and so on.)
- Could the author have used other types of graphics?

Prototype: Definitions

- Did the author use the basic concept of analysis by division?
- Are the divisions clear and appropriate?

* *For questions concerning math displays, please consult the* Writer's Handbook.

- Are other techniques—history, comparisons, contrasts—used successfully in the project?
- Are the graphics key players in defining by illustrating?

Prototype: Comparisons

- Is there a table to visualize the comparison in an effective way?
- Are the criteria (the measures or standards for analysis) clearly explained?
- Are the subjects more or less equal for the purposes of a fair comparison?
- Is the text developed by criteria or by subject?
- Did the author make the best choice for developing the project by criteria or by subject?

Fourth Readings

Editing Causal Analysis or Troubleshooting

The Technical Edit

- Are all the facts in order?
- Is all the math properly managed?
- Is the math (or chemistry) smoothly integrated into the language?*
- Are the graphics technically accurate?
- Are the schematics correct in every detail?

The General Edit

- Are there grammar and spelling errors?
- Are the page layouts properly managed?
- Is the introduction effective?
- Is the target audience clearly identified?
- Does the text language seem appropriate for the reader?

* *For questions concerning math displays, please consult the* Writer's Handbook.

- Are sentence complexity, paragraph length, and document length appropriate?
- Are the graphics appropriate for the readers?
- Are the graphics vivid?
- Are graphic features handled correctly? (Look at the callouts, legends, captions, figure numbers, credits, and so on.)
- Could the author have used other types of graphics?
- Are the graphics properly integrated?
- Does the text reference the graphics?
- Should locators (A,B,C) have been used to integrate the graphics and the text?
- If there are graphics in an appendix, are they appropriate for the location?

Prototypes: Causation and Troubleshooting

- In the opening paragraphs, does the author clearly explain whether the analysis is effect to cause or cause to effect?
- Are the causes clearly explained?
- Are the effects clearly explained?
- Are *all* causes and *all* effects accounted for, at least to your satisfaction?
- Is there a table or a flowchart to support a troubleshooting document?

Fifth Readings

Editing Academic Projects

The Technical Edit

- Are all the facts in order?
- Is all the math properly managed?
- Is the math (or chemistry) smoothly integrated into the language?*
- Are the graphics technically accurate?
- Are the schematics correct in every detail?

* *For questions concerning math displays, please consult the* Writer's Handbook.

The General Edit

- Are there grammar and spelling errors?
- Are the page layouts properly managed?
- Was the introduction effective?
- Is the target audience clearly identified?
- Does the text language seem appropriate for the reader?
- Are sentence complexity, paragraph length, and document length appropriate?
- Are the graphics appropriate for the readers?
- Are the graphics vivid?
- Are graphic features handled correctly? (Look at the callouts, legends, captions, figure numbers, credits, and so on.)
- Could the author have used other types of graphics?
- Are the graphics properly integrated?
- Does the text reference the graphics?
- Should locators (A,B,C) have been used to integrate the graphics and the text?
- If there are graphics in an appendix, are they appropriate for the location?

Citations

- Is the bibliography properly structured?
- Are the citation references uniform?
- Are text notes handled in an orderly fashion?

Prototype: Comparisons

- Is the format of the project quite clear to the reader?
- Was the document developed by subject ("sides") or by issues (criteria)?
- Was this choice appropriate?
- Does the introduction clearly distinguish the subjects and the issues?
- Is an opinion clearly stated by the author in the conclusion? Should there be one?
- If there is a table, is it a complement to the text?

* *For questions concerning math displays, please consult the* Writer's Handbook.

Prototype: Problem-Solution Analysis

- Is the project designed to maintain a balance between problems *and* solutions (50:50)?
- Is the project designed to primarily address problems *or* solutions (20:80)?
- Is the choice appropriate to the authors' goal?
- Do you see any oversights in the analysis?
- Does the introduction clearly highlight the problem and the solution?
- Does the conclusion sum up the analysis?
- Is there or should there be concluding recommendations?

Prototype: Risk Analysis

- Are the advantages clearly stated?
- Are the disadvantages clearly stated?
- If more than one subject was being discussed, was each subject clearly discussed in terms of advantages and disadvantages?
- If there is a table, is it a vivid complement to the text?

APPENDIX B

Templates and Tips

Document Templates

In all your writing activities it is extremely helpful to conserve your writing energy by using templates to construct your documents. A template is simply a mold in which you cast your essay or your business letter or memo. The layouts for projects such as we have seen throughout this text are time consuming, and it is helpful to be able to reach for a preconfigured document and fill in the blanks, so to speak.

In addition to the instructions for templates, other useful features are explained. You will find instructions for the automatic features for building a table of contents and a list of illustrations. Basic instructions for building a table are also provided. As a final note, there are directions for opening the equation editor.

The software corporations were slow to respond to this word processing need, but beginning with *Word 6* for Macs and *Word 7* for Windows, the template files began to take shape.

The templates are of several types. The "no-text" type of template marks the location for the parts of a document, of a business letter, for example. The "full text" varieties provide a completed document that can be altered as desired. A third variety of template serves as a "model" to demonstrate mock-ups of such productions as brochures. Most of the templates that will be of value to you are of the first two types—and templates of your own making. You should construct your own templates so that you can have an *outline* template and other types that are not in the existing program.

The explanations on the following pages describe the templates that are available on a number of popular software applications:

Word 98 or Office 98 for Macs

Word 7 for Windows

Office 97 for Windows

Office 2000 for Windows.

I selected these applications because they are evidently the most frequently installed word processing programs on most campuses. All the applications are similar but different. The instructions will also appear to be similar but there are important changes from program to program.

Note
The number of templates installed in your home computer or your computer lab systems can vary from installation to installation.

Word 98 or Office 98 for MAC

I. To Find Templates of Essay Components

These template are of the no-text variety.

1. Go to **File** on the menu bar.
2. Click on **New.**
3. In the dialog box select **Other Documents.**
4. Click on the **Thesis** icon.
   You will find the following:

   Title Page

   Table of Contents

   List of Illustrations

   Text Page

   Glossary

   Bibliography

II. To Save a Modified Template

The template can be altered to suit your needs or to conform to the suggestions of your instructors, your supervisor, or the recommendations in *Technical Document Basics.* These modifications can be saved as a *new* template.

1. Go to **File** on the menu bar.
2. Click on **Save As.**
3. In the dialog box, click on **Save File as Type** (it is near the bottom).
4. Select **Document Template.** This function automatically places the template in a template file.

5. Double-click on the **Other Documents** folder to place a template in the folder.
6. Important: Change the file name. Simply add a new date, for example.
7. Click on **Save.**

III. To Save an Original Template

If you build your own title page or any other new page layout, it can be used again and again. You can simply make a copy and modify it, or you can make a *template* and modify it. *Be sure to save the first outline you develop for your technical writing course. Save it as a template.*

1. Go to **File** on the menu bar.
2. Click on **Save As.**
3. In the dialog box, click on **Save File as Type** (it is near the bottom).
4. Select a location among the files.
5. Select **Document Template** (it is near the bottom).
6. In the dialog box window are the folders. Here you can file the template in any folder or make a folder (see Section IV). Move the cursor to the selected file and *double*-click.
7. Important: Change the file name. Simply add a new date, for example.
8. Click on **Save.**

IV. To Open a Personal Template Folder for Your Templates

1. Go to **File** on the menu bar.
2. Click on **Save As.**
3. Click on the **New** icon on the menu bar in the Save As dialog box.
4. Type a name for the folder on the **New Folder** screen.
5. Click on **Create.**
6. The new folder is created and opened automatically.
7. Important: Change the file name. Simply add a new date, for example.
8. Click on **Save.**

Locating Additional Features

Auto-Template for a Table of Contents

Auto-Template for a List of Illustrations

These two templates are *not* the fill-in-the-blank style and cannot be filled in. They are automatic features that construct the tables *from* the text if you used headings in the text body or figure numbers and captions for the illustrations.

1. Go to **Insert** on the menu bar.
2. Click on **Index and Tables.**
3. In the dialog box, select **Table of Contents** or **Table of Figures.**

To Build a Table

1. Go to **Table** on the menu bar.
2. Click on **Insert Table.**
3. In the dialog box, enter the number of rows and columns desired.
4. Click on **OK** for the basic format.
5. Or select **AutoFormats** for other styles.
6. There are many options in the autoformats. Select one that suits your needs.
7. See also the new feature, **Draw Table** when you pull down from **Table** on the menu bar.

To Use the Equation Editor

1. Go to **Insert** on the menu bar.
2. Click on **Object.**
3. To open, the program, scroll to **Microsoft Equation** and *double*-click.
4. For help with the editor, return to the menu bar and click on **Help.**
5. The help options list **Equation Editor: Help Topics.** Click to open.
6. Select from the choices in the dialog box with a *double* click to call up instructions.

For Help at Any Point

1. Click on **Help** in the menu bar.
2. Go to **Show Balloons** or **Microsoft Word Help.**
3. Activate the help program with a click.
4. Type in a question where indicated and click on **Search**.

Word 7 for Windows (Office 95)

I. To Find Templates of Essay Components

These templates are of the no-text variety.

1. Go to **File** on the menu bar.
2. Click on **New.**
3. In the dialog box select **Publications.**
4. Click on the **Thesis** icon.
   You will find the following:

   Title Page

   Table of Contents

   List of Illustrations

   Text Page

   Glossary

   Bibliography

II. To Save a Modified Template

The template can be altered to suit your needs or to conform to the suggestions of your instructors, your supervisor, or the recommendations in *Technical Document Basics*. These modifications can be saved as a *new* template.

1. Go to **File** on the menu bar.
2. Click on **Save As.**
3. In the dialog box, click on **Save File as Type** (it is near the bottom).
4. Select **Document Template** and click.
5. *Double*-click on the **Publications** folder to place a template in a folder.

6. Important: Change the file name. Simply add a new date, for example (do *not* use the slash).
7. Click on **Save.**

III. To Save an Original Template

If you build your own title page or any other page type, it can be used again and again. You can simply make a copy and modify it, or you can make a *template* and modify it. *Be sure to save the first outline you develop for your technical writing course. Save it as a template.*

1. Go to **File** on the menu bar.
2. Click on **Save As.**
3. In the dialog box, click on **Save File as Type** (it is near the bottom).
4. Select a location among the files.
5. Select **Document Template** and click (it is near the bottom).
6. In the dialog box window are the folders. Here you can file the template in any folder or make a folder (see Section IV). Move the cursor to the selected file and *double*-click.
7. Important: Change the file name. Simply add a new date, for example (do *not* use the slash).
8. Click on **Save.**

IV. To Open a Personal Template Folder for Your Templates

1. Go to **File** on the menu bar.
2. Click on **Save As.**
3. Click on the **New** icon on menu bar in the Save As dialog box. (To locate the proper icon, rest the cursor on each icon until the label pops up.)
4. Type a name for the folder on the New Folder screen.
5. Click on **Create.**
6. The new folder is created and saved automatically.
7. Important: Change the file name. Simply add a new date, for example (do *not* use the slash).

Locating Additional Features

Auto-Template for a Table of Contents

Auto-Template for a List of Illustrations

These two templates are *not* the fill-in-the-blank style and cannot be filled in. They are automatic features that construct the tables *from* the text if you used headings in the text body or figure numbers and captions for the illustrations.

1. Go to **Insert** on the menu bar.
2. Click on **Index and Tables.**
3. In the dialog box, select **Table of Contents** or **Table of Figures.**

To Build a Table

1. Go to **Table** on the menu bar.
2. Click on **Insert Table.**
3. In the dialog box, enter the number of rows and columns desired.
4. Click on **OK** for the basic format.
5. Or select **AutoFormats** for other styles.
6. There are many options in the autoformats. Select one that suits your needs.

To Use the Equation Editor

1. Go to **Insert** on the menu bar.
2. Click on **Object.**
3. To open the program, scroll to **Microsoft Equation** and *double*-click.
4. For help with the editor, return to the menu bar and click on **Help.**
5. The help options list "Equation Editor Help Topics." Click to open.
6. Select from the choices in the dialog box with a *double* click to call up instructions.

For Help at Any Point

1. Click on **Help** in the menu bar.
2. Go to **Answer Wizard.**
3. Activate the help program with a click.
4. Type in a question where indicated and click on **Search.**

Office 97 for Windows

I. To Find Templates of Essay Components

On one Office 97 version, you will be referred to a Web site for the templates. The templates have been placed on the Microsoft Web site, but you cannot get them unless you provide an e-mail address to download (which could have other purposes for the company). Your best bet is to move on to section III and save your own templates. Do *not* download on the computers in campus computer labs.

Other versions of Office 97 may or may not have the essay components templates. At least one of the versions removed any mention of essay templates.

II. To Save a Modified Template

The template can be altered to suit your needs or to conform to the suggestions of your instructors, your supervisor, or the recommendations in *Technical Document Basics*. These modifications can be saved as a *new* template.

1. Go to **File** on the menu bar.
2. Click on **Save As.**
3. In the Save-as-Type box, click on **Document Template.** This function automatically places the template in a template file.
4. *Double* click on the **Publications** folder to place a template in the folder.
5. Important: Change the file name. Simply add a new date, for example (do *not* use the slash).
6. Click on **Save.**

III. To Save an Original Template

If you build your own title page or any other page type, it can be used again and again. You can simply make a copy and modify it or you can make a *template* and modify it. *Be sure to save the first outline you develop. Save it as a template.*

1. Go to **File** on the menu bar.
2. Click on **Save As.**
3. In the Save-as-Type box, click on **Document Template,** which will take you to a template folder.
4. In the dialog box window are the folders. Here you can file the template in any folder or make a folder (see IV below). Move the cursor to the selected file and *double* click.
5. Important: Change the file name. Simply add a new date, for example (do *not* use the slash).
6. Click on **Save.**

IV. To Open a Personal Template Folder for Your Templates

1. Go to **File** on the menu bar.
2. Click on **Save As.**
3. In the Save-as-Type box, click on **Document Template,** which will take you to a template folder.
4. Click on the **Create New Folder** icon on the menu bar in the Save As dialog box. (To locate the proper icon, rest the cursor on each icon until the label pops up.)
5. Type a name for the folder on the **New Folder** screen.
6. Click on **OK.**
7. *Double* click on the folder you created to open it.
8. Important: Change the file name. Simply add a new date, for example (do *not* use the slash).
9. Click on **Save.**

Locating Additional Features

Auto-Template for a Table of Contents

Auto-Template for a List of Illustrations

These two templates are *not* the fill-in-the-blank style and cannot be filled in. They are automatic features that construct the tables *from* the text if you used headings in the text body or figure numbers and captions for the illustrations.

1. Go to **Insert** on the menu bar.
2. Click on **Index and Tables.**
3. In the dialog box, select **Table of Contents** or **Table of Figures.**

To Build a Table

1. Go to **Table** on the menu bar.
2. Click on **Insert Table.**
3. In the dialog box, enter the number of rows and columns desired.
4. Click on **OK** for the basic format.
5. Or select **AutoFormats** for other styles.
6. There are many options in the autoformats. Select one that suits your needs.
7. See also the new feature, **Draw Table,** when you pull down from **Table** on the menu bar.

To Use the Equation Editor

1. Go to **Insert** on the menu bar.
2. Click on **Object.**
3. To open, the program, scroll to **Microsoft Equation** and *double*-click.
4. For help with the editor, return to the menu bar and click on **Help.**
5. Click on **Help Topics: Equation Editor.**
6. Select from the choices in the dialog box with a *double* click to call up instructions.

For Help at Any Point

1. Click on **Help** in the menu bar.
2. Go to **Microsoft Word Help.**
3. Activate the help program with a click.
4. Type in a question where indicated and click on **Search.**
5. Select from options or use the Search button. Click on the **Option** or the **Search** button and a window will open.

Office 2000 for Windows

I. To Find Templates of Essay Components

1. Go to **File** on the menu bar.
2. Click on **New.**
3. In the dialog box select **Publications.**
4. Click on the **Thesis** icon.
   You will find the following:

   Title Page

   Table of Contents

   List of Figures

   Glossary

   Text Page

   Bibliography

III. To Save a Modified Template

The template can be altered to suit your needs or to conform to the suggestions of your instructors, your supervisor, or the recommendations in *Technical Document Basics.* These modifications can be saved as a *new* template.

1. Go to **File** on the menu bar.
2. Click on **Save As.**
3. In the Save-as-Type box, click on **Document Template,** which will take you to a template folder.

4. Important: Change the file name. Simply add a new date, for example (do *not* use the slash).
5. Click on **Save.**
6. Note the location where the template is being saved. It may save to different locations.

IV. To Save an Original Template

If you build your own title page or any other page type, it can be used again and again. You can simply make a copy and modify it or you can make a *template* and modify it. *Be sure to save the first outline you develop. Save it as a template.*

1. Go to **File** on the menu bar.
2. Click on **Save As.**
3. In the Save-as-Type box, click on **Document Template,** which will take you to a template folder.
4. Important: Give the file a name.
5. Click on **Save.**
6. Note the location where the template is being saved. It may save to different locations.

V. To Open a Personal Template Folder for Your Templates

1. Go to **File** on the menu bar.
2. Click on **Save As.**
3. In the Save-as-Type box, click on **Document Template,** which will take you to a template folder.
4. Click on the **Create New Folder** icon on the menu bar in the Save As dialog box. (To locate the proper icon, rest the cursor on each icon until the label pops up.)
5. Type a name for the folder on the New Folder screen.
6. Click on **OK** and the folder will automatically open.
7. Important: Change the file name. Simply add a new date, for example (do *not* use the slash).
8. Click on **Save.**

Locating Additional Features

Auto-Template for a Table of Contents

Auto-Template for a List of Illustrations

These two templates are *not* the fill-in-the-blank style and cannot be filled in. They are automatic features that construct the tables *from* the text if you used headings in the text body or figure numbers and captions for the illustrations.

1. Go to **Insert** on the menu bar.
2. Click on **Index And Tables.**
3. In the dialog box, select **Table Of Contents** or **Table Of Figures.**

To Build a Table

1. Go to **Table** on the menu bar.
2. Go to **Insert** and click on **Table.**
3. In the dialog box, enter the number of rows and columns desired.
4. Click on **OK** for the basic format.
5. Or select **Autoformats** for other styles.
6. There are many options in the autoformats. Select one that suits your needs.
7. See also the new feature, **Draw Table,** found under **Table** on the menu bar.

To Use the Equation Editor

1. Go to **Insert** on the menu bar.
2. Click on **Object.**
3. To open, the program, scroll to **Microsoft Equation** and *double* click.
4. For help with the editor, return to the menu bar and click on **Help.**
5. Click on **Help Topics: Equation Editor.**
6. Select from the choices in the dialog box with a *double* click to call up instructions.

VI. For Help at any Point

1. Click on **Help** in the menu bar.
2. Go to **Microsoft Word Help.**
3. Activate the help program with a click.
4. Type in a question where indicated and click on **Search**.
5. Select from the options above or use the Search button. Click on the **Option** or the **Search** button and a window will open.

Acknowledgments

There are four texts and three instructor's supplements in the *Wordworks*™ series. In order to properly acknowledge the many people and organizations that have contributed to this project, I have chosen to first extend my thanks to those who assisted the endeavor in a larger context. A great many more contributors helped shape the separate volumes.

The Wordworks Series

The *Wordworks*™ project would never have come about without the patience and generosity of my colleague and assistant Patricia Britz. There were 4000 pages of manuscript, endless keyboarding tasks, and elaborate page spreads. The project would have been impossible without Pat.

My colleague, Dr. Rita Smilkstein (published by Harcourt Brace), deserves very special thanks. Dr. Smilkstein has read every manuscript page of the project and contributed endless ideas and support.

To Stephen Helba, editor-in-chief at Prentice Hall, Pearson Education, I owe a special thanks. Some years ago I came home one day and found twenty-five pages of contracts spread all over the floor under an old fax machine. From the beginning Stephen took a personal interest in the *Wordworks*™ project and has been a source of encouragement for the many years it has taken to develop.

To Dr. David Mitchell, President of South Seattle Community College, I owe thanks for a special favor. When he was the former dean of my campus, I sought his help in finding a space where I could create a dedicated classroom for my program. His response was immediate. We surveyed a prospect and agreed to the experiment. He budgeted a fully provisioned room of tables and chairs and blackboards and other features I requested. That dedicated teaching and learning environment played a major role in developing the concepts embodied in the *Wordworks*™ series.

I would like to thank Marc Vassallo, Lee Anne White, Jeff Kolle, and other editors of the Taunton Press in Newtown, Connecticut. It was their faith in my ability to produce cover stories, feature articles, and shorter pieces for the nationally known Taunton magazines that give me the gumption to try to write the *Wordworks*™ series. It was a great boost to realize that each publication had a circulation of several hundred thousand readers and that, all told, I had reached several million readers thanks to the Taunton staff. I also discovered the excitement of writing copy that included photographic compositions and line art concepts.

It helps to have the support of a close friend when facing the misgivings involved in an enormous project. Rob Vinnedge, one of the Northwest's finest professional photographers, assisted me as I developed articles for the Taunton Press even when he was trying to meet his own book and magazine deadlines that were taking him as far away as Hong Kong. I regret that there was no time in our busy lives to share any shoots for *Wordworks*™, but the deadlines were too tight. What matters the most is that Rob said, again, again, and again, "You can do it."

Technical Document Basics for Engineering Technicians and Technologists

I am indebted to many former students for their enthusiasm and their willingness to appear in the *Wordworks*™ series. Many contributors developed projects that in some way appear in the *Wordworks*™ textbooks or the *Wordworks*™ supplements. A number of multi-page models appear in *Basic Composition Skills* and *Technical Document Basics.* Other models are located in the instructor's supplements for the two texts. The following contributors provided the selections that appear as multi-page models.

Richard Bigham
Janine Boyer Richards
Andrew Cameron
David Campbell
Corin Carper
Sean Caughlan
Jane Chateaubriand
Kathryn Chumbley
Ed Condon
Jennifer Dillard
Mark Eskridge
Edward Ferraro
Kenneth Tim Forman
David Hale
David Hilderbrand
Mike Kang
Allan Kellner
Ricky Keokitvon
Ansar Khalil
Eric Long
Dornie MacKenzie
Mike Meagher
Ben Minson
Janine Michelsons
Joe Mitchell
Susan Mutuc McCants
Greg Novlan
Michael Pitcher
Scott Schaper
Steve Schattenbild
Patricia Renderos
Troy Sewell
Dawn Shephard
David Stinson
Linda Strout
Michael Summers
Don Thornton
John Touliatos
Lucy Underwood
Anthony VanNorman
Mark Vansteenkiste
Jerald Yun

Damon Harrell
Paul Hefty
Carrie Pratt
Salim Rabaa

In order to properly develop two of the texts, I needed a wide variety of single-page or half-page samples that appear throughout *Basic Composition Skills* and *Technical Document Basics* and the instructor's supplements. As the books took shape, I found more and more material that was appropriate, and the list of contributors grew. Fortunately I had considerable success in contacting everyone involved, even though some of the people identified below are now in Texas, Arizona, California, and Oregon. To all of you, a special thanks.

James Affeld
Kayla Agan
Mitch Agan
Ben Andrews
Roman Ariri
Lynn Arnold
Erinn Barnett
Tyler Beam
Bob Bergstrom
John Bienick
Thomas Booze
Janine Boyer-Richards
Douglas Bradley
Dewi Cahyantari
David Campbell
Sean Caughlin
David Clemmons
Carol Collins
Wes Dang
Kevin Donnelly
Michael Elliston
Nadine Hamby
Brian Hanners
Art Hedley
Eric Hesselgesser
Mike Huckaby
Wayne Jarvimaki
Ernie Jean
Jeremiah Jester
Glen Johnson
Leonard Kannapell
Aaron King
Lawrence Kitchen
Steven Knobbs
Anna Komissarchik
Brian Kraft
Jay Voravong Kruise
Matt Laundroche
Ron Lewis
Clarence Lim
LeAnne Livingston
Eric Long
Peter Mikolajczyk
Kristy Moody
Jim Murray
Paul Nguyen
KimAnh Pham
Nick Pierce
Marianne Pinyuh
Michael Pitcher
Randy Rosen
Zijad Saric
Martin Saxer
Hassan Shirdavani
Robert Simpson
Mark Skullerud
Robert Smith
Greg Speers
Tim Stacey
Sandra McKay Stimson
Glenn Sudduth
Jamison Surquy
Earl Thompson

Diana Eng

Eugene Escarez

M.D. Esclavon

Randy Farlow

Mari Fortes

Chris Fraser

Bruce Fugere

Nazim Haji

Paul Luke

David MacDonald

Dornie MacKenzie

Sherry Sloan Manning

Blair Marshall

Ted Marier

Christine McCurdy

Janine Michelson

Tuan Tong

Elaine Tritt

Vu Quoc

Jeremy Watts

Chris Wiederhold

Karen Woodruff

Zuo Yan

To Shirley Bailey, a special thanks for taking the time out of her busy calendar to develop the casebook for the text. She chose a recent project that she developed for her company and documented its evolution from start to finish.

I decided to include shorter pieces that describe the writing experiences of other engineering technicians, and Todd Rendahl and Blair Jadwin promptly volunteered to discuss their work experiences. I appreciate their efforts to contribute to the project.

I would like to thank Andreas Brockhaus, specialist in software education, for acting as a resource on computer-related matters. Andreas developed the computer tips for the text and worked with me to make the appendix of templates as accurate as possible.

I would like to thank a number of my colleagues for their technical advice: Steve Anderson (physics), Lynn Arnold (network technology), Dale Cook (HVAC), Fred Edelman (CAD), Tom Griffith (chemistry), John Hagans (computer technology), Ralph Jenne (mathematics), Chris Sanders and Dennis Schaffer (electronics). Many of the diagrams in the text were first constructed with CAD by Stan Nelson, who generously donated his time to the project.

Publishers

I would like to thank the following publishing companies for granting me permission to use excerpts or illustrations from their publications, which are acknowledged in the text.

Addison-Wesley Publishing Co.

Delmar Publishers*

McGraw-Hill Publishing Co.*

Macmillan Magazines Ltd, UK

Macmillan USA*

Microcomputing Magazine

Plenum Publishing Corp.

Plumbing Engineering Magazine*

Prentice Hall/Pearson Education*

*Sources that are located in the teaching supplement are identified with an asterisk. Sources from McGraw-Hill appear in both the text and the supplement.

Web Sites

I would like to thank the following for permission to use images from their Web sites.

allelec.com

Hardware Central, internet.com Corp.

Bason

Arachnophilia at arachnoid.com

Microtell, UK

Shortcourses.com*

Software

Screen scans are a common feature of today's technical documents. The following companies kindly permitted me to use screen scans from their products.

America Online

Autodesk, Inc.

Intuit

Macromedia

Microsoft Inc.

Corporate Documents

I would like to thank the Tandy Corporation and the Xerox Corporation for their permission to use illustrations from their promotional literature. Additional documents appear through the courtesy of

Aiphone Communication Systems

Bardahl Corporation

Cruising Equipment Company

Digital Microwave Corporation

Hewlett-Packard Corporation

Pentec Corporation

Zetron Corporation

Professional Societies

The American Society for Testing Materials kindly permitted me to use several illustrations from the *Annual Book of ASTM Standards.*

Prentice-Hall/Pearson Education

Special thanks to the production staff at Prentice Hall in Columbus, Ohio, and the Clarinda Company. Two production teams saw to it that the *Wordworks Series* would go to press. At Prentice Hall, I would like to acknowledge the following staff:

Editor in Chief: Stephen Helba

Executive Editor: Debbie Yarnell

Associate Editor: Michelle Churma

Production Editor: Louise N. Sette

Design Coordinator: Robin G. Chukes

Cover Designer: Ceri Fitzgerald

Production Manager: Brian Fox

Marketing Manager: Jimmy Stephens

At Clarinda I would like to thank additional personnel for seeing *Wordworks* series through the production stages:

Manager of Publication Services: Cindy Miller

Account Manager: Jennifer Graham

For her patience and skills, a special thanks to Barbara Liguori for copyediting the *Wordworks*™ series.

I would also like to acknowledge the reviewers of this text:

Pamela S. Ecker, Cincinnati State Technical and Community College (OH)
Dr. Harold P. Erickson, Lake Superior College (MN)
Patricia Evenson, Northcentral Technical College (WI)
Anne Gervasi, North Lake College (TX)
Mary Francis Gibbons, Richland College (TX)
Charles F. Kemnitz, Penn College of Technology (PA)
Diane Minger, Cedar Valley College (TX)
Gerald Nix, San Juan College (NM)
M. Craig Sanders, Bellevue Community College (WA)
Laurie Shapiro, Miami-Dade Community College (FL)
Richard L. Steil, Southwest School of Electronics (TX)
David K. Vaughan, Air Force Institute of Technology (OH)

David W. Rigby

Index